FUNDAMENTALS OF AEROSPACE NAVIGATION AND GUIDANCE

This text covers fundamentals used in the navigation and guidance of modern aerospace vehicles, in both atmospheric and space flight. It can be used as a textbook supporting a graduate-level course on aerospace navigation and guidance, a guide for self-study, or a resource for practicing engineers and researchers.

It begins with an introduction that discusses why navigation and guidance ought to be considered together and delineates the class of systems of interest in navigation and guidance. The book then presents the necessary fundamentals in deterministic and stochastic systems theory and applies them to navigation. Next, the book considers guidance problems under a variety of assumptions, leading to the scenarios of homing, ballistic, and midcourse guidance. Then, the book treats optimization and optimal control for application in optimal guidance. In the final chapter, the book introduces problems in which two competing controls exercise authority over a system, leading to differential games. *Fundamentals of Aerospace Navigation and Guidance* features examples illustrating concepts and homework problems at the end of all chapters.

Pierre T. Kabamba is currently professor of aerospace engineering at the University of Michigan. He received a PhD in mechanical engineering from Columbia University in 1981 and joined the University of Michigan in 1983. His area of teaching and research is flight dynamics and control systems. His awards include the Class of 1938 E Distinguished Service Award (from the University of Michigan), a Best Paper Special Award (from the Society of Instrumentation and Control Engineers, Japan), and election to the rank of Fellow of IEEE. He is coauthor of the textbook *Quasilinear Control* (Cambridge University Press, 2011).

Anouck R. Girard is currently associate professor of aerospace engineering at the University of Michigan. She received a PhD in ocean engineering from the University of California, Berkeley, in 2002 and joined the University of Michigan in 2006. Her area of teaching and research is flight dynamics and control systems. Her awards include the Silver Shaft Teaching Award (from the University of Michigan) and a Best Student Paper Award (from the American Society of Mechanical Engineers).

Cambridge Aerospace Series

Editors:
Wei Shyy and Vigor Yang

1. J. M. Rolfe and K. J. Staples (eds.): *Flight Simulation*
2. P. Berlin: *The Geostationary Applications Satellite*
3. M. J. T. Smith: *Aircraft Noise*
4. N. X. Vinh: *Flight Mechanics of High-Performance Aircraft*
5. W. A. Mair and D. L. Birdsall: *Aircraft Performance*
6. M. J. Abzug and E. E. Larrabee: *Airplane Stability and Control*
7. M. J. Sidi: *Spacecraft Dynamics and Control*
8. J. D. Anderson: *A History of Aerodynamics*
9. A. M. Cruise, J. A. Bowles, C. V. Goodall, and T. J. Patrick: *Principles of Space Instrument Design*
10. G. A. Khoury (ed.): *Airship Technology*, Second Edition
11. J. P. Fielding: *Introduction to Aircraft Design*
12. J. G. Leishman: *Principles of Helicopter Aerodynamics*, Second Edition
13. J. Katz and A. Plotkin: *Low-Speed Aerodynamics*, Second Edition
14. M. J. Abzug and E. E. Larrabee: *Airplane Stability and Control: A History of the Technologies that Made Aviation Possible*, Second Edition
15. D. H. Hodges and G. A. Pierce: *Introduction to Structural Dynamics and Aeroelasticity*, Second Edition
16. W. Fehse: *Automatic Rendezvous and Docking of Spacecraft*
17. R. D. Flack: *Fundamentals of Jet Propulsion with Applications*
18. E. A. Baskharone: *Principles of Turbomachinery in Air-Breathing Engines*
19. D. D. Knight: *Numerical Methods for High-Speed Flows*
20. C. A. Wagner, T. Hüttl, and P. Sagaut (eds.): *Large-Eddy Simulation for Acoustics*
21. D. D. Joseph, T. Funada, and J. Wang: *Potential Flows of Viscous and Viscoelastic Fluids*
22. W. Shyy, Y. Lian, H. Liu, J. Tang, and D. Viieru: *Aerodynamics of Low Reynolds Number Flyers*
23. J. H. Saleh: *Analyses for Durability and System Design Lifetime*
24. B. K. Donaldson: *Analysis of Aircraft Structures*, Second Edition
25. C. Segal: *The Scramjet Engine: Processes and Characteristics*
26. J. F. Doyle: *Guided Explorations of the Mechanics of Solids and Structures*
27. A. K. Kundu: *Aircraft Design*
28. M. I. Friswell, J. E. T. Penny, S. D. Garvey, and A. W. Lees: *Dynamics of Rotating Machines*
29. B. A. Conway (ed): *Spacecraft Trajectory Optimization*
30. R. J. Adrian and J. Westerweel: *Particle Image Velocimetry*
31. G. A. Flandro, H. M. McMahon, and R. L. Roach: *Basic Aerodynamics*
32. H. Babinsky and J. K. Harvey: *Shock Wave–Boundary-Layer Interactions*
33. C. K. W. Tam: *Computational Aeroacoustics: A Wave Number Approach*
34. A. Filippone: *Advanced Aircraft Flight Performance*
35. I. Chopra and J. Sirohi: *Smart Structures Theory*
36. W. Johnson: *Rotorcraft Aeromechanics*
37. W. Shyy, H. Aono, C. K. Kang, and H. Liu: *An Introduction to Flapping Wing Aerodynamics*
38. T. C. Lieuwen and V. Yang: *Gas Turbine Emissions*
39. P. Kabamba and A. Girard: *Fundamentals of Aerospace Navigation and Guidance*
40. R. M. Cummings, S. A. Morton, W. H. Mason and, D. R. McDaniel: *Applied Computational Aerodynamics*

Fundamentals of Aerospace Navigation and Guidance

Pierre T. Kabamba
University of Michigan at Ann Arbor

Anouck R. Girard
University of Michigan at Ann Arbor

CAMBRIDGE
UNIVERSITY PRESS

CAMBRIDGE
UNIVERSITY PRESS

University Printing House, Cambridge CB2 8BS, United Kingdom

One Liberty Plaza, 20th Floor, New York, NY 10006, USA

477 Williamstown Road, Port Melbourne, VIC 3207, Australia

314-321, 3rd Floor, Plot 3, Splendor Forum, Jasola District Centre, New Delhi - 110025, India

79 Anson Road, #06-04/06, Singapore 079906

Cambridge University Press is part of the University of Cambridge.

It furthers the University's mission by disseminating knowledge in the pursuit of education, learning and research at the highest international levels of excellence.

www.cambridge.org
Information on this title: www.cambridge.org/9781107070943

First published 2014

A catalogue record for this publication is available from the British Library

Library of Congress Cataloging in Publication data
Kabamba, Pierre T., author.
Fundamentals of aerospace navigation and guidance / Pierre T. Kabamba, University of Michigan, Anouck R. Girard, University of Michigan.
pages cm
Includes bibliographical references and index.
ISBN 978-1-107-07094-3 (hardback)
1. Navigation (Astronautics) 2. Space vehicles. 3. Trajectory optimization. 4. Space flight. I. Girard, Anouck. II. Title.
TL1065.K33 2014
629.1–dc23 2014009748

ISBN 978-1-107-07094-3 Hardback

To our families, Joséphine, Monique, Orianne, and Louis,
Michael, Leo, and Mia.
With a special commendation for Mia, who, at age three months,
listened to the formal read-through of the whole book.

Contents

Preface

This book arose out of lecture notes for the graduate-level course AE584 Guidance, Navigation, and Avionics, taught in the Department of Aerospace Engineering at the University of Michigan at Ann Arbor. The course was originally introduced by Professor Emeritus Robert M. Howe and was taught for a number of years by the authors.

This book can be used as a textbook supporting a graduate-level formal course on aerospace navigation and guidance. It can also be used as a guide for self-study on the subject. Finally, it can be used as a reference text. To enhance its usability, the book contains numerous examples illustrating concepts and homework problems at the end of all chapters. A partial solution manual for these homework problems is available with limited circulation.

Although this book uses many mathematical results, it is not a mathematics text. The results are stated with some rigor, but only a few proofs of special interest are given. The interested reader may consult the bibliography listed at the end of each chapter for proofs and additional information.

The prerequisite knowledge includes undergraduate instruction in elementary linear algebra, differential equations, and flight mechanics, which are standard in aerospace engineering curricula. Key definitions and mathematical results are given in the appendix.

The intended audience for this book encompasses graduate students in engineering specializing in flight dynamics, guidance, navigation, and control, together with practicing engineers and researchers in the field. In the authors' experience, this text maps to a two-semester graduate sequence, with the first semester covering roughly the first six chapters and the second semester covering roughly the last four chapters.

The authors acknowledge and are thankful for the intellectual contributions of their teachers, colleagues, and especially several generations of students who have learned this material at Michigan. The authors also acknowledge and are thankful for financial support from the Air Force Research Laboratory, the Air Force Office of Scientific Research, the Boeing Company, the Office of Naval Research, and the National Science Foundation.

Pierre T. Kabamba and Anouck R. Girard
Ann Arbor, Michigan

1 Introduction

1.1 Purpose and Motivation

The purpose of this book is to present fundamentals used in the navigation and guidance of aerospace vehicles. By **fundamentals**, we mean a body of knowledge that does not change with advances in technology. **Aerospace** vehicles encompass all kinds of craft flying in the atmosphere or in space. **Navigation** is concerned with the questions: Where is the vehicle? What is its velocity? And what are its angular orientation and angular rates? **Guidance** is concerned with the question: What maneuvers should we perform to cause the vehicle to go where we want, while meeting the specifications of the mission?

For centuries, devices such as charts, magnetic compasses, drafting compasses, sextants, and clocks were sufficient to solve the navigation and guidance problems of the time. In those days, velocities over land and sea were relatively slow, navigators could count on a variety of visible references (such as landmarks, beacons, or stars), and a human pilot controlled the vehicle during the entirety of the trip. However, the advent of air and space travel has posed substantial new challenges to navigators. First, aerospace vehicles generally travel much faster and farther than their land and sea counterparts. Second, whereas navigation on land and sea can be based on visible references, an aerospace navigator may be deprived of such resources. Finally, many aerospace missions require a high level of automation, either because they are unmanned or to minimize crew fatigue.

In response to these new challenges, the aerospace community has standardized the use of avionic equipment such as radio transmitters, laser velocimeters, laser gyroscopes, and digital computers. However, despite these remarkable technological advances, the underlying mathematical principles that govern the use of data for navigation and guidance have remained the same. For instance, the fifteenth-century sea navigator would determine the location of a ship by measuring the apparent elevation of known stars using a sextant. The ship would then be located at the intersection of loci drawn on a map. In this century, the navigation of an unmanned interplanetary spacecraft is typically performed by measuring the angles between lines of sight to known heavenly bodies, such as the Sun, planets, and stars. The spacecraft is similarly found at the intersection of several loci drawn in three-dimensional space.

This book aims at an exposition of the mathematical methods used in aerospace navigation and guidance, with a strong emphasis on fundamentals rather than any particular hardware implementation. Such an exposition is possible because these methods, which use geometry, linear algebra, differential calculus, and stochastic error analysis, are hardware-independent. This book does not attempt to cover all the fundamentals; however, it is the authors' experience that the fundamentals it presents constitute a good initial body of knowledge for an engineer or researcher in the field of navigation and guidance.

Before delving earnestly into the technical material, it is worthwhile to answer the question: Why study fundamentals at all? We can give at least two compelling answers to this question. The first is purely utilitarian: by their very nature, fundamentals have a value that transcends time and technology. Hence, studying fundamentals is a better investment than studying ephemeral knowledge that may soon become obsolete due to technological advances. The second answer is foundational: fundamentals are a sound basis on which to build knowledge. As a practical consequence, fundamentals typically provide excellent guidance in the solution of engineering analysis, synthesis, and design problems.

1.2 Problem Statement

Throughout this book, we consider the following three vector quantities:

1. $x(t)$ denotes the actual state vector of the vehicle. This vector typically contains the variables that specify the position and angular orientation of the vehicle, and their time derivatives.
2. $x_r(t)$ denotes the reference state vector of the vehicle, that is, the desired value of the state vector.
3. $\hat{x}(t)$ denotes the estimated state vector, that is, the value of the state vector, as estimated by the navigator, based on the sensor measurements.

The three preceding vectors generate the following:

1. The vector $\delta x(t) = x(t) - x_r(t)$ is called the **guidance error**.
2. The vector $\tilde{x}(t) = x(t) - \hat{x}(t)$ is called the **navigation error**.

Thus, the guidance error is, roughly speaking, the difference between where the vehicle is located and where it is supposed to be. It is generally caused by disturbances acting on the vehicle, such as wind gusts, currents, and thrusting misalignment. The navigation error is, roughly speaking, the difference between where the vehicle is located and where the navigator believes it is. This error is generally due to uncertainties and errors in the readings of the sensors. It is important to realize that, even without disturbances or measurement errors, the guidance and navigation errors interact. For instance, if the vehicle is on its nominal course at some initial time t_0 (that is, $\delta x(t_0) = 0$), but the navigator believes that the vehicle is off-course (that is, $\tilde{x}(t_0) \neq 0$), this will prompt the navigator to "mis-correct" the course of the vehicle, causing $\delta x(t_0 + \tau) \neq 0$ for some small positive τ.

The problem of combined navigation and guidance can now, in broad terms, be stated as that of *causing the vehicle to follow its desired path, despite disturbances and navigation uncertainties*. In other words, we want to go where we are supposed to go, despite the facts that we don't know exactly where we are and that our motion is undergoing unpredictable disturbances.

We present the following argument to show how such an achievement is, in principle, possible and to justify why we consider the problems of navigation and guidance together. Assume that we guide the vehicle based on navigation, that is, we take guidance decisions assuming that our estimate of $x(t)$ is indeed $x(t)$ itself. This results in an **estimated guidance error** $\hat{x}(t) - x_r(t)$, that is, the difference between where we believe we are and where we are supposed to be. Note that the estimated guidance error can also be viewed as the guidance error we would incur if the navigation error happened to be zero. Assume that we have a bound on the estimated guidance error, that is, we have found a positive number ϵ_g such that $||\hat{x}(t) - x_r(t)|| \leq \epsilon_g$. Also assume that we have a bound on the navigation error, that is, we have found a positive number ϵ_n such that $||x(t) - \hat{x}(t)|| \leq \epsilon_n$. Then, a simple application of the triangle inequality yields

$$\begin{aligned} ||\delta x(t)|| &= ||x(t) - x_r(t)|| \\ &= ||(x(t) - \hat{x}(t)) + (\hat{x}(t) - x_r(t))|| \\ &\leq ||x(t) - \hat{x}(t)|| + ||\hat{x}(t) - x_r(t)|| \end{aligned} \tag{1.1}$$

and therefore

$$||\delta x(t)|| \leq \epsilon_n + \epsilon_g, \tag{1.2}$$

which implies that we can guarantee a bound on the true guidance error. In other words, small navigation error and small estimated guidance error imply good guidance.

Therefore, in practice, we proceed as follows. First, we develop navigation strategies and quantify their errors. Then, we obtain guidance laws and analyze their performance in the absence of navigation errors. Finally, we assess the degradation of performance of the guidance laws due to the navigation errors. Note that this procedure assumes that a guidance law that is designed based on perfect knowledge of $x(t)$ would work similarly well if, in its implementation, $x(t)$ were replaced by $\hat{x}(t)$. This **separation property** is formally justified in Chapter 7.

1.3 Scope of the Book

This book applies systems and control theory to the motion of aerospace vehicles. In broad terms, systems theory deals with signals and cause-effect relationships between them, whereas control theory seeks to achieve performance of systems in the presence of uncertainty. Let us clarify what we mean.

1.3.1 Systems Theory

For now, let us define a signal as a vector function of time – in Chapter 2, we give a more precise definition of the term *signal*, but here vector functions of time suffice. Then, a system is simply a portion of the universe where we postulate that some signals cause other signals. We call the cause-signals inputs and the effect-signals outputs, and we represent their cause-effect relationship in a block diagram such

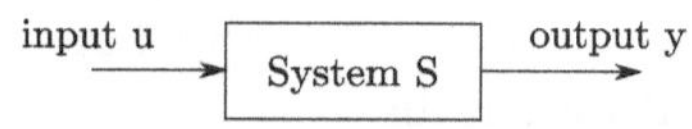

Figure 1.1. Block diagram representation of a system.

as in Figure 1.1, where the inputs enter the system and the outputs exit the system. Formally, we write

$$y = S(u). \tag{1.3}$$

For instance, in a spring system, such as in Figure 1.2, we typically postulate that the force is the input and the deformation is the output.

The following taxonomy is used to characterize systems.

1. **Linear versus nonlinear.** In a linear system, the superposition principle holds, whereby a linear combination of inputs produces as output the same linear combination of the respective outputs. In other words, for all inputs u_1, u_2 and real numbers α_1, α_2, we have

   $$S(\alpha_1 u_1 + \alpha_2 u_2) = \alpha_1 S(u_1) + \alpha_2 S(u_2). \tag{1.4}$$

   Otherwise, the system is called nonlinear. For instance, a spring system is often modeled as linear.
2. **Static versus dynamic.** In a static, or **memoryless**, system, the output at any time, $y(t)$, depends only on the input at the same time, $u(t)$, but does not depend on the past time history of input $u(\tau)$, $\tau < t$. Otherwise, the system is called dynamic. Typically, the mathematical description of a static system is in terms of algebraic equations, whereas that of a dynamic system is in terms of differential equations. For instance, the spring system of Figure 1.2 is static, whereas a mass-spring system, such as in Figure 1.3, is dynamic.
3. **Time invariant versus time varying.** In a time invariant system, an arbitrary time shift of the input produces exactly the same time shift of the output. In other words, let the system Δ_τ represent the time shift of magnitude τ, so that $(\Delta_\tau(u))(t) = u(t - \tau)$. Then, for a time invariant system, for all signals u and time shift magnitudes τ, we have

   $$S(\Delta_\tau(u)) = \Delta_\tau(S(u)). \tag{1.5}$$

   Otherwise, the system is called time varying. For instance, a spring system where the spring stiffness is constant is a time invariant system.
4. **Causal versus anticipatory.** In a dynamic, strictly causal (resp. causal) system, $y(t)$ is determined by $u(\tau)$, $\tau < t$ (resp. $\tau \leq t$), but does not depend on $u(\tau)$, $\tau \geq t$ (resp. $\tau > t$). Otherwise, the system is called **noncausal** or anticipatory. For instance, the mass-spring system of Figure 1.3 is strictly causal.
5. **Lumped parameter versus distributed parameter.** In a lumped parameter system, the mathematical description is in terms of a finite number of ordinary differential equations (ODEs), whereas in a distributed parameter system, it is

Figure 1.2. Spring system.

Figure 1.3. Mass-spring system.

in terms of an infinite number of ODEs or a partial differential equation. For instance, the mass-spring system of Figure 1.3 is lumped parameter.

6. **Deterministic versus stochastic.** In a stochastic system, some signals are random processes, that is, they are unpredictable but exhibit statistical regularity – these terms are defined precisely in Chapter 3. For instance, an aircraft flying through stochastic turbulence is a stochastic system.

In view of the preceding taxonomy, the systems considered in this book are mostly linear, dynamic, time varying, strictly causal, lumped parameter, and stochastic. This is the class of systems that is most relevant to the study of navigation and guidance.

1.3.2 Control Theory

A generic control system is represented in Figure 1.4, where:

1. The system P, called the **plant**, is given and is to be controlled
2. The system C, called the **controller**, is unknown and is used to control the plant
3. The input u_1, called **exogenous**, is not under the authority of the designer
4. The input u_2, called **endogenous**, is under the authority of the designer
5. The output y_1, called the **performance output**, is used to quantify the performance of the controller
6. The output y_2, called the **measured output**, is available to the designer for feedback to achieve performance

The goal of control, then, is to find a system C that causes the signal y_1 to be small despite uncertainties in the signal u_1.

REMARK 1.1 *Note that in this book, nonlinearities in dynamics do not play a significant role, which is justified in Chapter 2. For this reason, this book is separate from the literature on nonlinear dynamic systems (see, e.g., [44], [69], [41], [64]). Also note that in this book, the uncertainties are in the signals rather than the dynamics of plants. This is because, owing to sufficient instrumentation, dynamic models of aerospace systems are relatively accurate, even though the reading of sensors may be corrupted by noise. In that respect, this book is separate from the literature on adaptive control systems (see, e.g., [55], [3], [48]) and model predictive control systems (see, e.g., [16], [20], [79]).*

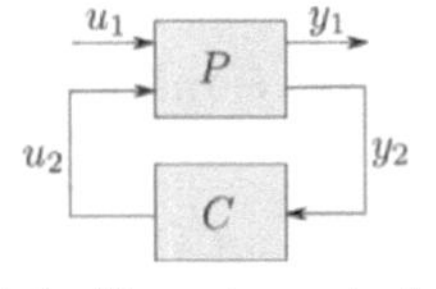

Figure 1.4. Generic control system.

1.3.3 Aerospace Applications

In general, the application of systems and control theory to the motion of aerospace vehicles may cover four areas:

1. **Path planning.** The determination of a nominal flight path for a vehicle to accomplish specified objectives subject to specified constraints. This corresponds to choosing $x_r(t)$ in (1.1).
2. **Navigation.** The estimation of the state of the vehicle, from outputs of specified imperfect sensors, and using principles of geometry and astronomy. This corresponds to computing $\hat{x}(t)$ in (1.1).
3. **Guidance.** A strategy for following the nominal flight path, given off-nominal conditions, and despite disturbances and navigation uncertainties.
4. **Attitude control.** A strategy for maintaining an angular orientation consistent with the guidance law and the constraints, despite disturbances.

Note that these four areas may overlap. For instance, guidance and attitude control interact in airplane flight where, typically, a change of trajectory requires a change of attitude (for example, in a bank-to-turn aircraft). Another instance is that of guidance and path planning, which are coupled in homing guidance: when a disturbance perturbs the pursuer's trajectory, rather than returning to the original nominal path, the pursuer plans a new path to achieve intercept.

The subject of this book is navigation and guidance, as defined earlier.

1.4 Examples

In this section, we outline how navigation and guidance are used to help accomplish a number of prototypical aerospace missions. Technical details are treated in subsequent chapters; the purpose here is simply to give the reader an appreciation of the importance of navigation and guidance throughout the spectrum of aerospace missions.

1.4.1 Transoceanic Jetliner Flight

In a modern jetliner, navigation is typically accomplished using the Global Positioning System (GPS) and other electronic aids. Path planning and guidance are typically accomplished by specifying a sequence of waypoints over which the aircraft is to fly. Path planning, navigation, guidance, and control are typically integrated through the Flight Management System (FMS), with an ergonomic display for use by the pilots, such as in Figure 1.5.

In such missions, navigation errors may, for instance, take the jetliner over forbidden territory, with tragic results. Guidance errors may, for instance, cause a landing aircraft to overshoot or undershoot the runway, also with tragic results.

1.4.2 Intelligence, Surveillance, and Reconnaissance with Unmanned Aerial Vehicle

A large proportion of military unmanned aerial vehicle (UAV) sorties are tasked with collecting information through intelligence, surveillance, and reconnaissance

Figure 1.5. A control display unit (CDU) used to control the first flight management system (FMS), UNS-1. Image courtesy of FAA Instrument Flying Handbook, FAA-H-8083-15B.

(ISR) missions [59]. In these missions, navigation is typically accomplished by combining GPS estimates, inertial navigation system (INS) estimates, and radar telemetry. Path planning and guidance are typically accomplished by a human remote-operator specifying a sequence of waypoints over which the UAV is to fly. Path planning, navigation, guidance, and control are typically integrated in a command center, with ergonomic displays for remote-operators, such as in Figure 1.6.

In such missions, navigation errors may compromise the UAV, especially when combined with a loss of radio link. Of special concern in adversarial situations is the possibility of GPS spoofing that, through a maliciously generated navigation error, may deliver an autonomous UAV to the enemy [22]. Guidance errors may also occur and jeopardize the UAV.

1.4.3 Homing Guidance of Heat-Seeking Missile

Homing missiles are capable of intercepting a target despite its maneuvers, as follows. Some missiles carry a heat seeker, such as in Figure 1.7, a device that can determine the line of sight from the missile to a heat source, for example, the exhaust of an enemy aircraft engine. An effective way to guide a homing missile is then to steer it so that its turn rate is proportional to the turn rate of the line of sight. Hence, in this situation, navigation consists of estimating the turn rate of the line of sight, and guidance consists of steering the missile so that its turn rate is proportional to that of the line of sight. This process can be automated through an autopilot such as in Figure 1.8.

Figure 1.6. General Atomics MQ-1 Predator control station. Image courtesy of U.S. Air Force, Master Sgt. Steve Horton.

Figure 1.7. U.S. Marine Corps Lance Cpl. Leander Pickens arms an AIM-9 Sidewinder missile on a FA-18C Hornet. Image courtesy of U.S. Navy ID 980220-N-0507F-003.

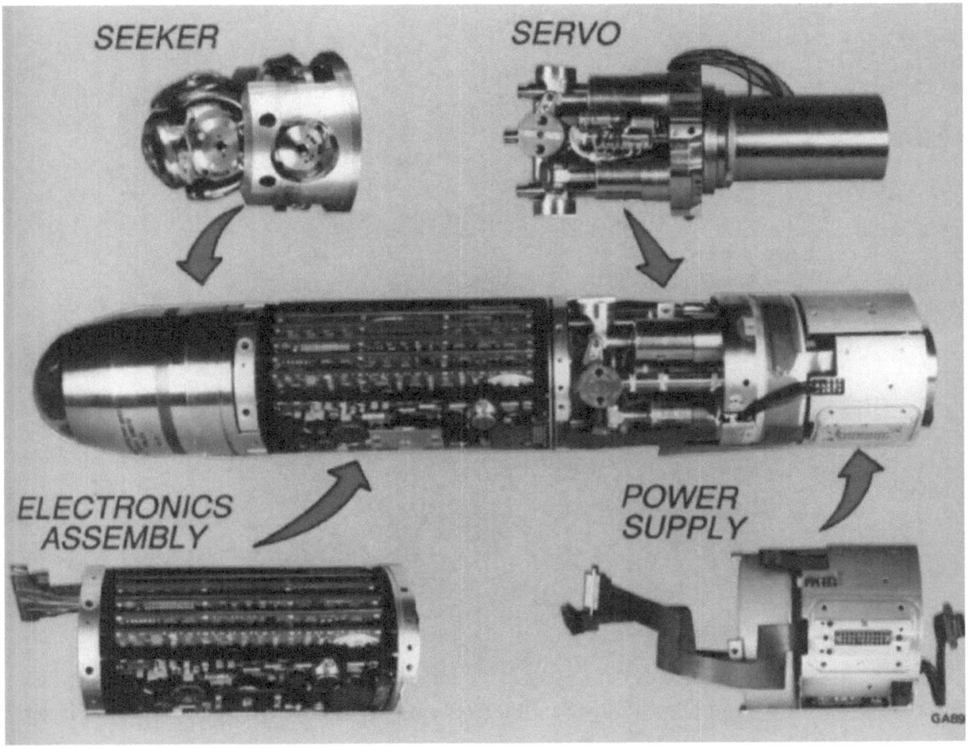

Figure 1.8. AIM-9R Sidewinder subsystems. The autopilot is in the Electronics Assembly. Image courtesy of the U.S. Navy.

For homing missiles, navigation and guidance errors may cause the missile to miss the target, with possible collateral damage. The problem is aggravated when the missile autopilot has a sluggish dynamic response.

1.4.4 Spacecraft Orbital Maneuvers

For spacecraft orbiting Earth, navigation is typically accomplished using radar telemetry from Earth-based tracking stations, such as in Figure 1.9. The outcome of navigation is the orbit determination, that is, the computation of the six orbital elements of the spacecraft. Note that the particular mission of each spacecraft typically puts stringent requirements on its orbit, for example, geosynchronous or Sun synchronous [18]. Guidance is typically achieved through changes of orbit via impulsive burns, as in Figure 1.10. Navigation and guidance are typically integrated and monitored in mission control centers, such as in Figure 1.11.

In space missions, both navigation and guidance errors may leave the spacecraft on an orbit that is useless for the mission at hand.

1.4.5 Interplanetary Travel

For interplanetary travel, navigation is typically accomplished using measurements of angles between lines of sight to bright heavenly objects and radar telemetry from

Figure 1.9. Tracking and Data Relay Satellite System (TDRSS) – Guam remote site. Image courtesy of NASA.

Earth-based tracking stations, as in Figure 1.12. Here also the outcome of navigation is the orbit determination, guidance consists of orbit changes via impulsive burns, and these activities are integrated in mission control centers.

The "final frontier" of interplanetary travel is fraught with significant technical challenges. Here also, both navigation and guidance errors may doom the spacecraft and the mission.

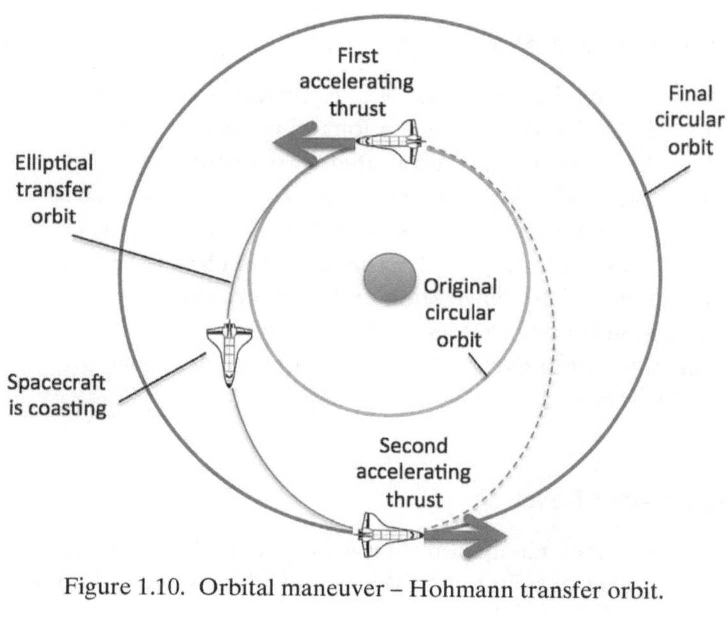

Figure 1.10. Orbital maneuver – Hohmann transfer orbit.

Figure 1.11. U.S. International Space Station ISS Flight Control Room, 2006. Image courtesy of NASA.

Figure 1.12. Deep Space Network (DSN) Goldstone 70 m antenna. Image courtesy of NASA.

1.5 Content of the Book

The subject of this book is navigation and guidance, as defined in Section 1.3. Chapter 2 presents the elements of deterministic systems theory that we need to analyze and synthesize navigation and guidance systems, with emphasis on linear time varying dynamic systems. Chapter 3 presents the elements of stochastic systems theory that we need, with a focus on how linear dynamic systems respond to random inputs. Chapter 4 applies the results of Chapters 2 and 3 to the study of navigation systems and avionics hardware. Chapters 5, 6, and 7 study terminal guidance under various assumptions, leading to the scenarios of homing guidance, ballistic guidance, and midcourse guidance, respectively. Chapters 8 and 9 study optimization and optimal guidance strategies. While Chapters 5, 6, 7, and 9 essentially assume guidance against a target whose maneuvers are independent of the pursuer's, Chapter 10 considers targets that respond to maneuvers of the pursuer, leading to the formalism of differential games.

1.6 Summary of Key Results

The key result in Chapter 1 is inequality (1.2). This result justifies why we consider the problems of navigation and guidance together. It also proves that good navigation together with good navigation-based guidance imply good mission performance.

1.7 Bibliographic Notes for Further Reading

As stated earlier, this book emphasizes the fundamental aspects of aerospace guidance and navigation, as opposed to their technological and hardware aspects. In that respect, this book is significantly different from the majority of books on aerospace navigation and guidance. The publication that is closest in its emphasis to this book is [38], from which much of Chapters 5 and 6 originates. Several books are available on the subject of aerospace navigation and guidance, including [49], [10], [66], [31], [23].

In Section 1.3, we stated that systems and control theory could be applied to the motion of aerospace vehicles in four areas: path planning, navigation, guidance, and attitude control. An early statement of the idea of partitioning the control system of a vehicle into layers is in [14]. Partitioning the architecture of a vehicle control system into three layers (strategic layer, tactical layer, and execution layer) is suggested in [34].

1.8 Homework Problems

For this chapter, the homework problems consist of finding documented mishaps because of failure of navigation or guidance. In a formal course, we suggest that the instructor specify what the acceptable sources are (e.g., general public news, technical magazines, military magazines, NASA news releases, FAA accident reports). The instructor may also specify how recent these mishaps must be.

PROBLEM 1.1 *Find a documented instance of failure of navigation in a transoceanic jetliner flight. Be specific as to what went wrong and how this affected the mission.*

PROBLEM 1.2 *Find a documented instance of failure of guidance in a transoceanic jetliner flight. Be specific as to what went wrong and how this affected the mission.*

PROBLEM 1.3 *Find a documented instance of failure of navigation in a remotely operated UAV. Be specific as to what went wrong and how this affected the mission.*

PROBLEM 1.4 *Find a documented instance of failure of guidance in a remotely operated UAV. Be specific as to what went wrong and how this affected the mission.*

PROBLEM 1.5 *Find a documented instance of failure of navigation or guidance in a homing missile. Be specific as to what went wrong and how this affected the mission.*

PROBLEM 1.6 *Find a documented instance of failure of navigation or guidance in an Earth-orbital mission. Be specific as to what went wrong and how this affected the mission.*

PROBLEM 1.7 *Find a documented instance of failure of navigation or guidance in an interplanetary mission. Be specific as to what went wrong and how this affected the mission.*

2 Deterministic Systems Theory

This chapter presents the fundamentals of deterministic systems theory that we use in the analysis and synthesis of aerospace navigation and guidance systems. The emphasis is on linear, time varying dynamic systems. This emphasis is justified in Section 2.1, where it is shown that when the trajectory of a nonlinear system undergoes a small perturbation, the time history of that perturbation can be approximated using a linear system. Thus, this linear approach is appropriate when the navigation and guidance errors are small. This, of course, implicitly assumes that we use successful navigation and guidance laws to ensure small errors.

In addition to system linearization, Section 2.1 surveys properties of linear dynamic systems: the state transition matrix and its features, various notions of stability and their criteria, the variation of constants formula, and the impulse response. Section 2.2 studies observability, which is a central concept in navigation. Section 2.3 considers the particular case of linear, time invariant dynamic systems. Section 2.4 presents the method of adjoints, which is invaluable in terminal guidance because it yields the miss distance due to a perturbation of the trajectory as a function of the time at which the perturbation is applied. Finally, Section 2.5 discusses controllability and the duality between controllability and observability. Sections 2.6, 2.7, and 2.8 present a summary of the key results in the chapter, bibliographic notes for further reading, and homework problems, respectively.

2.1 Linear Dynamic Systems

2.1.1 System Linearization

Let us start by considering the following general nonlinear dynamic system, representing the equations of motion of a vehicle in some coordinate system:

$$\begin{aligned}\dot{x}(t) &= f(x(t), u(t), t)\\ y(t) &= g(x(t), t),\end{aligned} \tag{2.1}$$

where $x(t) \in \mathbb{R}^n$ is called the state vector, $u(t) \in \mathbb{R}^m$ is called the input vector, and $y(t) \in \mathbb{R}^p$ is called the output vector. When (2.1) represents the equations of motion

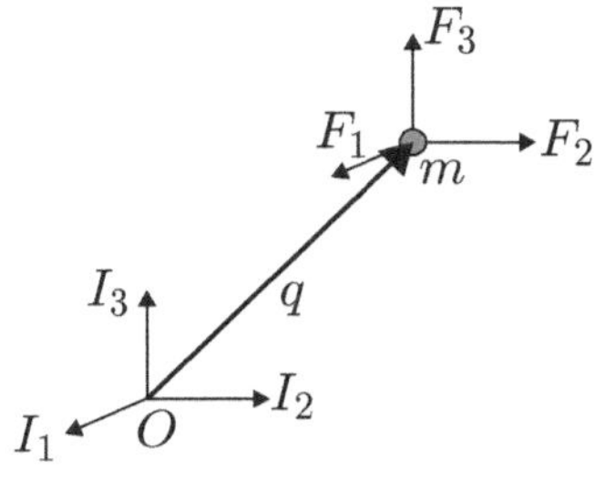

Figure 2.1. Particle in Cartesian coordinates.

of a vehicle, typically $x(t)$ contains positions and velocities; $u(t)$ contains control actions, such as thrusting forces, torques, or accelerations, as read by accelerometers; and $y(t)$ contains the readings of position and velocity sensors.

EXAMPLE 2.1 *Consider the motion of a particle in three-dimensional Cartesian coordinates (see Figure 2.1). Let* (q_1, q_2, q_3) *be the Cartesian coordinate vector of the particle of mass* m *in an inertial frame* $\{O, I_1, I_2, I_3\}$, *let* (v_1, v_2, v_3) *be the Cartesian components of velocity, and let* (F_1, F_2, F_3) *be the Cartesian components of applied force. Newton's equations of motion have the form*

$$\ddot{q}_i = F_i/m, \quad i = 1, 2, 3 \tag{2.2}$$

and can be developed into

$$\begin{aligned}
\dot{q}_1 &= v_1 \\
\dot{v}_1 &= F_1/m \\
\dot{q}_2 &= v_2 \\
\dot{v}_2 &= F_2/m \\
\dot{q}_3 &= v_3 \\
\dot{v}_3 &= F_3/m
\end{aligned} \tag{2.3}$$

which have the form (2.1), *with* $x = (q_1, v_1, q_2, v_2, q_3, v_3)^T$ *and* $u = (F_1/m, F_2/m, F_3/m)^T$.

EXAMPLE 2.2 *Consider the motion of a particle in a plane with polar coordinates* (r, θ) *(see Figure 2.2). Let* F_r *and* F_θ *be the polar components of the applied force. The*

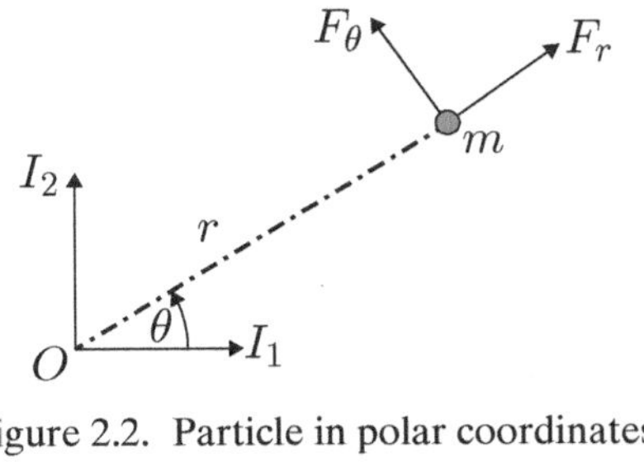

Figure 2.2. Particle in polar coordinates.

equations of motion are derived in Chapter 6 as

$$\ddot{r} - r\dot{\theta}^2 = F_r/m$$
$$r\ddot{\theta} + 2\dot{r}\dot{\theta} = F_\theta/m. \tag{2.4}$$

Define

$$v_r = \dot{r}$$
$$v_\theta = r\dot{\theta}, \tag{2.5}$$

which represent the radial and tangential components of velocity, respectively. The equations of motion take the form

$$\dot{r} = v_r$$
$$\dot{\theta} = v_\theta/r$$
$$\dot{v}_r = v_\theta^2/r + F_r/m$$
$$\dot{v}_\theta = -v_r v_\theta/r + F_\theta/m \tag{2.6}$$

which also have the form (2.1), *with* $x = (r, \theta, v_r, v_\theta)^T$ *and* $u = (F_r/m, F_\theta/m)^T$.

Now assume that for system (2.1), we have a nominal trajectory ($x^0(t)$, $u^0(t)$, $y^0(t)$). Let

$$\delta x(t) = x(t) - x^0(t)$$
$$\delta u(t) = u(t) - u^0(t),$$
$$\delta y(t) = y(t) - y^0(t) \tag{2.7}$$

represent the excursions of $x(t)$, $u(t)$, and $y(t)$ away from their nominal values, respectively. Equation (2.1) implies that

$$\dot{x}^0(t) + \delta\dot{x}(t) = f(x^0(t) + \delta x(t), u^0(t) + \delta u(t), t)$$
$$y^0(t) + \delta y(t) = g(x^0(t) + \delta x(t), t). \tag{2.8}$$

Develop each entry of $f(.,.,.)$ in a Taylor series expansion to first order (see Section A.3) of the form

$$f_i(x^0 + \delta x, u^0 + \delta u, t) = f_i(x^0, u^0, t) + \sum_{j=1}^{n} \frac{\partial f_i}{\partial x_j}\delta x_j + \sum_{k=1}^{m} \frac{\partial f_i}{\partial u_k}\delta u_k + \text{H.O.T.},$$
$$i = 1, \ldots, n, \tag{2.9}$$

where H.O.T. stands for "higher-order terms" and where the partial derivatives are computed along the nominal trajectory. These n scalar equations can be written in compact form as a single matrix equation:

$$f(x^0 + \delta x, u^0 + \delta u, t) = f(x^0, u^0, t) + \left(\frac{\partial f}{\partial x}\right)^T_{(x^0,u^0,t)} \delta x + \left(\frac{\partial f}{\partial u}\right)^T_{(x^0,u^0,t)} \delta u + \text{H.O.T.}, \tag{2.10}$$

where the $n \times n$ and $m \times n$ matrices $(\partial f/\partial x)$ and $(\partial f/\partial u)$ are, respectively, defined by

$$
\begin{aligned}
\left(\frac{\partial f}{\partial x}\right)_{ij} &= \frac{\partial f_j}{\partial x_i}, \ 1 \le i \le n, 1 \le j \le n \\
\left(\frac{\partial f}{\partial u}\right)_{ki} &= \frac{\partial f_i}{\partial u_k}, \ 1 \le i \le n, 1 \le k \le m.
\end{aligned} \tag{2.11}
$$

Assume that δx, δu, δy are small and ignore the higher-order terms to obtain

$$
\begin{aligned}
\dot{x}^0(t) + \delta\dot{x}(t) &= f(x^0, u^0, t) + \left(\frac{\partial f}{\partial x}\right)^T_{(x^0,u^0,t)} \delta x(t) + \left(\frac{\partial f}{\partial u}\right)^T_{(x^0,u^0,t)} \delta u(t) \\
y^0(t) + \delta y(t) &= g\left(x^0(t), t\right) + \left(\frac{\partial g}{\partial x}\right)^T_{(x^0,t)} \delta x(t).
\end{aligned} \tag{2.12}
$$

Now, because $x^0(t)$, $u^0(t)$, and $y^0(t)$ are a nominal solution, they satisfy

$$
\begin{aligned}
\dot{x}^0(t) &= f(x^0(t), u^0(t), t) \\
y^0(t) &= g(x^0(t), t).
\end{aligned} \tag{2.13}
$$

Let us therefore define

$$
\begin{aligned}
A(t) &= \left(\frac{\partial f}{\partial x}\right)^T_{(x^0,u^0,t)}, \\
B(t) &= \left(\frac{\partial f}{\partial u}\right)^T_{(x^0,u^0,t)}, \\
C(t) &= \left(\frac{\partial g}{\partial x}\right)^T_{(x^0,t)}.
\end{aligned} \tag{2.14}
$$

In other words, $A(t)$, $B(t)$, and $C(t)$ represent the Jacobian matrices of f and g computed along the nominal trajectory. Subtract (2.13) from (2.12), and use (2.14) to obtain a linear dynamic system in standard form:

$$
\begin{aligned}
\delta\dot{x}(t) &= A(t)\delta x(t) + B(t)\delta u(t) \\
\delta y(t) &= C(t)\delta x(t).
\end{aligned} \tag{2.15}
$$

In (2.15), $A(t)$, $B(t)$, and $C(t)$ are referred to as the **state matrix**, **input matrix**, and **output matrix**, respectively.

EXAMPLE 2.3 (Double Integrator) *Here we introduce an important class of single-input–single-output systems where the second derivative of the output is the input, or conversely, the output is the second integral of the input, hence the name* ***double integrator****. We use this class of systems throughout the book to illustrate technical concepts. Double integrators are ubiquitous in navigation and guidance because they model basic physical laws. Instances include the following:*

1. *Consider the one-dimensional version of the system in Example 2.1: motion of a particle along a line, which can be used to model the motion of a spaceship along a line, with application to docking (see Figure 2.3). Using the force per unit mass*

Figure 2.3. Docking of the SpaceX Dragon onto the International Space Station. Image courtesy of NASA.

as an input and the translational displacement as an output, Newton's law implies that this system is a double integrator.

2. *Consider the rotational motion of a spaceship around a principal axis of inertia under the influence of a torque actuator (see Figure 2.4). Using the torque per unit moment of inertia as input and the rotational displacement as output, the law of angular momentum implies that this system is also a double integrator.*
3. *Consider the slewing of a star tracker around an axis (see Figure 2.5). Here also, using the torque per unit moment of inertia as input and the rotational displacement as output, the law of angular momentum implies that this system is a double integrator.*
4. *Consider small, slow rolling motions of a conventional aircraft (see Figure 2.6). Using aileron deflection as input and roll angle as output, the law of angular momentum implies that this system is, with appropriate scaling, a double integrator.*

There are many other examples of double integrators in aerospace navigation and guidance. They all have in common the model shown in Figure 2.7 and are

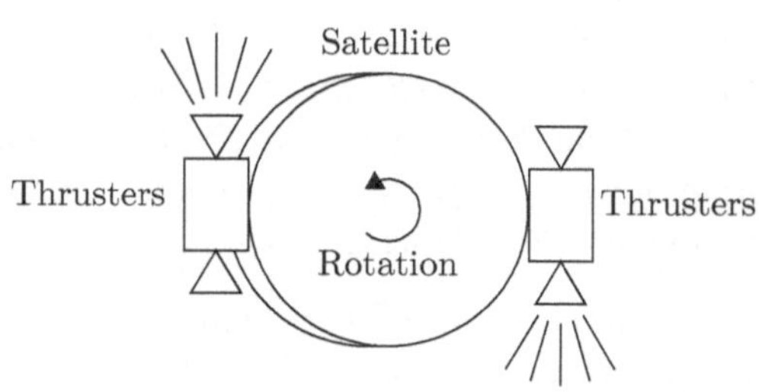

Figure 2.4. Satellite attitude control using thrusters.

Figure 2.5. Star tracker of Space Shuttle Columbia. Image courtesy of Smithsonian Institution, National Air and Space Museum.

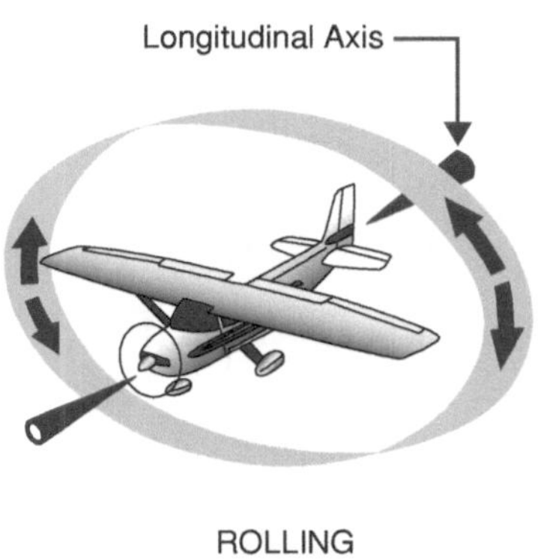

Figure 2.6. Rolling motion of a conventional aircraft. Image adapted from *Pilot's Handbook of Aeronautical Knowledge*, FAA-H-8083-25.

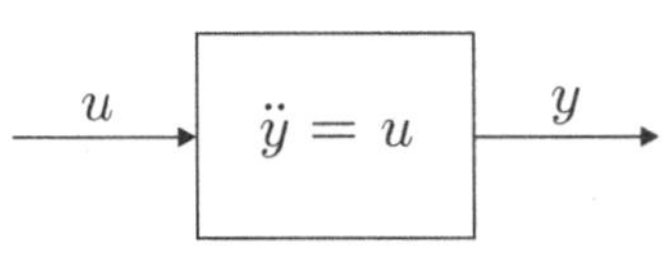

Figure 2.7. Double integrator.

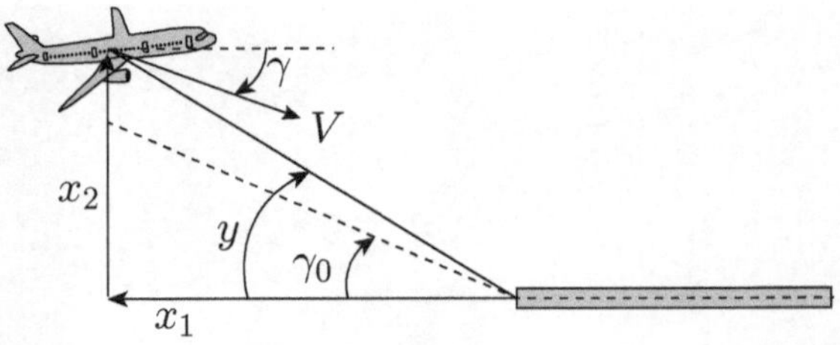

Figure 2.8. Guided approach to landing of a conventional aircraft.

described by the equation

$$\ddot{y}(t) = u(t). \tag{2.16}$$

Now, defining

$$\begin{aligned} x_1(t) &= y(t) \\ x_2(t) &= \dot{y}(t), \end{aligned} \tag{2.17}$$

we obtain linear dynamic equations in standard form:

$$\begin{aligned} \begin{bmatrix} \dot{x}_1 \\ \dot{x}_2 \end{bmatrix} &= \begin{bmatrix} 0 & 1 \\ 0 & 0 \end{bmatrix} \begin{bmatrix} x_1 \\ x_2 \end{bmatrix} + \begin{bmatrix} 0 \\ 1 \end{bmatrix} u \\ y &= \begin{bmatrix} 1 & 0 \end{bmatrix} \begin{bmatrix} x_1 \\ x_2 \end{bmatrix}, \end{aligned} \tag{2.18}$$

where the output y represents the reading of an appropriate sensor. Note that this example did not require linearization because the equation of motion (2.16) *is already linear.*

EXAMPLE 2.4 (Guided Approach to Landing) *Consider a longitudinal model for the guided approach to landing of a conventional aircraft (see Figure 2.8). This model is simplified in the sense that it assumes constant pitch angle and constant speed. The equations of motion are*

$$\begin{aligned} \dot{x}_1 &= V \cos \gamma, \\ \dot{x}_2 &= V \sin \gamma, \\ y &= \arctan\left(\frac{x_2}{x_1}\right), \end{aligned} \tag{2.19}$$

where x_1 is the horizontal distance traveled, x_2 is the altitude, V is the speed, γ is the flight path angle and control input, and y is the elevation of the aircraft, as seen from the runway. The nominal solution is given by

$$\begin{aligned} \dot{x}_1^0 &= V \cos \gamma^0, \\ \dot{x}_2^0 &= V \sin \gamma^0, \\ x_1^0(t_f) &= 0, \\ x_2^0(t_f) &= 0, \end{aligned} \tag{2.20}$$

where γ^0 is the ideal flight path angle for approach, t_f is the time of touchdown, and the runway starts at $(x_1, x_2) = (0, 0)$. The nominal solution is, therefore,

$$\begin{aligned} x_1^0(t) &= (t - t_f)V \cos\gamma^0, \\ x_2^0(t) &= (t - t_f)V \sin\gamma^0, \\ y^0(t) &= \gamma^0. \end{aligned} \tag{2.21}$$

Linearization about the nominal yields

$$\begin{aligned} \begin{bmatrix} \delta\dot{x}_1 \\ \delta\dot{x}_2 \end{bmatrix} &= \begin{bmatrix} 0 & 0 \\ 0 & 0 \end{bmatrix} \begin{bmatrix} \delta x_1 \\ \delta x_2 \end{bmatrix} + \begin{bmatrix} -V\sin\gamma^0 \\ V\cos\gamma^0 \end{bmatrix} \delta\gamma \\ \delta y &= \begin{bmatrix} -\frac{\sin\gamma^0}{(t-t_f)V} & \frac{\cos\gamma^0}{(t-t_f)V} \end{bmatrix} \begin{bmatrix} \delta x_1 \\ \delta x_2 \end{bmatrix}. \end{aligned} \tag{2.22}$$

Because it is relatively easy for a pilot to see changes in elevation, we use the simple guidance law

$$\delta\dot{\gamma} = \lambda\delta\dot{y}, \tag{2.23}$$

where λ is a constant. The last equation can be integrated as

$$\delta\gamma = \lambda\delta y + \delta u, \tag{2.24}$$

where δu is a constant of integration. The closed loop system is then

$$\begin{aligned} \begin{bmatrix} \delta\dot{x}_1 \\ \delta\dot{x}_2 \end{bmatrix} &= \begin{bmatrix} \lambda\frac{\sin^2\gamma^0}{t-t_f} & -\lambda\frac{\sin\gamma^0\cos\gamma^0}{t-t_f} \\ -\lambda\frac{\sin^2\gamma^0}{t-t_f} & \lambda\frac{\cos^2\gamma^0}{t-t_f} \end{bmatrix} \begin{bmatrix} \delta x_1 \\ \delta x_2 \end{bmatrix} + \begin{bmatrix} -V\sin\gamma^0 \\ V\cos\gamma^0 \end{bmatrix} \delta u \\ \delta y &= \begin{bmatrix} -\frac{\sin\gamma^0}{(t-t_f)V} & \frac{\cos\gamma^0}{(t-t_f)V} \end{bmatrix} \begin{bmatrix} \delta x_1 \\ \delta x_2 \end{bmatrix}, \end{aligned} \tag{2.25}$$

which is time varying.

2.1.2 Properties of Linear Dynamic Systems

Here we survey properties of systems of the form (2.15) that are important for later use. First, consider the homogeneous n-dimensional vector equation with boundary condition

$$\begin{aligned} \dot{x}(t) &= A(t)x(t) \\ x(t_0) &= x_0. \end{aligned} \tag{2.26}$$

PROPOSITION 2.1 *For every initial condition $x_0 \in \mathbb{R}^n$ and initial time $t_0 \in \mathbb{R}$,* (2.26) *possesses a unique solution, defined for all time t, and satisfying $x(t_0) = x_0$.*

PROPOSITION 2.2 *All the solutions of $\dot{x}(t) = A(t)x(t)$ form a real n-dimensional vector space.*

The practical implication of Propositions 2.1 and 2.2 is that it is possible to find n linearly independent solutions of $\dot{x}(t) = A(t)x(t)$ and express every solution of (2.26) (line 1) as a linear combination of these n solutions. This is the basic idea behind the definition of the state transition matrix.

PROPOSITION 2.3 *The equation $\dot{x}(t) = A(t)x(t)$ possesses a unique* ***state transition matrix***, *$\Phi(t, \tau) \in \mathbb{R}^{n\times n}$, defined for all real t and τ, and satisfying*

$$\frac{\partial \Phi(t,\tau)}{\partial t} = A(t)\Phi(t,\tau)$$

$$\Phi(\tau,\tau) = I_n = \begin{bmatrix} 1 & 0 & \dots & 0 \\ 0 & 1 & \dots & 0 \\ \dots & & & \\ 0 & 0 & \dots & 1 \end{bmatrix}. \tag{2.27}$$

PROPOSITION 2.4 *The state transition matrix $\Phi(t, \tau)$ of $\dot{x}(t) = A(t)x(t)$ is nonsingular for all real t and τ.*

PROPOSITION 2.5 *For all real t_1, t_2, and t_3, the state transition matrix $\Phi(t, \tau)$ of $\dot{x}(t) = A(t)x(t)$ satisfies*

$$\Phi(t_1,t_3) = \Phi(t_1,t_2)\Phi(t_2,t_3)$$
$$\Phi^{-1}(t_1,t_2) = \Phi(t_2,t_1). \tag{2.28}$$

PROPOSITION 2.6 *Let $x(t; t_0, x_0)$ denote the unique solution of (2.26) satisfying $x(t_0) = x_0$. Then, this solution is given in terms of the state transition matrix by*

$$x(t; t_0, x_0) = \Phi(t, t_0)x_0. \tag{2.29}$$

We now introduce the notions of stability and asymptotic stability, which play an important role in some of our future applications.

DEFINITION 2.1 *The system (2.26) is called* ***stable*** *if, for all initial times t_0 and all initial conditions x_0, the solution $x(t; t_0, x_0)$ satisfying $x(t_0) = x_0$ is bounded for all $t \geq t_0$.*

DEFINITION 2.2 *The system (2.26) is called* ***asymptotically stable*** *if it is stable and every solution decays to zero, that is, for all t_0, x_0, the solution $x(t; t_0, x_0)$ satisfying $x(t_0) = x_0$ is such that*

$$\lim_{t\to\infty} x(t; t_0, x_0) = 0. \tag{2.30}$$

Stability and asymptotic stability of a general linear time varying dynamic system such as (2.26) can be checked using the following criteria.

PROPOSITION 2.7 *System (2.26) is stable if and only if its state transition matrix is uniformly bounded, by which we mean that there exists a constant $k > 0$ such that*

$$\forall t_0 \in \mathbb{R}, \forall t \geq t_0, |\Phi_{ij}(t, t_0)| \leq k < \infty, \; 1 \leq i \leq n, 1 \leq j \leq n. \tag{2.31}$$

It is important to realize that in Proposition 2.7, the constant k must be independent of t_0 and t. For given t_0 and t, it is generally easy to find bounds for the magnitudes of the entries of the state transition matrix. However, in Proposition 2.7, these bounds must be uniform, that is, they must hold **for all** t_0 and t.

PROPOSITION 2.8 *System (2.26) is asymptotically stable if and only if its state transition matrix is uniformly exponentially bounded, by which we mean that there exists a pair of constants $k_1 > 0$ and $k_2 > 0$ such that*

$$\forall t_0 \in \mathbb{R}, \forall t \geq t_0, |\Phi_{ij}(t, t_0)| \leq k_1 e^{-k_2(t-t_0)} < \infty, \; 1 \leq i \leq n, 1 \leq j \leq n. \quad (2.32)$$

It is also important to realize that in Proposition 2.8, the two constants k_1 and k_2 must be independent of the arguments of the state transition matrix.

We are now ready to consider the full linear time varying dynamic system with boundary condition

$$\begin{aligned} \dot{x}(t) &= A(t)x(t) + B(t)u(t), \\ y(t) &= C(t)x(t), \\ x(t_0) &= x_0, \end{aligned} \quad (2.33)$$

which differs from (2.15) only by removal of the δ symbols for notational convenience.

PROPOSITION 2.9 *For system (2.33), the solution $x(t; t_0, x_0)$, $y(t; t_0, x_0, u)$ that satisfies $x(t_0) = x_0$ and has been driven by u is*

$$x(t) = x(t; t_0, x_0, u) = \Phi(t, t_0)x_0 + \int_{t_0}^{t} \Phi(t, \tau)B(\tau)u(\tau)\,\mathrm{d}\tau$$

$$y(t) = y(t; t_0, x_0, u) = C(t)\Phi(t, t_0)x_0 + \int_{t_0}^{t} C(t)\Phi(t, \tau)B(\tau)u(\tau)\,\mathrm{d}\tau. \quad (2.34)$$

The equations (2.34) are commonly called the **variation of constants formulas**. They contain two terms. The first term is the contribution of the initial condition x_0 and is called the **free response**. The second term is the contribution of the input u and is called the **forced response**.

In (2.34), we define the matrix

$$G(t, \tau) = C(t)\Phi(t, \tau)B(\tau), \; t \geq \tau \quad (2.35)$$

as the **impulse response** or weighting pattern of system (2.33). In practice, $G_{ij}(t, \tau)$ represents the response at time t of the ith output due to an impulse applied at time τ in the jth input.

The presence of an input vector u in system (2.33) allows the introduction of another type of stability, which requires the definition of signals and their norms, as follows.

DEFINITION 2.3 *Two vector functions of time are said to be **essentially equal**, or **equal almost everywhere**, if the set of times at which they differ has measure zero.*

DEFINITION 2.4 *Let $u : \mathbb{R} \to \mathbb{R}^n$ be a vector function of time. The* ***signal*** *u is the equivalence class of all the vector functions of time that are essentially equal to the function u.*

DEFINITION 2.5 *The* ***norm*** *of a signal u, denoted $||u||$, is the quantity*

$$||u|| = \sqrt{\int_{-\infty}^{\infty} u^T(t)u(t)\,\mathrm{d}t}. \tag{2.36}$$

The signal u is called ***bounded*** *if its norm* (2.36) *is finite.*

REMARK 2.1 *The notion of norm in Definition 2.5 is a generalization to signals of the familiar notion of Euclidian norm in $\mathbb{R}^n$, that is, square root of the sum of the squares of the coordinates. The norm* (2.36) *satisfies the standard properties of norms; that is, for all real number α and signals u, v,*

$$\begin{aligned} ||u|| = 0 \Leftrightarrow u = 0,& \\ ||\alpha u|| &= |\alpha|||u||, \\ ||u + v|| &\leq ||u|| + ||v||. \end{aligned} \tag{2.37}$$

We can now introduce a new definition of system stability, as follows.

DEFINITION 2.6 *System* (2.33) *is called* ***bounded-input, bounded-output (BIBO) stable*** *if, whenever $x_0 = 0$, the forced output response $y(t)$ to every bounded input $u(t)$ is itself bounded.*

DEFINITION 2.7 *Let system* (2.33) *with impulse response* (2.35) *be BIBO stable. Then its* ***induced norm****, denoted $||G||$, is the quantity*

$$||G|| = \sup_{||u||<\infty, x_0=0} \frac{||y||}{||u||}. \tag{2.38}$$

Note, from the definition and adherence property of the supremum given in Appendix A.1, that if system (2.33), (2.35) is BIBO stable, then the norm of its forced response satisfies the tight bound:

$$||y|| \leq ||G||||u||. \tag{2.39}$$

BIBO stability of a system can therefore be ascertained by computing its induced norm (2.38). However, the underlying optimization problem is generally difficult. Fortunately, BIBO stability can also be checked using the following criterion.

PROPOSITION 2.10 *System* (2.33) *is BIBO stable if and only if its impulse response matrix is uniformly absolutely integrable, by which we mean that there exists a constant $k > 0$ such that for all real t, we have*

$$\int_{-\infty}^{t} |G_{ij}(t, \tau)|\,\mathrm{d}\tau \leq k < \infty, 1 \leq i \leq p, 1 \leq j \leq m. \tag{2.40}$$

It is again important to realize that in Proposition 2.10, the same constant must be a bound to the integral in (2.40) *for all t*. It is in this sense that k is a uniform bound.

EXAMPLE 2.5 *Consider again the double integrator, introduced and motivated in Example 2.3, namely,*

$$\begin{bmatrix} \dot{x}_1 \\ \dot{x}_2 \end{bmatrix} = \begin{bmatrix} 0 & 1 \\ 0 & 0 \end{bmatrix} \begin{bmatrix} x_1 \\ x_2 \end{bmatrix} + \begin{bmatrix} 0 \\ 1 \end{bmatrix} u$$

$$y = \begin{bmatrix} 1 & 0 \end{bmatrix} \begin{bmatrix} x_1 \\ x_2 \end{bmatrix}. \tag{2.41}$$

For this system, the state transition matrix is the 2×2 *matrix satisfying*

$$\frac{\partial \Phi(t,\tau)}{\partial t} = \begin{bmatrix} 0 & 1 \\ 0 & 0 \end{bmatrix} \Phi(t,\tau)$$

$$\Phi(\tau,\tau) = \begin{bmatrix} 1 & 0 \\ 0 & 1 \end{bmatrix}. \tag{2.42}$$

A little algebra yields the state transition matrix:

$$\Phi(t,\tau) = \begin{bmatrix} 1 & t-\tau \\ 0 & 1 \end{bmatrix}. \tag{2.43}$$

The free response, which is the solution of the homogeneous equation, is therefore

$$x_{\text{free}}(t) = \Phi(t,t_0)x_0 = \begin{bmatrix} 1 & t-t_0 \\ 0 & 1 \end{bmatrix} \begin{bmatrix} x_{10} \\ x_{20} \end{bmatrix}, \tag{2.44}$$

or equivalently,

$$x_{1,\text{free}}(t) = x_{10} + (t-t_0)x_{20}$$

$$x_{2,\text{free}}(t) = x_{20}, \tag{2.45}$$

where x_{10} *and* x_{20} *represent the initial position and velocity, respectively. Equation* (2.45) *is therefore consistent with our expectation that, for a free particle, the velocity be a constant equal to the initial velocity, and the displacement increase linearly, proportional to the velocity and the elapsed time.*

The state transition matrix (2.43) *is such that*

$$\lim_{t\to\infty} \Phi_{12}(t,\tau) = \infty, \tag{2.46}$$

which violates the requirement of Proposition 2.7, implying that the homogeneous equation is unstable.

The variation of constants formula (2.34) *(line 2) yields the following solution for the forced problem:*

$$\begin{aligned} y(t) &= \begin{bmatrix} 1 & 0 \end{bmatrix} \begin{bmatrix} 1 & t-t_0 \\ 0 & 1 \end{bmatrix} \begin{bmatrix} x_{10} \\ x_{20} \end{bmatrix} + \int_{t_0}^{t} \begin{bmatrix} 1 & 0 \end{bmatrix} \begin{bmatrix} 1 & t-\tau \\ 0 & 1 \end{bmatrix} \begin{bmatrix} 0 \\ 1 \end{bmatrix} u(\tau)\, d\tau \\ &= x_{10} + \int_{t_0}^{t} (t-\tau)u(\tau)\, d\tau + x_{20}(t-t_0). \end{aligned} \tag{2.47}$$

Note that differentiating equation (2.47) *twice with respect to t using Leibniz's rule yields successively*

$$\dot{y}(t) = \int_{t_0}^{t} u(\tau)\,\mathrm{d}\tau + x_{20}$$
$$\ddot{y}(t) = u(t), \tag{2.48}$$

which is exactly what we expect.

Also, the impulse response is

$$G(t, \tau) = t - \tau \tag{2.49}$$

and is not uniformly absolutely integrable. Therefore, as per Proposition 2.10, the system (2.41) *is not BIBO stable.*

REMINDER 2.1 *Leibniz's rule for differentiation under the integral sign, named after Gottfried Wilhem von Leibniz, tells us that if we consider a function $f(x, \theta)$ such that $\frac{\partial}{\partial \theta} f(x, \theta)$ exists and is continuous, then*

$$\frac{\mathrm{d}}{\mathrm{d}\theta}\left(\int_{a(\theta)}^{b(\theta)} f(x, \theta)\,\mathrm{d}x\right) = \left(\int_{a(\theta)}^{b(\theta)} \frac{\partial}{\partial \theta} f(x, \theta)\,\mathrm{d}x\right) + f(b(\theta), \theta)b'(\theta) - f(a(\theta), \theta)a'(\theta), \tag{2.50}$$

where ′ denotes derivative.

REMARK 2.2 *Although stability is generally desirable in navigation and guidance, several practical guidance systems have double integrator dynamics and are therefore, as per Example 2.5, unstable. These systems, which include bullets and other ballistic projectiles, still perform acceptably because the duration of the engagement is short enough. As a consequence, the instability does not have enough time to degrade accuracy beyond tolerance.*

2.2 Observability

Observability generally refers to our ability to compute the state vector of a system knowing the time history of the input and output vectors. Consider our standard linear time varying dynamic system:

$$\dot{x}(t) = A(t)x(t) + B(t)u(t)$$
$$y(t) = C(t)x(t). \tag{2.51}$$

In this section, we are concerned with the following type of question: Assume we know $u(t)$; can the knowledge of $y(t)$ tell us something about $x(t)$?

The importance of this question should be clear in the context of navigation. Typically, the vector $u(t)$ contains the difference between the nominal and actual control actions applied to the vehicle and is known; the vector $y(t)$ contains the difference between the nominal and actual readings of the position and velocity sensors and is also known. The vector $x(t)$ contains the difference between the nominal and actual position and velocity variables of the vehicle and has to be estimated by the navigator based on the available information.

Recall the variation of constants formula:

$$y(t) = y(t; t_0, x_0, u) = C(t)\Phi(t, t_0)x_0 + \int_{t_0}^{t} C(t)\Phi(t, \tau)B(\tau)u(\tau)\,d\tau, \qquad (2.52)$$

where $\Phi(t, \tau)$ is the state transition matrix.

DEFINITION 2.8 *For system* (2.51), *a state x_0 is* ***unobservable at time*** *t_0 if, for all times $t_1 \geq t_0$, we have $y(t_1; t_0, x_0, 0) = 0$, that is, the free response ($u \equiv 0$) due to initial condition x_0 at t_0 is zero. A state x_0 is* ***observable*** *if it is not unobservable. System* (2.51) *is* ***completely observable*** *if the only unobservable state is the origin 0.*

In Definition 2.8, it should be clear that the origin of the state space is always unobservable in a linear time varying dynamic system. Also, it is easily seen that the set of unobservable states is a vector space. Observability at a given time t_0 can be checked using the following result.

PROPOSITION 2.11 *For system* (2.51), *the state x_0 is unobservable at t_0 if and only if, for all $t \geq t_0$, we have*

$$C(t)\Phi(t, t_0)x_0 \equiv 0. \qquad (2.53)$$

PROOF. This is a direct consequence of the variation of constants formula (2.52) and Definition 2.8.

As a corollary to Proposition 2.11, we have the following.

PROPOSITION 2.12 *For system* (2.51), *the state x_0 is unobservable at t_0 if and only if, for all $t \geq t_0$, we have $M(t_0, t_1)x_0 = 0$, where*

$$M(t_0, t_1) = \int_{t_0}^{t_1} \Phi^T(t, t_0)C^T(t)C(t)\Phi(t, t_0)\,dt. \qquad (2.54)$$

In particular, system (2.51) *is completely observable at t_0 if and only if there exists $t_1 > t_0$ such that $M(t_0, t_1)$ is nonsingular.*

REMARK 2.3 *The symmetric matrix $M(t_0, t_1)$ defined in Proposition 2.12 is always at least positive semidefinite, by which we mean that*

$$\forall x_0 \in \mathbb{R}^n, \;\; x_0^T M(t_0, t_1)x_0 \geq 0.$$

PROOF. Assume that the state x_0 is unobservable at t_0. Equation (2.53) implies that for all $t \geq t_0$, we have

$$\Phi^T(t, t_0)C^T(t)C(t)\Phi(t, t_0)x_0 = 0. \qquad (2.55)$$

Integrating between t_0 and an arbitrary $t_1 \geq t_0$ yields

$$M(t_0, t_1)x_0 = 0. \qquad (2.56)$$

Conversely, assume that the state x_0 is such that for all $t_1 \geq t_0$, we have $M(t_0, t_1)x_0 = 0$. Premultiplying by x_0^T and using the definition of M from

Proposition 2.12, we obtain

$$\int_{t_0}^{t_1} ||C(t)\Phi(t, t_0)x_0||^2 \, \mathrm{d}t = 0, \tag{2.57}$$

which implies (2.53). Hence x_0 is unobservable, and the proof is complete.

The matrix defined in Proposition 2.12 is called the **observability Gramian**, and the practical implication of Proposition 2.12 is that a criterion for observability at time t_0 is that the observability Gramian $M(t, t_0)$ be nonsingular (and consequently positive definite) for some $t \geq t_0$.

We now elucidate the link between observability and our ability to determine the initial condition x_0 and the current value of the state $x(t)$ from knowledge of the inputs and outputs.

PROPOSITION 2.13 *Consider the linear dynamic system* (2.51) *over the interval* $t \in [t_0, t_1]$. *The initial condition* x_0 *is determined from the time history of the input* $u(t)$ *and output* $y(t)$ *to within an additive constant vector belonging to the null space of the matrix* $M(t_0, t_1)$ *of Proposition 2.12. The initial condition* x_0 *is uniquely determined from the time history of the input and output if* $M(t_0, t_1)$ *is nonsingular. Moreover, if two initial conditions* x_0 *and* $\bar{x}_0$ *are such that* $x_0 - \bar{x}_0$ *is unobservable at* t_0, *then, for all* $t \geq t_0$, $y(t; t_0, x_0, 0) = y(t; t_0, \bar{x}_0, 0)$.

PROOF. Consider the variation of constants formula (2.52). Premultiply the left-hand side and right-hand side by $\Phi^T(t, t_0)C^T(t)$ and integrate with respect to t on the interval $[t_0, t_1]$ to obtain

$$M(t_0, t_1)x_0 = \int_{t_0}^{t_1} \Phi^T(t, t_0)C^T(t)\hat{y}(t) \, \mathrm{d}t, \tag{2.58}$$

where

$$\hat{y}(t) = y(t) - \int_{t_0}^{t} C(t)\Phi(t, \tau)B(\tau)u(\tau) \, \mathrm{d}\tau. \tag{2.59}$$

Because in (2.58) the right-hand side is known, this is a system of simultaneous linear equations for the vector unknown x_0, with as many equations as unknowns. The solution is determined to within an additive vector in the null space of the matrix $M(t_0, t_1)$. If this matrix is nonsingular, the solution of (2.58) exists, is unique, and is given by

$$x_0 = M^{-1}(t_0, t_1) \int_{t_0}^{t_1} \Phi^T(t, t_0)C^T(t)\hat{y}(t) \, \mathrm{d}t. \tag{2.60}$$

Finally, if $x_0 - \bar{x}_0$ is unobservable at t_0, then from Proposition 2.11, for all $t \geq t_0$, we have

$$C(t)\Phi(t, t_0)(x_0 - \bar{x}_0) = 0, \tag{2.61}$$

which implies

$$C(t)\Phi(t,t_0) = C(t)\Phi(t,t_0)\bar{x}_0 \tag{2.62}$$

and proves that $y(t;t_0,x_0,0) = y(t;t_0,\bar{x}_0,0)$.

Proposition 2.13 implies that, to determine without ambiguity the initial condition of a linear dynamic system, this system must be observable. Moreover, when that is the case, Proposition 2.13 suggests a method for computing the current state of the linear system. Indeed, using (2.60) for the initial condition in the variation of constants formula (2.34) (line 1), we obtain

$$x(t) = \Phi(t,t_0)M^{-1}(t_0,t_1)\int_{t_0}^{t_1}\Phi^T(t,t_0)C^T(t)\hat{y}(t)\,\mathrm{d}t + \int_{t_0}^{t}\Phi(t,\tau)B(\tau)u(\tau)\,\mathrm{d}\tau, \tag{2.63}$$

where

$$\hat{y}(t) = y(t) - \int_{t_0}^{t} C(t)\Phi(t,\tau)B(\tau)u(\tau)\,\mathrm{d}\tau. \tag{2.64}$$

Notice that the right-hand side of (2.63) is known if we know the input and output vectors over the interval $[t_0, t]$. Therefore, in principle, (2.63) and (2.64) can be used to compute the current value of the state vector $x(t)$. This yields an observer that is nonrecursive in the sense that knowing the state vector at time t does not help us compute it at time $t+\sigma$, no matter how small σ is: the integrals in (2.63) and (2.64) must be recomputed over the interval $[t_0, t+\sigma]$. In Chapter 4, we introduce an observation method that is better, because it is recursive in the sense outlined earlier.

EXAMPLE 2.6 *Consider again the double integrator introduced and motivated in Example 2.3, namely,*

$$\begin{bmatrix}\dot{x}_1\\ \dot{x}_2\end{bmatrix} = \begin{bmatrix}0 & 1\\ 0 & 0\end{bmatrix}\begin{bmatrix}x_1\\ x_2\end{bmatrix} + \begin{bmatrix}0\\ 1\end{bmatrix}u$$

$$y = \begin{bmatrix}1 & 0\end{bmatrix}\begin{bmatrix}x_1\\ x_2\end{bmatrix}. \tag{2.65}$$

Here we assume that we know the range y and the acceleration u. In practice, this is achieved by using a range sensor and an accelerometer. The state transition matrix of system (2.65) *is*

$$\Phi(t,0) = \begin{bmatrix}1 & t\\ 0 & 1\end{bmatrix}. \tag{2.66}$$

Therefore, the observability Gramian of system (2.65) *over an interval* $[0, t_1]$ *is*

$$\begin{aligned} M(0,t_1) &= \int_0^{t_1}\begin{bmatrix}1\\ t\end{bmatrix}\begin{bmatrix}1 & t\end{bmatrix}\mathrm{d}t \\ &= \int_0^{t_1}\begin{bmatrix}1 & t\\ t & t^2\end{bmatrix}\mathrm{d}t \\ &= \begin{bmatrix}t_1 & t_1^2/2\\ t_1^2/2 & t_1^3/3\end{bmatrix} \end{aligned} \tag{2.67}$$

and is nonsingular for all $t_1 > 0$. Therefore, all initial conditions are observable at time 0. If, however, we consider the same dynamic system as in (2.15) (line 1), but with a velocity sensor instead of a range sensor, the output equation becomes

$$y = \begin{bmatrix} 0 & 1 \end{bmatrix} x. \tag{2.68}$$

With the same state transition matrix as in (2.66), the observability Gramian is now

$$\begin{aligned} M(0, t_1) &= \int_0^{t_1} \begin{bmatrix} 0 \\ 1 \end{bmatrix} \begin{bmatrix} 0 & 1 \end{bmatrix} \mathrm{d}t \\ &= \int_0^{t_1} \begin{bmatrix} 0 & 0 \\ 0 & 1 \end{bmatrix} \mathrm{d}t \\ &= \begin{bmatrix} 0 & 0 \\ 0 & t_1 \end{bmatrix} \end{aligned} \tag{2.69}$$

and is singular for all $t_1 > 0$. Therefore, the system is unobservable at time 0. Specifically, the unobservable states (which must belong to the null space of the observability Gramian) are along the direction

$$x_0 = \begin{bmatrix} 1 \\ 0 \end{bmatrix} \tag{2.70}$$

and represent initial conditions with zero velocity. This tells us the well-expected fact that for a double integrator, ***it is impossible to determine the initial position with just a velocity sensor****.*

2.3 Time Invariant Systems

In this section, we consider the particular case of linear time invariant systems, to which all the notions introduced earlier in the chapter apply, and for which the criteria established there take particularly simple and elegant forms. Consider the time invariant version of system (2.51), that is,

$$\begin{aligned} \dot{x}(t) &= Ax(t) + Bu(t) \\ y(t) &= Cx(t) \\ A, B, C &\equiv \text{constant}. \end{aligned} \tag{2.71}$$

Because of the time invariance of system (2.71), it can be argued that the state transition matrix must be shift invariant. In other words, for all time shifts σ, the effect at any time t due to an initial condition applied at time τ must be the same as the effect at time $t + \sigma$ of the same initial condition if it were applied at time $\tau + \sigma$. Recalling (2.29), it is clear that, mathematically, this means that the state transition matrix must satisfy

$$\forall t, \tau, \sigma \in \mathbb{R}, \ \Phi(t + \sigma, \tau + \sigma) = \Phi(t, \tau). \tag{2.72}$$

Equation (2.72) holds, in particular, for $\sigma = -\tau$, yielding

$$\forall t, \tau \in \mathbb{R}, \ \Phi(t, \tau) = \Phi(t - \tau, 0). \tag{2.73}$$

Now, recalling (2.27), the matrix $\Phi(t, 0)$ satisfies the matrix differential equation with initial condition

$$\begin{aligned} \dot{\Phi}(t, 0) &= A\Phi(t, 0) \\ \Phi(0, 0) &= I. \end{aligned} \tag{2.74}$$

Equation (2.74) can be solved using the method of Laplace transforms. Let $\Psi(s)$ be the Laplace transform of $\Phi(t, 0)$, that is,

$$\Psi(s) = \int_0^{\infty} \Phi(t, 0)e^{-st}\, dt. \tag{2.75}$$

Applying the differentiation rule for Laplace transforms to (2.74), we have

$$s\Psi(s) - I = A\Psi(s), \tag{2.76}$$

or equivalently,

$$\Psi(s) = (sI - A)^{-1}. \tag{2.77}$$

Note that the matrix $(sI - A)$ is a rational function of s, that is, a ratio of polynomials in s. As a consequence, the right-hand side of (2.77) is a matrix where every entry is a rational function of s. We can therefore use tables of Laplace transforms to compute $\Phi(t, 0)$ by computing the inverse Laplace transform of each entry of the matrix $\Psi(s)$ in (2.77).

Also, applying the power series solution method to the matrix differential equation (2.74) yields

$$\Phi(t, 0) = I + At + \frac{A^2t^2}{2!} + \frac{A^3t^3}{3!} + \cdots + \frac{A^kt^k}{k!} + \cdots, \tag{2.78}$$

which leads to the following definition that generalizes the familiar notion of exponential function.

DEFINITION 2.9 *The **exponential** of a real $n \times n$ matrix A is the real $n \times n$ matrix*

$$e^A = \exp(A) = I + A + \frac{A^2}{2!} + \frac{A^3}{3!} + \cdots + \frac{A^k}{k!} + \cdots \tag{2.79}$$

REMARK 2.4 *By identifying the inverse Laplace transform of* (2.77) *with* (2.78) *using* (2.79), *we obtain the formula*

$$e^{At} = \mathcal{L}^{-1}\left((sI - A)^{-1}\right), \tag{2.80}$$

where $\mathcal{L}^{-1}$ represents inverse Laplace transform. This formula can be used to compute the exponential of a matrix; it involves explicitly obtaining the rational matrix $(sI - A)^{-1}$ and then computing the inverse Laplace transform of each of its entries.

REMARK 2.5 *For a linear time invariant system,* (2.73) *and* (2.78) *imply that the state transition matrix satisfies*

$$\Phi(t, \tau) = e^{A(t-\tau)}. \tag{2.81}$$

The variation of constants formula (2.34) then becomes

$$y(t) = Ce^{A(t-t_0)}x_0 + \int_{t_0}^{t} Ce^{A(t-\tau)}Bu(\tau)\,\mathrm{d}\tau. \tag{2.82}$$

EXAMPLE 2.7 *Consider again the double integrator introduced and motivated in Example 2.3, namely,*

$$\begin{bmatrix}\dot{x}_1\\ \dot{x}_2\end{bmatrix} = \begin{bmatrix}0 & 1\\ 0 & 0\end{bmatrix}\begin{bmatrix}x_1\\ x_2\end{bmatrix} + \begin{bmatrix}0\\ 1\end{bmatrix}u$$
$$y = \begin{bmatrix}1 & 0\end{bmatrix}\begin{bmatrix}x_1\\ x_2\end{bmatrix}. \tag{2.83}$$

The exponential function of the state matrix is computed using the method outlined in Remark 2.4, as follows:

$$\begin{aligned} sI - A &= \begin{bmatrix}s & -1\\ 0 & s\end{bmatrix},\\ (sI - A)^{-1} &= \frac{1}{s^2}\begin{bmatrix}s & 1\\ 0 & s\end{bmatrix}\\ &= \begin{bmatrix}1/s & 1/s^2\\ 0 & 1/s\end{bmatrix},\\ e^{At} &= \mathcal{L}^{-1}\left((sI - A)^{-1}\right)\\ &= \begin{bmatrix}1 & t\\ 0 & 1\end{bmatrix}. \end{aligned} \tag{2.84}$$

Also, note that in this particular case, the state matrix A is nilpotent because

$$A^2 = \begin{bmatrix}0 & 1\\ 0 & 0\end{bmatrix}\begin{bmatrix}0 & 1\\ 0 & 0\end{bmatrix} = 0_2, \tag{2.85}$$

and $A^k = 0_2$, for all $k \geq 2$. As a consequence, we can also easily use Definition 2.9 to obtain this exponential as

$$\begin{aligned} e^{At} &= I + At + \frac{A^2t^2}{2!} + \frac{A^3t^3}{3!} + \cdots\\ &= I + At\\ &= \begin{bmatrix}1 & t\\ 0 & 1\end{bmatrix}. \end{aligned} \tag{2.86}$$

Then, the state transition matrix is

$$\Phi(t,\tau) = e^{A(t-\tau)} = \begin{bmatrix}1 & t-\tau\\ 0 & 1\end{bmatrix}, \tag{2.87}$$

in agreement with Example 2.5.

2.3.1 Stability of Linear Time Invariant Systems

First, note that from (2.80), the entries of e^{At} are combinations of exponentials and polynomials because the entries of $(sI - A)^{-1}$ are rational functions.

DEFINITION 2.10 *If $A \in \mathbb{R}^{n\times n}$, $\lambda \in \mathbb{C}$ is called an **eigenvalue** of A if*

$$\det(\lambda I - A) = 0. \tag{2.88}$$

If λ is an eigenvalue of A, any vector $x \in \mathbb{C}^n$ that satisfies

$$\begin{aligned}(\lambda I - A)x &= 0\\ x &\neq 0\end{aligned} \tag{2.89}$$

*is called an **eigenvector associated with** λ. The polynomial*

$$p(\lambda) = \det(\lambda I - A) \tag{2.90}$$

*is called the **characteristic polynomial** of A.*

We note that in Definition 2.10, the characteristic polynomial is a polynomial of degree n, whose roots are precisely the eigenvalues.

DEFINITION 2.11 *The **algebraic multiplicity** of an eigenvalue is the multiplicity of this eigenvalue as a root of the characteristic polynomial $p(\lambda)$. Specifically, we say that λ is a root of algebraic multiplicity k for polynomial $p(.)$ if $p(\lambda) = 0$, $p'(\lambda) = 0$, $p''(\lambda) = 0, \ldots, p^{(k-1)}(\lambda) = 0$, and $p^{(k)}(\lambda) \neq 0$. The **geometric multiplicity** of an eigenvalue λ is $\dim \mathcal{N}(\lambda I - A) = n - \operatorname{rank}(\lambda I - A)$.*

In Definition 2.11, the geometric multiplicity also represents the number of linearly independent eigenvectors associated with λ. The algebraic and geometric multiplicities are related as follows.

PROPOSITION 2.14 *If λ is an eigenvalue of A, then its geometric multiplicity is smaller than or equal to its algebraic multiplicity.*

This property leads to the following definition.

DEFINITION 2.12 *An eigenvalue λ is called **defective** if its geometric multiplicity is strictly smaller than its algebraic multiplicity.*

We are now ready to state the following results on stability of linear time invariant dynamic systems.

PROPOSITION 2.15 *Let $A \in \mathbb{R}^{n\times n}$. The linear time invariant homogeneous equation*

$$\dot{x}(t) = Ax(t) \tag{2.91}$$

is stable if and only if all the eigenvalues of A are in the closed left half of the complex plane ($\mathrm{Re}(\lambda) \leq 0$)*, and those on the imaginary axis are not defective.*

PROPOSITION 2.16 *Let* $A \in \mathbb{R}^{n \times n}$*. The linear time invariant homogeneous equation*

$$\dot{x}(t) = Ax(t) \tag{2.92}$$

is asymptotically stable if and only if all the eigenvalues of A are in the open left half of the complex plane ($\mathrm{Re}(\lambda) < 0$)*.*

EXAMPLE 2.8 *Consider the homogeneous part of the linear system in Example 2.5, that is,*

$$\dot{x} = \begin{bmatrix} 0 & 1 \\ 0 & 0 \end{bmatrix} x. \tag{2.93}$$

The characteristic polynomial is

$$\det(\lambda I - A) = \det \begin{bmatrix} \lambda & -1 \\ 0 & \lambda \end{bmatrix} = \lambda^2, \tag{2.94}$$

implying from Definition 2.11 that $\lambda = 0$ *is an eigenvalue with algebraic multiplicity 2. Now, the only eigenvector corresponding to* $\lambda = 0$ *is along*

$$\begin{bmatrix} 0 \\ 1 \end{bmatrix}. \tag{2.95}$$

This implies that the geometric multiplicity of the eigenvalue $\lambda = 0$ *is 1. Therefore, from Definition 2.12, the eigenvalue* $\lambda = 0$ *is defective. Now 0 is on the imaginary axis of the complex plane. Therefore, by Proposition 2.15, the system* (2.93) *is unstable. This agrees with the conclusion of Example 2.5.*

2.3.2 BIBO Stability of Linear Time Invariant Systems

Consider again the linear time invariant system

$$\begin{aligned} \dot{x}(t) &= Ax(t) + Bu(t) \\ y(t) &= Cx(t) \\ A, B, C &\equiv \text{constant} \end{aligned} \tag{2.96}$$

with transfer function

$$G(s) = C(sI - A)^{-1}B. \tag{2.97}$$

We have the following.

PROPOSITION 2.17 *The linear time invariant system* (2.96) *is BIBO stable if and only if all the poles of the transfer function G(s) in* (2.97) *are in the open left half of the complex plane* ($Re(\lambda) < 0$).

EXAMPLE 2.9 *Consider again the double integrator, introduced and motivated in Example 2.3:*

$$\begin{bmatrix}\dot{x}_1\\ \dot{x}_2\end{bmatrix} = \begin{bmatrix}0 & 1\\ 0 & 0\end{bmatrix}\begin{bmatrix}x_1\\ x_2\end{bmatrix} + \begin{bmatrix}0\\ 1\end{bmatrix}u$$

$$y = \begin{bmatrix}1 & 0\end{bmatrix}\begin{bmatrix}x_1\\ x_2\end{bmatrix}. \tag{2.98}$$

The transfer function is

$$\begin{aligned} G(s) &= \begin{bmatrix}1 & 0\end{bmatrix}\begin{bmatrix}s & -1\\ 0 & s\end{bmatrix}^{-1}\begin{bmatrix}0\\ 1\end{bmatrix} \\ &= \frac{1}{s^2}\begin{bmatrix}1 & 0\end{bmatrix}\begin{bmatrix}s & 1\\ 0 & s\end{bmatrix}\begin{bmatrix}0\\ 1\end{bmatrix} \\ &= \frac{1}{s^2}. \end{aligned} \tag{2.99}$$

Because $G(s)$ has a (double) pole on the imaginary axis, Proposition 2.17 implies that system (2.98) *is not BIBO stable. This agrees with the conclusion of Example 2.5.*

2.3.3 Observability of Linear Time Invariant Systems

Consider again the linear time invariant system

$$\begin{aligned} \dot{x}(t) &= Ax(t) + Bu(t) \\ y(t) &= Cx(t) \\ A, B, C &\equiv \text{constant}. \end{aligned} \tag{2.100}$$

We have the following.

PROPOSITION 2.18 *For the linear time invariant system* (2.100), *the following five statements are equivalent:*

1. *System* (2.100) *is observable.*
2. *For all $\tau > 0$, the symmetric matrix*

$$M(\tau) = \int_0^\tau e^{A^T t} C^T C e^{At}\, dt \tag{2.101}$$

   is nonsingular (i.e., positive definite in this particular case).
3. *The matrix*

$$O = \begin{bmatrix} C \\ CA \\ CA^2 \\ \cdots \\ CA^{n-1} \end{bmatrix} \tag{2.102}$$

   has full rank n.

4. *There exists no eigenvector of A that is orthogonal to the rows of C; that is,*

$$Ax = \lambda x \text{ and } Cx = 0 \tag{2.103}$$

imply

$$x = 0. \tag{2.104}$$

5. *The matrix*

$$\begin{bmatrix} sI - A \\ C \end{bmatrix} \tag{2.105}$$

has full rank for every complex number s.

EXAMPLE 2.10 *Consider again the double integrator introduced and motivated in Example 2.3:*

$$\begin{aligned} \dot{x} &= \begin{bmatrix} 0 & 1 \\ 0 & 0 \end{bmatrix} x \\ y &= \begin{bmatrix} 1 & 0 \end{bmatrix} x. \end{aligned} \tag{2.106}$$

In Example 2.6, we used criterion 2 of Proposition 2.18 to show that the system (2.106) *is observable. Also here the matrix O of criterion 3 is*

$$\begin{aligned} O &= \begin{bmatrix} C \\ CA \end{bmatrix} \\ &= \begin{bmatrix} 1 & 0 \\ 0 & 1 \end{bmatrix}, \end{aligned} \tag{2.107}$$

which has full rank 2. Therefore, from criterion 3, we also conclude that system (2.106) *is observable. To apply criterion 4, we first compute the only eigenvector of the state matrix A, which is along*

$$\begin{bmatrix} 1 \\ 0 \end{bmatrix} \tag{2.108}$$

and is not orthogonal to the output matrix $\begin{bmatrix} 1 & 0 \end{bmatrix}$*. Therefore, from criterion 4, we also conclude that system* (2.106) *is observable.*

EXAMPLE 2.11 *Now, conversely, consider the linear time invariant system*

$$\begin{aligned} \dot{x} &= \begin{bmatrix} 0 & 1 \\ 0 & 0 \end{bmatrix} x \\ y &= \begin{bmatrix} 0 & 1 \end{bmatrix} x, \end{aligned} \tag{2.109}$$

which is different from the system (2.106) *only by the fact that we measure the velocity instead of the position. In Example 2.6, we had also used criterion 2 to show that the system* (2.109) *is unobservable. The matrix O of criterion 3 is now*

$$O = \begin{bmatrix} 0 & 1 \\ 0 & 1 \end{bmatrix}, \tag{2.110}$$

which has only rank 1. Therefore, by criterion 3, system (2.109) *is unobservable. Also, the vector*

$$x = \begin{bmatrix} 1 \\ 0 \end{bmatrix} \tag{2.111}$$

is an eigenvector of A associated with the eigenvalue 0, and

$$Cx = \begin{bmatrix} 0 & 1 \end{bmatrix} \begin{bmatrix} 1 \\ 0 \end{bmatrix} = 0. \tag{2.112}$$

Therefore, criterion 4 implies that system (2.109) *is unobservable. Also, for* $s = 0$, *we have*

$$\text{rank} \begin{bmatrix} sI - A \\ C \end{bmatrix} = \text{rank} \begin{bmatrix} 0 & -1 \\ 0 & 0 \\ 0 & 1 \end{bmatrix} = 1 < 2. \tag{2.113}$$

Therefore, by criterion 5, we also conclude that system (2.109) *is unobservable.*

2.4 The Method of Adjoints

At first sight, the variation of constants formula (2.34) together with the definition of the state transition matrix from Proposition 2.3 look quite appropriate for computing the output, at a given final time, of a linear dynamic system. On further examination, however, they are seen to contain formidable difficulties indeed. The problem can be understood as follows. Consider a linear dynamic system with zero initial condition:

$$\begin{aligned} \dot{x}(t) &= A(t)x(t) + B(t)u(t) \\ y(t) &= C(t)x(t) \\ x(t_0) &= 0 \\ & \quad t \in [t_0, t_f]. \end{aligned} \tag{2.114}$$

Assume that we know the time history of the input $u(t)$ on the given interval, and we want to compute the output at the final time. A casual application of the variation of constants formula yields

$$\begin{aligned} y(t_f, t_0, x_0, u) &= y(t_f) \\ &= \int_{t_0}^{t_f} C(t_f)\Phi(t_f, \tau)B(\tau)u(\tau)\, d\tau \\ &= \int_{t_0}^{t_f} G(t_f, \tau)u(\tau)\, d\tau, \end{aligned} \tag{2.115}$$

where $G(t, \tau)$ is the impulse response matrix. The kth column of $G(t_f, \tau)$ represents the response at time t_f due to an impulse at time τ in the kth component of $u(t)$. It should be realized that applying (2.115) requires computing

$$\Phi(t_f, \tau) \text{ or } G(t_f, \tau) \text{ for all } \tau \in [t_0, t_f]. \tag{2.116}$$

Moreover, according to the definition of the state transition matrix from Proposition 2.3, each value of τ requires integrating the differential equation with boundary condition

$$\begin{aligned} \frac{\partial \Phi(t, \tau)}{\partial t} &= A(t)\Phi(t, \tau) \\ \Phi(\tau, \tau) &= I_n. \end{aligned} \tag{2.117}$$

Equivalently, for each τ, one would need to simulate the system to determine the response at the final time due to an impulse applied at time τ. This clearly implies a prohibitive amount of computations.

The difficulty outlined earlier can be overcome by using the method of adjoints presented subsequently. Associated with system (2.114), consider the **adjoint equation**

$$\dot{p}(t) = -A^T(t)p(t). \tag{2.118}$$

We have

$$\begin{aligned} \frac{\mathrm{d}}{\mathrm{d}t}(p^T x) &= \dot{p}^T x + p^T \dot{x} \\ &= -p^T A x + p^T A x + p^T B u \\ &= p^T B u. \end{aligned} \tag{2.119}$$

Integrating (2.119) between t_0 and t_f, we obtain

$$p^T(t_f)x(t_f) = p^T(t_0)x(t_0) + \int_{t_0}^{t_f} p^T(\tau)B(\tau)u(\tau)\,\mathrm{d}\tau. \tag{2.120}$$

Therefore, if we choose the final value of the adjoint vector to satisfy

$$p(t_f) = C^T(t_f), \tag{2.121}$$

the final value of the output will be given by

$$y(t_f) = C(t_f)x(t_f) = p^T(t_f)x(t_f) = p^T(t_0)x(t_0) + \int_{t_0}^{t_f} p^T(\tau)B(\tau)u(\tau)\,\mathrm{d}\tau. \tag{2.122}$$

Remarkably, (2.122) requires only two integrations: one integration backward to obtain the time history of the adjoint vector $p(t)$, and the other forward to compute $y(t_f)$. This compares quite favorably with the infinite number of integrations required by (2.115).

An alternative way of deriving the adjoint method is to obtain a differential equation that generates the state transition matrix as a function of its *second* argument instead of its *first* argument, as in Proposition 2.3. This equation turns out to be precisely the adjoint equation (2.118). This idea is developed in more detail in Problem 2.4.

EXAMPLE 2.12 *Consider the first-order linear time varying system*

$$\begin{aligned}\dot{x}(t) &= tx(t) + u(t)\\ y(t) &= x(t)\\ x(t_0) &= 0.\end{aligned} \tag{2.123}$$

Assume that we know the time history of the input on the interval $[t_0, t_f]$ *and that we want to compute the resulting output at time* t_f. *Here the state transition matrix is easily computed as*

$$\Phi(t, \tau) = e^{\frac{t^2-\tau^2}{2}}, \tag{2.124}$$

so that, from the variation of constants formula, the final output is given by

$$y(t_f) = \int_{t_0}^{t_f} e^{\frac{t_f^2-\tau^2}{2}} u(\tau)\, \mathrm{d}\tau. \tag{2.125}$$

It is important to note that, in practice, it is quite exceptional to have at one's disposal a closed-form expression of the state transition matrix such as that of (2.124). *In general, in the absence of such a closed-form expression,* (2.125) *would require one integration for every value of* τ.

Now, consider the adjoint equation for (2.123):

$$\begin{aligned}\dot{p} &= -tp\\ p(t_f) &= 1.\end{aligned} \tag{2.126}$$

A simple backward integration yields

$$p(t) = e^{\frac{t_f^2-t^2}{2}}, \tag{2.127}$$

so that the final output is also given by

$$y(t_f) = \int_{t_0}^{t_f} p(\tau)u(\tau)\, \mathrm{d}\tau, \tag{2.128}$$

where (2.126) *and* (2.128) *require only two integrations.*

2.5 Controllability and Duality

Consider again the standard linear time varying system, repeated here for convenience:

$$\begin{aligned}\dot{x}(t) &= A(t)x(t) + B(t)u(t)\\ x(t_0) &= x_0.\end{aligned} \tag{2.129}$$

Here we consider the question: Given t_0 and x_0, can we find the time history of the input, $u(t)$, that drives the state from x_0 at time t_0 to 0 in finite time t_f? This question is important in Chapter 7.

Recall the variation of constants formula:

$$\begin{aligned} x(t) &= \Phi(t, t_0)x_0 + \int_{t_0}^{t} \Phi(t, \tau)B(\tau)u(\tau)\,\mathrm{d}\tau \\ &= \Phi(t, t_0)\left(x_0 + \int_{t_0}^{t} \Phi(t_0, \tau)B(\tau)u(\tau)\,\mathrm{d}\tau\right). \end{aligned} \tag{2.130}$$

DEFINITION 2.13 *A state x_0 is **controllable** at time t_0 if there exist a finite time $t_f > t_0$ and $u_0(t)$, $t \in [t_0, t_f]$, such that $x(t_f; t_0, x_0, u_0(.)) = 0$. (Therefore, the state at time t_f resulting from initial conditions x_0, t_0, and input u_0 is equal to zero.) A state x_0 is **uncontrollable** at time t_0 if it is not controllable. System (2.129) is **completely controllable** at t_0 if all states are controllable at t_0.*

REMARK 2.6 *The set of controllable states forms a vector space.*

DEFINITION 2.14 *The **range space** of a functional map $L(u) = \int_{t_0}^{t_1} G(t)u(t)\,\mathrm{d}t$ is given by*

$$\mathcal{R}(L) = \left\{x : \exists u(.) : x = \int_{t_0}^{t_1} G(t)u(t)\,\mathrm{d}t\right\}. \tag{2.131}$$

PROPOSITION 2.19 *Consider the vector differential equation $\dot{z}(t) = B(t)u(t)$. Then z_0 is controllable at t_0 if and only if $\exists t_1 : z_o \in \mathcal{R}(L)$, where $L(u) = \int_{t_0}^{t_1} B(t)u(t)\,\mathrm{d}t$.*

PROOF. Use the fact that z_0 is controllable if and only if $\exists t_1, u_0 : z(t_1) = 0 = z_0 + \int_{t_0}^{t_1} B(\tau)u_0(\tau)\,\mathrm{d}\tau$.

PROPOSITION 2.20 *Let $G(t)$ be an $n \times m$ real integrable matrix. Define the linear functional operator $L(u) = \int_{t_0}^{t_1} G(t)u(t)\,\mathrm{d}t$ and the real $n \times n$ matrix $W = \int_{t_0}^{t_1} G(t)G^T(t)\,\mathrm{d}t$. Then $\mathcal{R}(L) = \mathcal{R}(W)$.*

PROOF. Consider first the case where $x \in \mathcal{R}(W)$. Then, there exists an η such that $x \in W\eta$. Let $u(t) = G^T(t)\eta$. Then,

$$\begin{aligned} \int_{t_0}^{t_1} G(\tau)u(\tau)\,\mathrm{d}\tau &= W\eta = x \\ &\Rightarrow x \in \mathcal{R}(L). \end{aligned} \tag{2.132}$$

Let us now consider the case where $x \notin \mathcal{R}(W)$. Then, there exists an x_n such that $Wx_n = 0$ and $x^T x_n \neq 0$. Indeed, recall that $\mathbb{R}^n = \mathcal{R}(W) \oplus \mathcal{N}(W^T)$, with $\mathcal{R}(W) \perp \mathcal{N}(W^T)$, that is,

$$\exists! x_r \in \mathcal{R}(W), \exists! x_n \in \mathcal{N}(W^T) : x = x_r + x_n; x_r \perp x_n. \tag{2.133}$$

Assume that there exists a u such that $x = \int_{t_0}^{t_1} G(\tau)u(\tau)\,\mathrm{d}\tau$. Then,

$$x_n^T x = \int_{t_0}^{t_1} x_n^T G(\tau)u(\tau)\,\mathrm{d}\tau \neq 0. \tag{2.134}$$

Now,

$$x_n^T W x_n = \int_{t_0}^{t_1} (x_n^T G(\tau))(G^T(\tau)x_n)\,\mathrm{d}\tau = 0. \tag{2.135}$$

This implies

$$G^T(t)x_n \equiv 0, \tag{2.136}$$

which contradicts (2.134). This contradiction completes the proof.

PROPOSITION 2.21 *Consider the vector differential equation* $\dot{z}(t) = B(t)u(t)$; z_0 *is controllable at* t_0 *if and only if*

$$\exists t_1 : z_0 \in \mathcal{R}\left(\int_{t_0}^{t_1} B(\tau)B^T(\tau)\,\mathrm{d}\tau\right). \tag{2.137}$$

PROPOSITION 2.22 *Consider the system* $\dot{x}(t) = A(t)x(t) + B(t)u(t)$. *A state* x_0 *is controllable at* t_0 *if and only if there exists* $t_1 > t_0$ *such that* $x_0 \in \mathcal{R}(W(t_0, t_1))$, *where*

$$W(t_0, t_1) = \int_{t_0}^{t_1} \Phi(t_0, \tau)B(\tau)B^T(\tau)\Phi^T(t_0, \tau)\,\mathrm{d}\tau. \tag{2.138}$$

PROOF. Let $z(t) = \Phi(t_0, t)x(t)$. Then, $x(t) = \Phi(t, t_0)z(t)$. In addition, x_0 is controllable, which is equivalent to the fact that z_0 $(= x_0)$ is also controllable. We have

$$\begin{aligned}\dot{x}(t) &= A(t)\Phi(t, t_0)z(t) + \Phi(t, t_0)\dot{z}(t)\\ &= A(t)\Phi(t, t_0)z(t) + B(t)u(t),\end{aligned} \tag{2.139}$$

implying that

$$\dot{z}(t) = \Phi(t_0, t)B(t)u(t). \tag{2.140}$$

Then, use Proposition 2.21 to obtain the result.

REMARK 2.7 *The matrix* $W(t_0, t_1)$ *in* (2.138) *is called the* ***controllability Gramian***. *It is always symmetric and positive semidefinite.*

REMARK 2.8 *System* (2.129) *is completely controllable at* t_0 *if* $\exists t_1 > t_0 : W(t_0, t_1) > 0$.

REMARK 2.9 *If system* (2.129) *is controllable at* t_0, *then one control that drives the state of the system from initial condition* x_0 *at time* t_0 *to the origin at final time* t_1 *is*

$$u_0(t) = -B^T(t)\Phi^T(t_0, t)W^{-1}(t_0, t_1)x_0. \tag{2.141}$$

REMARK 2.10 *For the system*

$$\begin{aligned}\dot{x}(t) &= A(t)x(t)\\ y(t) &= C(t)x(t),\end{aligned} \tag{2.142}$$

we have introduced the observability Gramian:

$$M(t_0, t_1) = \int_{t_0}^{t_1} \Phi^T(t, t_0)C^T(t)C(t)\Phi(t, t_0)\,\mathrm{d}t. \tag{2.143}$$

Our results on controllability look suspiciously similar to those on observability. This similarity is clarified in the next result.

PROPOSITION 2.23 (Duality) *Consider the triples*

$$(A(t), B(t), C(t)) \text{ and } (-A^T(t), C^T(t), B^T(t)). \tag{2.144}$$

Then the controllable (respectively, uncontrollable) states of one are the observable (respectively, unobservable) states of the other.

PROOF. The proof is left as an optional exercise. Compute the Gramians, and see that the controllability (observability) Gramian of one is the observability (controllability) Gramian of the other.

As a consequence, all of our results on observability from Section 2.2 extend to controllability by duality, including the criteria for time invariant systems presented in Section 2.3.

2.6 Summary of Key Results

The key results in Chapter 2 are as follows:

1. Equation (2.27), which defines the state transition matrix of a linear dynamic system
2. Propositions 2.7 and 2.8, which provide criteria for stability and asymptotic stability of a linear dynamic system
3. Equations (2.34), the variation of constants formulas, which provide the state and output of a linear dynamic system in terms of initial state and input
4. Proposition 2.10, which provides a criterion for BIBO stability of a linear dynamic system
5. Proposition 2.12, which provides a criterion for observability of a linear dynamic system
6. Equations (2.80) and (2.81), which provide the state transition matrix for a linear time invariant system
7. Equations (2.118), (2.121), and (2.122) which provide the output of a linear dynamic system using the method of adjoints
8. Proposition 2.23, which states the duality between controllability and observability

2.7 Bibliographic Notes for Further Reading

The material in Chapter 2 is standard and is well covered in many texts, including [13], [17], [43], and [35].

2.8 Homework Problems

PROBLEM 2.1 *Compute the state transition matrices corresponding to the following state matrices:*

1. *Time invariant case*

$$A = \begin{bmatrix} 0 & -\omega \\ \omega & 0 \end{bmatrix}$$

2. *Time varying case*

$$A(t) = \begin{bmatrix} 1 & t \\ 0 & 1 \end{bmatrix}$$

PROBLEM 2.2 *Show that if the state matrix $A(t)$ has the form*

$$A = \begin{bmatrix} A_{11}(t) & A_{12}(t) \\ 0 & A_{22}(t) \end{bmatrix},$$

then the state transition matrix corresponding to $A(t)$ has the form

$$\Phi(t, \tau) = \begin{bmatrix} \Phi_{11}(t, \tau) & \Phi_{12}(t, \tau) \\ 0 & \Phi_{22}(t, \tau) \end{bmatrix},$$

where $\Phi_{11}(t, \tau)$ and $\Phi_{22}(t, \tau)$ are the state transition matrices corresponding to $A_{11}(t)$ and $A_{22}(t)$, respectively.

PROBLEM 2.3 *Let $P(t)$ be a square, differentiable, and nonsingular matrix. Show that*

$$\frac{\mathrm{d}}{\mathrm{d}t}\left(P^{-1}(t)\right) = -P^{-1}(t)\dot{P}(t)P^{-1}(t).$$

Hint: *Differentiate the identity $P(t)P^{-1}(t) = I$ with respect to time.*

PROBLEM 2.4 *If $\Phi(t, \tau)$ is the state transition matrix corresponding to the state matrix $A(t)$, show that*

$$\frac{\partial}{\partial \tau}\Phi(t, \tau) = -\Phi(t, \tau)A(\tau).$$

PROBLEM 2.5 *Let $\Phi(t, \tau)$ and $\Psi(t, \tau)$ be the state transition matrices associated with the state matrices $A(t)$ and $-A^T(t)$, respectively. Show that*

$$\begin{aligned} \Phi^T(t, \tau)\Psi(t, \tau) &= \Psi^T(t, \tau)\Phi(t, \tau) \\ &= \Psi(t, \tau)\Phi^T(t, \tau) \\ &= \Phi(t, \tau)\Psi^T(t, \tau) \\ &= I. \end{aligned}$$

Deduce that if $A(t)$ is skew-symmetric for all t, then $\Phi(t, \tau)$ is orthogonal for all t and τ.

PROBLEM 2.6 *The equations of motion for an orbiting satellite are*

$$\dot{r} = v_r$$
$$\dot{\theta} = \frac{v_\theta}{r}$$
$$\dot{v}_r = \frac{v_\theta^2}{r} - \frac{k}{r^2} + u_r$$
$$\dot{v}_\theta = -\frac{v_r v_\theta}{r} + u_\theta$$

where r is the radius of the orbit, θ is the anomaly angle, v_r is the radial velocity, v_θ is the tangential velocity, u_r and u_θ are the radial and tangential components of the acceleration due to thrust, and k is a constant.

Assume that we have a nominal trajectory $r^0, \theta^0, v_r^0, v_\theta^0, u_r^0, u_\theta^0$. Give the linearized equations for $\delta r, \delta\theta, \delta v_r, \delta v_\theta, \delta u_r, \delta u_\theta$.

PROBLEM 2.7 *With the same notation as in Problem 2.6, consider the particular case of a circular orbit, that is, $u_r^0 = 0$, $u_\theta^0 = 0$, $r = r^0 = \text{constant}$, $\dot{\theta}^0 = \omega^0 = \text{constant}$. Obtain the linearized equations for $\delta r, \delta\theta, \delta v_r, \delta v_\theta, \delta u_r, \delta u_\theta$.*

PROBLEM 2.8 *For what values of the real parameter a are the following systems completely observable? When appropriate, indicate the unobservable directions:*

1.

$$\dot{x}_1 = -x_1 + x_2$$
$$\dot{x}_2 = -x_2$$
$$\dot{x}_3 = -x_3$$
$$y = 2ax_1 + ax_2 + (a-1)x_3$$

2.

$$\dot{x}_1 = -x_1 + x_2$$
$$\dot{x}_2 = -x_2$$
$$\dot{x}_3 = ax_3$$
$$y = x_1 + x_3$$

PROBLEM 2.9 *Consider the linear system*

$$\begin{bmatrix}\dot{x}_1 \\ \dot{x}_2\end{bmatrix} = \begin{bmatrix}A_{11}(t) & A_{12}(t) \\ 0 & A_{22}(t)\end{bmatrix}\begin{bmatrix}x_1 \\ x_2\end{bmatrix} + \begin{bmatrix}B_1(t) \\ B_2(t)\end{bmatrix} u(t)$$

$$y = \begin{bmatrix}0 & C_2(t)\end{bmatrix}\begin{bmatrix}x_1 \\ x_2\end{bmatrix},$$

where the submatrices $A_{11}, A_{12}, A_{22}, B_1, B_2, C_2$ are of compatible dimensions.

1. *Show that initial conditions of the form* $\begin{bmatrix} x_{10} \\ 0 \end{bmatrix}$ *are unobservable for all* t_0, *that is, that* x_1 *is unobservable.*
2. *Show that the impulse response of the system is independent of* $A_{11}(t)$ *and* $B_1(t)$.

PROBLEM 2.10 *For the linear observable system*

$$\dot{x} = A(t)x$$
$$y = C(t)x,$$

we have seen that a formula for a nonrecursive observer is

$$\hat{x}(t) = \Phi(t, t_0)M^{-1}(t_0, t)\int_{t_0}^{t} \Phi(\tau, t_0)C^T(\tau)y(\tau)\,\mathrm{d}\tau.$$

Generalize this formula to the nonhomogeneous system

$$\dot{x} = A(t)x + B(t)u$$
$$y = C(t)x,$$

where u is known.

PROBLEM 2.11 *Consider a linear time invariant system with input–output relation*

$$Y(s) = \frac{b_{n-1}s^{n-1} + b_{n-2}s^{n-2} + \cdots + b_1 s + b_0}{s^n + a_{n-1}s^{n-1} + a_{n-2}s^{n-2} + \cdots + a_1 s + a_0}U(s), \tag{2.145}$$

where $Y = \mathcal{L}(y)$, $U = \mathcal{L}(u)$ *are the Laplace transforms of the signals y and u, respectively. Find an nth-order realization for this system, that is, find matrices A, b, c with* $A \in \mathbb{R}^{n\times n}$ *such that* (2.145) *is implied by*

$$\dot{x} = Ax + bu$$
$$y = cx.$$

Hint: *Write* (2.145) *as an nth-order differential equation, and then define*

$$x_1 = y, x_2 = \dot{y}, x_3 = \ddot{y}, \ldots, x_n = y^{(n-1)}.$$

First, realize

$$\frac{b_0}{s^n + a_{n-1}s^{n-1} + a_{n-2}s^{n-2} + \cdots + a_1 s + a_0},$$
$$\frac{b_1 s}{s^n + a_{n-1}s^{n-1} + a_{n-2}s^{n-2} + \cdots + a_1 s + a_0},$$
$$\cdots$$
$$\frac{b_{n-1}s^{n-1}}{s^n + a_{n-1}s^{n-1} + a_{n-2}s^{n-2} + \cdots + a_1 s + a_0}$$

separately, then use superposition.

PROBLEM 2.12 *Consider the linear system*

$$\dot{x} = A(t)x, \; t \in [t_0, t_f]$$
$$y = C(t)x$$
$$x(t_0) = x_0.$$

Assume that t_f is fixed and that we want to compute the sensitivities $\frac{\partial y(t_f)}{\partial x_0}$ and $\frac{\partial y(t_f)}{\partial t_0}$ for ***all*** *values of $t_0 \leq t_f$. Show that the method of adjoints solves this problem with just* ***one*** *backward integration.*

PROBLEM 2.13 *We are given a nonlinear differential equation and a function J of the final state, that is,*

$$\dot{x} = f(x, u, t), \; t \in [t_o, t_f] \tag{2.146}$$

$$x(t_0) = x_0 \tag{2.147}$$

$$J = g(x(t_f)), \tag{2.148}$$

where t_0, t_f, x_0 are given and the functions f and g are differentiable. Assume that we are given a nominal trajectory $x^0(t), u^0(t), \; t \in [t_0, t_f]$, and assume that the nominal input is perturbed by a small $\delta u(t), \; t \in [t_0, t_f]$. Use the method of adjoints to compute δJ, the first-order perturbation of J due to δu.

Hint: *Write*

$$\delta J = \left(\frac{\partial g}{\partial x}\right)^T_{x^0(t_f)} = c\delta x(t_f),$$

where c is a constant row vector. Start by linearizing equation (2.146) around the nominal trajectory to obtain a linear differential equation relating $\delta u(t)$ with $\delta x(t)$, $t \in [t_0, t_f]$.

PROBLEM 2.14 *The equations for the lateral dynamics of a* ***bank-to-turn*** *aircraft are given as*

$$\dot{\psi} = k_1\phi$$
$$\ddot{\phi} = k_2 u,$$

where ψ is the heading angle, ϕ is the roll angle, u is the aileron deflection, and the constants k_1 and k_2 are nonzero. Note that this system is a triple integrator.

1. *Is this system observable through ψ?*
2. *Is this system observable through ϕ?*

In both cases, specify the unobservable states.

PROBLEM 2.15 ***Inertial navigation*** *is a navigation procedure that consists of measuring the acceleration of a vehicle and integrating from measured initial conditions to obtain the current position.* ***Dead-reckoning*** *does the same based on velocity measurements.*

Consider the motion of a vehicle along a straight line. The equation of motion is

$$\ddot{x} = u,$$

where x is the position and u is the acceleration. Assume that inertial navigation is used. Let $\epsilon_1 > 0$, $\epsilon_2 > 0$ be bounds on the measurement errors of initial position and velocity, respectively. Assume that the accelerometer measures u with an error bounded by $\epsilon_a > 0$.

1. *Obtain a bound for the position error at the end of a trip of duration T.*
2. *Show that inertial navigation, when used alone, is impractical for trips of long duration.*

3 Stochastic Systems Theory

This chapter presents the elements of stochastic systems theory that we use in the analysis and synthesis of navigation and guidance systems. Roughly speaking, a stochastic phenomenon is randomly unpredictable but exhibits "statistical regularity," in a sense defined subsequently. Consistent with our emphasis on linear systems, justified in Chapter 2, the purpose of this chapter is to quantify how linear systems respond to uncertain input signals.

We use stochastic models to account for uncertainty for at least two compelling reasons. The first one is purely pragmatic: stochastic models have proven extremely effective in modeling uncertainty in science and engineering and provide a rational basis for taking and optimizing decisions in the presence of the unknown. The second, deeper reason, appeals to the **fundamental postulate of natural science** [24], which states that "nature is governed by laws that are uniform." Although this postulate leads us to expect that natural phenomena be highly and exactly repeatable, the uncertainty we observe in practice seems to violate this expectation. A way to resolve this dilemma is to relax the notion of exact repeatability into that of statistical regularity, leading seamlessly to stochastic systems theory. Hence stochastic systems theory affords a provision for uncertainty within the context of the fundamental postulate of natural science.

The motivation of this chapter is that many phenomena in navigation and guidance can be modeled as stochastic. They include measurement errors in components of acceleration, velocity and position, drift of gyroscopes, wind gusts, and so on, and the navigation and guidance errors that the aforementioned generate. In fact, it can be claimed that without such phenomena, there would really be no substantive navigation or guidance problem. Hence, in practical terms, this chapter quantifies the response of linear dynamic systems to stochastic inputs.

Sections 3.1–3.5 focus on the theory of random variables and Sections 3.6–3.8 on the theory of random processes. Specifically, Section 3.1 presents the basic building blocks of probability theory, and Section 3.2 utilizes these blocks to define random variables and their probability distribution. Section 3.3 presents the notions of expected value and characteristic function. Section 3.4 discusses independence and correlation. Section 3.5 presents and motivates the Gaussian distribution. Sections 3.6 and 3.7 introduce random and Gauss–Markov processes, respectively. Section 3.8 introduces linear Gauss–Markov models, which we use in navigation and guidance. Sections 3.9, 3.10, and 3.11 present a summary of the key results in the chapter, bibliographic notes for further reading, and homework problems, respectively.

3.1 Probability Spaces

Let us start by defining an **event**, roughly speaking, as a possible outcome of an uncertain experiment. An example of an event could be "to measure the wind velocity v at a specified location and time and have 1 m/s $\leq v \leq$ 2 m/s." Let A be an event. Assume that we set up a number of runs, indexed by i, and that in run i we perform the experiment of which A is a possible outcome N_i times (e.g., we measure the wind velocity N_i times). Assume that, in those N_i trials, event A occurs $n_i(A)$ times. We say that event A is **statistically regular** if there exists a number $P(A)$ such that

$$\forall i, \lim_{N_i \to \infty} \frac{n_i(A)}{N_i} = P(A). \tag{3.1}$$

In other words, from run to run, as the number of trials increases, the proportion of occurrences of A tends to the same limit, independent of the run.

DEFINITION 3.1 *Let event A be statistically regular. The* ***probability*** *of A is the limit in* (3.1). *If $P(A) = 1$, we say that event A is* ***certain****.*

Definition 3.1 points to a notion of repeatability that is weaker than the familiar exact repeatability: what is repeatable is the proportion of occurrences of A over a number of trials that is large enough, rather than A itself over each trial.

Definition 3.1 has at least two possible deficiencies. First, it requires an infinite (or at least "large") number of experiments. Second, one may not be able, or willing, to actually perform those experiments – for instance, one may quite understandably be reluctant to determine experimentally the probability of being shot in a game of Russian roulette. In view of these limitations, probability theory can be viewed as a necessary formalization of this intuitive notion of "frequency of occurrence." However, Definition 3.1 may be quite sufficient for the technical purpose at hand. This is the case in Monte Carlo simulations, for instance.

DEFINITION 3.2 *A* ***sample space*** *Ω is the set of all possible outcomes of an uncertain experiment. The elements of the sample space are called the* ***samples****.*

EXAMPLE 3.1 *Assume that we measure the wind velocity once, as earlier. Then, for this measurement, a reasonable sample space is the set of real numbers $\mathbb{R}$. Assume we measure the wind velocity at the same location, in the same direction, but at two different given times. In this case, a reasonable sample space is $\mathbb{R}^2$, the set of couples of real numbers. Assume that, at a given location, we measure simultaneously the wind velocity, the atmospheric pressure, and the temperature. Because in this case we know that the wind velocity is a real number, the atmospheric pressure is a real positive number, the temperature (on the Celsius scale) is a number greater than -273, a reasonable sample space is the Cartesian product $\mathbb{R} \times (0, \infty) \times (-273, \infty)$.*

DEFINITION 3.3 *Given a sample space Ω, an* **event** *A is a subset of Ω.*

EXAMPLE 3.2 *Assume that the experiment is to measure wind velocity once, as earlier. The sample space is $\mathbb{R}$. Then, the interval $[1, 2] = \{v | 1 \leq v \leq 2\}$ is an event. Similarly, the union of two intervals $(-\infty, 1) \cup (2, \infty) = \{v | v < 1 \text{ or } v > 2\}$ is also an event.*

DEFINITION 3.4 *Given a sample space Ω, a* **σ-algebra** *F is a class of events, closed under complementation and countable union. By this, we mean that*

$$A \in F \Rightarrow A^* \in F, \text{ where } A^* = \{\omega \in \Omega : \omega \notin A\}$$

$$A_1, A_2, \ldots \in F \Rightarrow \bigcup_{i=1}^{\infty} A_i \in F. \tag{3.2}$$

EXAMPLE 3.3 *Given any sample space Ω, the set $F_1 = \{\Omega, \emptyset\}$ where $\emptyset$ represents the empty set is a σ-algebra. In fact, it is the "smallest" σ-algebra in the sense that every σ-algebra must be a superset of F_1. Similarly, the set of all subsets of Ω, $F_2 = 2^{\Omega}$, is also a σ-algebra. In fact, it is the "largest" σ-algebra in the sense that every σ-algebra must be a subset of F_2.*

DEFINITION 3.5 *Given a sample space Ω and a σ-algebra F, a* **probability function** *is a function $P : F \to \mathbb{R} : A \mapsto P(A)$ satisfying the three conditions*

$$\forall A, P(A) \geq 0,$$

$$P(\Omega) = 1 \text{ (that is, the event } \Omega \text{ is certain)},$$

$$\forall A_1, A_2, \ldots \in F \text{ such that } A_i \cap A_j = \emptyset \text{ for } i \neq j, P\left(\bigcup_{i=1}^{\infty} A_i\right) = \sum_{i=1}^{\infty} P(A_i). \tag{3.3}$$

REMARK 3.1 *Definition 3.5 looks abstract; however, the reader is already familiar with several functions that have properties (3.3). Indeed, consider, for instance, a three-dimensional object of volume 1 and the volume of its sub-objects. It is clear that the volume of every sub-object is always greater than or equal to zero. Also, the volume of the object is 1 by assumption. Finally, whenever sub-objects are disjoint, the volume of their union is exactly the sum of their volumes. Therefore, the function "volume," which to a sub-object associates its volume, behaves exactly as specified by (3.3). Another similar example is obtained by considering an object of mass 1 and the masses of all of its sub-objects. The same argument shows that the function "mass," which to a sub-object associates its mass, behaves exactly as specified by (3.3). The preceding analogies lead to the following question: If one pursues the analogy between probability function and mass, what concepts are analogous to those of mass density, center of mass, and inertia matrix? The answers consist of the concepts of probability density function, expected value, and covariance matrix, respectively, which are developed in the next three sections.*

When two events A_i and A_j satisfy $A_i \cap A_j = \emptyset$, we say that the events A_i and A_j are **mutually exclusive.** From Definition 3.5, it should be clear that probability functions are "nondecreasing" in the sense that if $A \subset B$, then $P(A) \leq P(B)$. Indeed, $B = A \cup (A^* \cap B)$ and $A \cap (A^* \cap B) = \emptyset$ imply that $P(B) = P(A) + P(A^* \cap B)$. Also, it should be clear that a probability function is always between 0 and 1. Indeed, if A is an event, then $A \subset \Omega$, implying $P(A) \leq P(\Omega) = 1$. Hence, $0 \leq P(A) \leq 1$.

DEFINITION 3.6 *A **probability space** is a triple (Ω, F, P) where Ω is a sample space, F is a σ-algebra, and P is a probability function as defined earlier.*

DEFINITION 3.7 *If A and B are events of a probability space, the **conditional probability** of A given B is defined as*

$$P(A|B) = \frac{P(A \cap B)}{P(B)} \quad \text{if } P(B) \neq 0. \tag{3.4}$$

3.2 Random Variables and Distributions

The next step in our formalization of random phenomena is to introduce some "structure" in the sample space. Roughly speaking, we would like to perform on samples the same operations as on vectors of real numbers, such as sum, difference, and multiplication by a scalar. We would also like to account for the fact that samples pertaining to the same experiment may be functionally related, just like a real vector may be a function of another real vector. Clearly the simplest way to do this is to define a function that, to a sample, associates a vector in $\mathbb{R}^n$. This procedure is similar to that in analytic geometry, which, using the coordinate map, associates to a point in the plane the vector in $\mathbb{R}^2$ containing its Cartesian coordinates. This leads to the classical notion of a random vector, as the image of this map. Before giving the formal definition of a random vector, we need to define the following notation.

DEFINITION 3.8 *For $a \in \mathbb{R}^n$ and $b \in \mathbb{R}^n$, we say that*

$$a \leq b \quad \text{if} \quad a_1 \leq b_1, a_2 \leq b_2, \ldots, a_n \leq b_n. \tag{3.5}$$

We can now give the following definition.

DEFINITION 3.9 *Given a probability space (Ω, F, P), a **random vector** is a function $x : \Omega \to \mathbb{R}^n : \omega \mapsto x(\omega)$ satisfying the two conditions*

$$\begin{aligned} \forall a \in \mathbb{R}^n, \{\omega : x(\omega) \leq a\} &\in F \\ P\{\omega : x(\omega) = \pm\infty\} &= 0. \end{aligned} \tag{3.6}$$

*The components of a random vector are called **random variables**.*

REMARK 3.2 *We often consider that the random vector is the image of the map in (3.6) rather than the map itself. This slight abuse of language is exactly the same as that which, in analytic geometry, causes us to say "the coordinates of a point" rather than "the image of the point by the coordinate map." This type of abuse – identifying a map with its image – is quite common in mathematics and should not cause any confusion.*

EXAMPLE 3.4 *Assume that, like in Example 3.1, we measure simultaneously at the same location the wind velocity v, the atmospheric pressure p, and the temperature T. The sample space and the σ-algebra are*

$$\begin{aligned} \Omega &= \mathbb{R} \times \mathbb{R}^+ \times (-273, +\infty), \\ F &= 2^\Omega. \end{aligned} \tag{3.7}$$

Samples have the form $\omega = (v, p, T)$ where v is the wind velocity, p is the atmospheric pressure, and T is the temperature. Events are sets of samples, or equivalently subsets of Ω. To every sample ω, we associate $x(\omega) = \omega$ by the identity. Hence,

$$x : \Omega \to \mathbb{R}^3 : \omega \mapsto x(\omega) = \omega. \tag{3.8}$$

Then, for all v_0, p_0, and T_0, sets of samples of the form

$$\{\omega = (v, p, T) : v \leq v_0, p \leq p_0, T \leq T_0\} \tag{3.9}$$

are events. Moreover, samples are always finite, implying that ω is a random vector.

In our work, the sample space Ω has the form $\Omega = \mathbb{R}^n$ and the events are subsets of $\mathbb{R}^n$. The samples, that is, the elements of $\mathbb{R}^n$, are considered the random vectors. In other words, we choose the random vector of Definition 3.9 as the identity function, and we identify the random vectors with the samples. Events are then subsets of $\mathbb{R}^n$, and a probability function associates a real number to every event according to Definition 3.5.

In Remark 3.1, we argued that a probability density function behaves like the mass function. Following this analogy, we now introduce the concept that plays a similar role to that of "mass density."

DEFINITION 3.10 *Let $x \in \mathbb{R}^n$ be a random vector. The* ***probability density function*** *of x is any (possibly generalized) function f_x such that, for every $a \in \mathbb{R}^n$, the probability of the event $x \leq a$ is given by the multiple integral*

$$P(x \leq a) = \int_{-\infty}^{a} f_x(b)\, \mathrm{d}b. \tag{3.10}$$

If $x = \left(x_1^T, x_2^T\right)^T$, then f_x is called the ***joint probability density function*** *of x_1 and x_2.*

An important consequence of Definition 3.10 is that, whenever $D \subset \mathbb{R}^n$ is a quadrable domain (i.e., its boundary, ∂D, has measure zero), its probability is given by the multiple integral

$$P(x \in D) = \int_D f(\xi)\, \mathrm{d}\xi. \tag{3.11}$$

In particular, because every probability function must satisfy (3.3) (line 2), the probability density function of every random vector must satisfy

$$\int_{\mathbb{R}^n} f(\xi)\, \mathrm{d}\xi = 1. \tag{3.12}$$

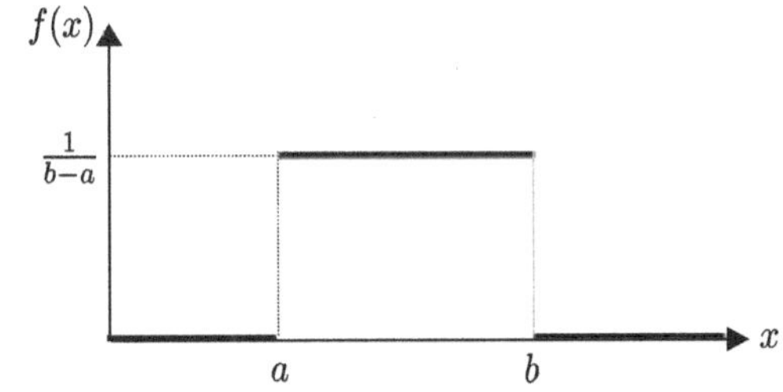

Figure 3.1. Uniform distribution.

In (3.11), if we consider a domain of differentially small size, we obtain the interpretation

$$f(\xi_1, \xi_2, \ldots, \xi_n)\, \mathrm{d}\xi_1\, \mathrm{d}\xi_2 \ldots \mathrm{d}\xi_n \text{ represents } P(\xi \leq x \leq \xi + \mathrm{d}\xi), \tag{3.13}$$

which justifies the use of the word "density" in the phrase "probability density function." The interpretation (3.13) together with the definition in (3.3) (line 1) imply that every probability density function satisfies

$$\forall x,\ f_x(x) \geq 0. \tag{3.14}$$

EXAMPLE 3.5 *Note that in Definition 3.10, generalized functions allow for the possibility of a Dirac delta as a probability density function. To illustrate this, let $a \in \mathbb{R}$ be a constant, and let $x \in \mathbb{R}$ be a random variable. To enforce the equality $x = a$ with certainty, we ascribe to x the probability density function*

$$f_x(x) = \delta\,(x - a)\,. \tag{3.15}$$

This is similar to, in physics, saying that the mass density function of a point mass is proportional to a Dirac delta function.

EXAMPLE 3.6 *A random vector is said to be* ***uniformly distributed*** *if its probability density function is constant. The value of that constant is specified by the requirement (3.12). Consider a random variable that is uniformly distributed over an interval $[a, b]$. The probability density function is given by*

$$f(x) = \begin{cases} \frac{1}{b-a} & \text{if } x \in [a, b] \\ 0 & \text{if } x \notin [a, b] \end{cases} \tag{3.16}$$

and is shown in Figure 3.1.

For a random vector that is uniformly distributed over the domain

$$D = [a_1, b_1] \times [a_2, b_2] \times \cdots \times [a_n, b_n], \tag{3.17}$$

the probability density function is

$$f(x) = \begin{cases} \prod_{i=1}^{n} \frac{1}{b_i - a_i} & \text{if } x \in D \\ 0 & \text{if } x \notin D. \end{cases} \tag{3.18}$$

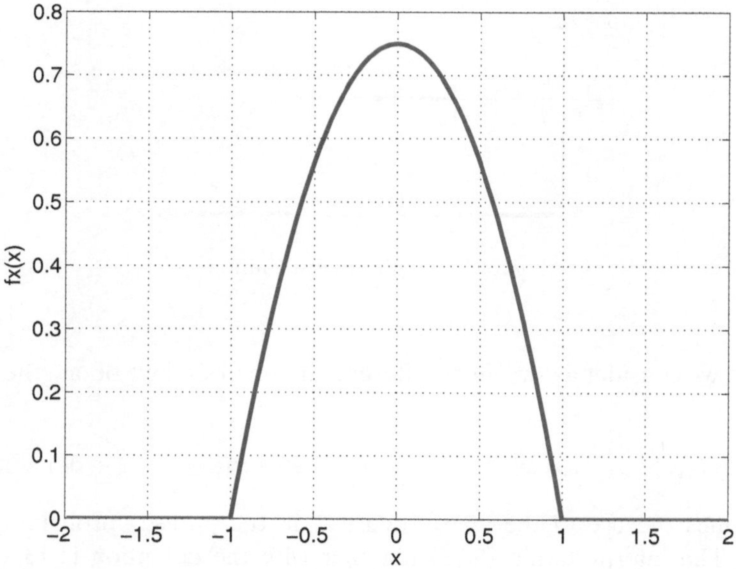

Figure 3.2. f_x, probability density function of the parabolic distribution with radius R (plotted for $R = 1$).

EXAMPLE 3.7 *Consider the* ***parabolic distribution with radius*** *R. Its probability density function, represented in Figure 3.2, is*

$$f_x(x) = \begin{cases} \frac{3}{4R^3}(R^2 - x^2) & \text{if } x^2 \le R^2 \\ 0 & \text{if } x^2 > R^2, \end{cases} \tag{3.19}$$

which satisfies (3.12).

The parabolic distribution can be generalized to several dimensions. For instance, the ***two-dimensional parabolic distribution of radius 1*** *has probability density function*

$$f_{xy}(x, y) = \begin{cases} \frac{2}{\pi}(1 - x^2 - y^2) & \text{if } x^2 + y^2 \le 1 \\ 0 & \text{if } x^2 + y^2 > 1, \end{cases} \tag{3.20}$$

which satisfies (3.12) and is represented in Figure 3.3.

EXAMPLE 3.8 *Consider again a random variable that is uniformly distributed over the interval $[a, b]$, as given by (3.16). Let $[c, d]$ be a subinterval of $[a, b]$. Then the*

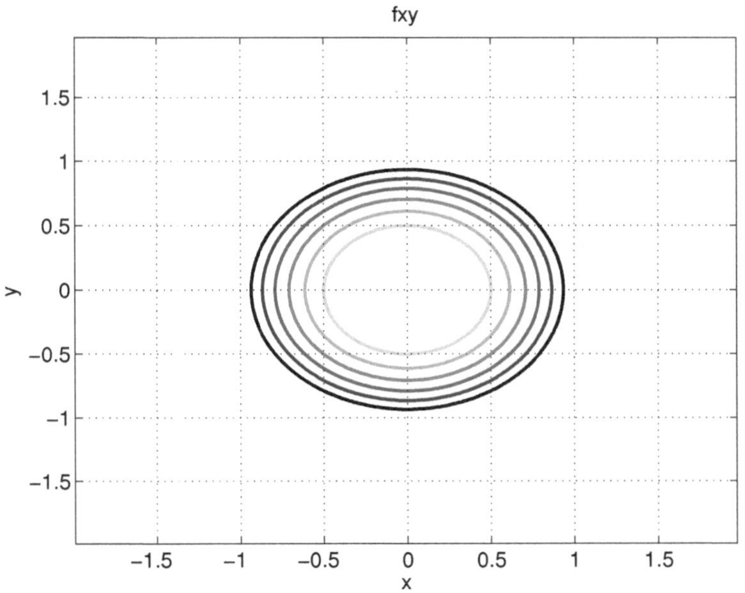

Figure 3.3. Level curves of f_{xy}, probability density function for two-dimensional parabolic distribution of radius 1.

probability of the subinterval is given by

$$P(x \in [c, d]) = \int_{[c,d]} f(\xi)\, \mathrm{d}\xi$$
$$= \frac{d-c}{b-a}. \tag{3.21}$$

For the parabolic distribution with radius R (3.19), let $[a, b]$ be a subinterval of $[-R, R]$. Then, the probability of the subinterval is

$$P(x \in [a, b]) = \int_a^b \frac{3}{4R^3}(R^2 - x^2)\, \mathrm{d}x \tag{3.22}$$
$$= \frac{3}{4R^3}\left(R^2(b-a) - \frac{b^3 - a^3}{3}\right).$$

DEFINITION 3.11 *If $x = (x_1^T, x_2^T)^T$ is a random vector, x_1, x_2 are subvectors of x, and x has joint probability density function $f_x(x_1, x_2)$, then the* ***marginal probability density function*** *of x_1 is*

$$f_{x_1}(x_1) = \int_{-\infty}^{\infty} f_x(x_1, x_2)\, \mathrm{d}x_2. \tag{3.23}$$

DEFINITION 3.12 *If $x = (x_1^T, x_2^T)^T$ is a random vector, x_1, x_2 are subvectors of x, and x has joint probability density function $f_x(x_1, x_2)$, then the* ***conditional probability density function*** *of x_1 given x_2 is, from* ***Bayes's rule****,*

$$f(x_1|x_2) = \frac{f_x(x_1, x_2)}{f_x(x_2)}. \tag{3.24}$$

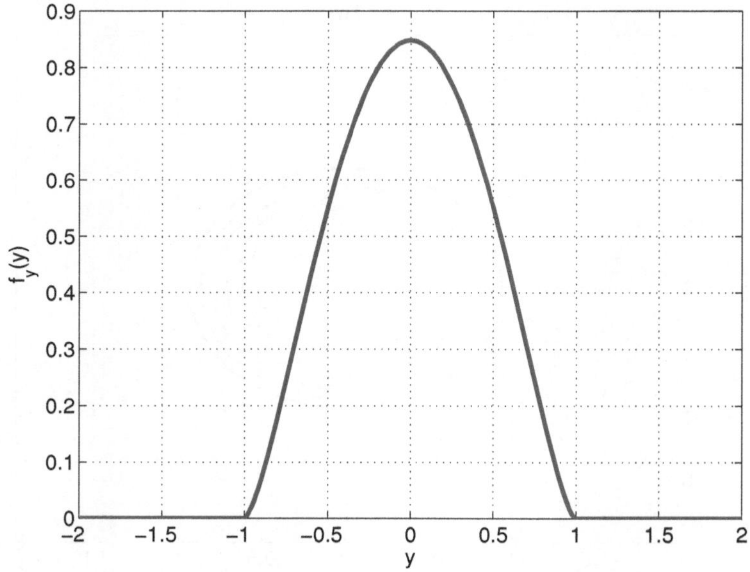

Figure 3.4. f_y, marginal probability density function of a subvector of a vector that is parabolically distributed over the unit disk.

REMARK 3.3 *The marginal and conditional probability functions are probability functions in their own right; that is, they satisfy properties* (3.10), (3.11), *and* (3.12).

EXAMPLE 3.9 *Consider the two-dimensional parabolic distribution with radius 1* (3.20). *The marginal probability density function of y is*

$$f_y(y) = \int_{-\infty}^{\infty} f_{xy}(x, y)\, \mathrm{d}y \tag{3.25}$$

$$= \frac{2}{\pi} \int_{-\sqrt{1-y^2}}^{\sqrt{1-y^2}} (1 - x^2 - y^2)\, \mathrm{d}x$$

$$= \begin{cases} \frac{8}{3\pi}(1 - y^2)\sqrt{1 - y^2} & \textit{if } y^2 \leq 1 \\ 0 & \textit{if } y^2 > 1 \end{cases},$$

which satisfies (3.16) *and is represented in Figure 3.4.*

The conditional distribution of x given y is

$$f(x|y) = \frac{f_{xy}(x, y)}{f_y(y)}$$

$$= \begin{cases} \frac{3}{4} \frac{(1-x^2-y^2)}{(1-y^2)\sqrt{1-y^2}} & \textit{if } x^2 \leq 1 - y^2 \text{ and } y^2 \leq 1 \\ 0 & \textit{otherwise,} \end{cases} \tag{3.26}$$

which satisfies (3.16) *and is represented in Figure 3.5 for several values of y. Note that, as per* (3.26), *the conditional distribution of x given y is itself parabolic with radius* $\sqrt{1 - y^2}$.

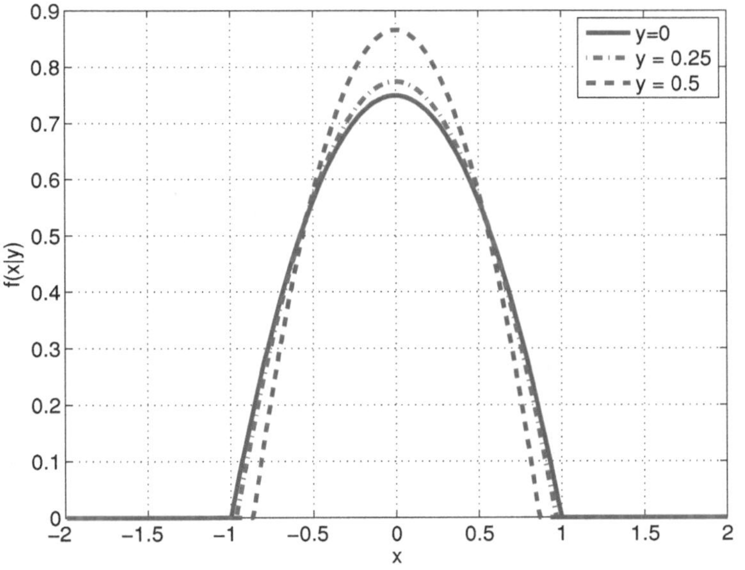

Figure 3.5. $f(x|y)$, conditional probability density function of a subvector of a vector that is parabolically distributed over the unit disk.

3.3 Expected Value and Characteristic Function

DEFINITION 3.13 *If $x \in \mathbb{R}^n$ is a random vector with probability density function $f(x)$ and $g : \mathbb{R}^n \to \mathbb{R}^m$ is a function of x, the* ***expected value*** *of g is the multiple integral*

$$E[g(x)] = \int_{\mathbb{R}^n} g(x)f(x)\,\mathrm{d}x \in \mathbb{R}^m. \tag{3.27}$$

The ***mean value*** *of x is the expected value of x, that is, the multiple integral*

$$\bar{x} = E[x] = \int_{R^n} xf(x)\,\mathrm{d}x \in \mathbb{R}^n. \tag{3.28}$$

The ***covariance matrix*** *of x is the multiple integral*

$$\begin{aligned} P_{xx} &= E[(x-\bar{x})(x-\bar{x})^T] \\ &= \int_{R^n} (x-\bar{x})(x-\bar{x})^T f(x)\,\mathrm{d}x \in \mathbb{R}^{n\times n}. \end{aligned} \tag{3.29}$$

If x_1, x_2 are subvectors of x, the ***cross-covariance matrix*** *of x_1 and x_2, $P_{x_1x_2}$, is the multiple integral*

$$E[(x_1-\bar{x}_1)(x_2-\bar{x}_2)^T] = \int_{R^n} (x_1-\bar{x}_1)(x_2-\bar{x}_2)^T f(x)\,\mathrm{d}x \in \mathbb{R}^{n_1\times n_2}. \tag{3.30}$$

In Definition 3.13, the mean value $\bar{x}$ denotes the average value of the random vector and plays a role similar to that of the center of mass, where the probability density function plays the role of mass density. The covariance matrix P_{xx} contains the second central moments and plays a role similar to that of the inertia matrix for mass distribution. The covariance matrix is obviously symmetric. It is also positive semidefinite in the sense that

$$\forall v \in \mathbb{R}^n, v^T P_x v \geq 0. \tag{3.31}$$

Indeed,

$$\begin{aligned} v^T P_x v &= \int_{\mathbb{R}^n} v^T (x - \bar{x})(x - \bar{x})^T v f(x)\,\mathrm{d}x \\ &= \int_{\mathbb{R}^n} [(x - \bar{x})]^2 f(x)\,\mathrm{d}x \geq 0. \end{aligned} \tag{3.32}$$

The nondiagonal elements of the covariance matrix have the form

$$\begin{aligned} P_{ij} &= \int_{\mathbb{R}^n} (x_i - \bar{x}_i)(x_j - \bar{x}_j)^T f(x)\,\mathrm{d}x \\ &= \int_{\mathbb{R}^2} (x_i - \bar{x}_i)(x_j - \bar{x}_j)^T f_{x_i x_j}(x_i, x_j)\,\mathrm{d}x_i\,\mathrm{d}x_j \end{aligned} \tag{3.33}$$

and denote the coupling between the ith and jth components of x. The diagonal elements of the covariance matrix have the form

$$P_{ii} = \int_{\mathbb{R}^2} (x_i - \bar{x}_i)^2 f_{x_i}(x_i)\,\mathrm{d}x_i\,\mathrm{d}x_i \geq 0 \tag{3.34}$$

and denote the "spread" of the ith component of x around its mean value. The **standard deviation** of the ith component of x is defined as $\sqrt{P_{ii}}$. Also note that the covariance matrix can be computed as

$$P_x = \int_{\mathbb{R}^n} xx^T f(x)\,\mathrm{d}x - \bar{x}\bar{x}^T. \tag{3.35}$$

DEFINITION 3.14 *If x is a random vector of order n with probability density function $f(x)$, the* ***characteristic function*** *of x is*

$$\phi_x(s) = E(e^{jx^T s}) = \int_{\mathbb{R}^n} (e^{jx^T s}) f(x)\,\mathrm{d}x, \tag{3.36}$$

where $j^2 = -1$ and s is a complex vector of order n.

Definition 3.14 is very reminiscent of Fourier transforms, for which there is a well-known inversion formula. Similarly, here we have the inversion formula

$$f(x) = \frac{1}{(2\pi)^n} \int_{\mathbb{R}^n} e^{-jx^T s} \phi_x(s)\,\mathrm{d}s. \tag{3.37}$$

The statistical properties of a random vector x are equivalently specified by its probability density function $f(x)$ or by its characteristic function $\phi_x(s)$. The characteristic function $\phi_x(s)$ is useful to compute the probability density functions of functions of x (see Problem 3.15) and to define the Gaussian distribution.

3.4 Independence and Correlation

DEFINITION 3.15 *Two random vectors x and y are* ***independent*** *if their joint probability density function (or equivalently their joint characteristic function) can be factored as*

$$f(x, y) = f_x(x)f_y(y)$$
$$\phi_{xy}(s, r) = \phi_x(s)\phi_y(r), \tag{3.38}$$

respectively.

If x and y are independent, then the conditional density function and conditional mean of x given y satisfy

$$f(x|y) = \frac{f_x(x)f_y(y)}{f_y(y)} = f_x(x)$$
$$E(x|y) = \int_{R^n} xf(x|y)\,\mathrm{d}x = \int_{R^n} xf_x(x)\,\mathrm{d}x = E(x) = \bar{x}. \tag{3.39}$$

DEFINITION 3.16 *Two random vectors x and y are* ***uncorrelated*** *if*

$$E[xy^T] = E[x]E[y^T]. \tag{3.40}$$

If x and y are uncorrelated, then their cross-correlation matrix P_{xy} must be zero. Indeed,

$$\begin{aligned} P_{xy} &= E[(x - \bar{x})(y - \bar{y})^T] \\ &= E[xy^T] + E[\bar{x}\bar{y}^T] - E[x\bar{y}^T] - E[\bar{x}y^T] \\ &= 0. \end{aligned} \tag{3.41}$$

For x and y scalar, it is customary to define

$$P_{xy} = \frac{E[(x - \bar{x})(y - \bar{y})]}{\sqrt{E[(x - \bar{x})^2]}\sqrt{E[(y - \bar{y})^2]}} \tag{3.42}$$

as their **correlation coefficient**. Then, x and y are uncorrelated if and only if their correlation coefficient is zero. Moreover, we always have

$$|P_{xy}| \leq 1. \tag{3.43}$$

In general, when x and y are independent vectors, they are necessarily uncorrelated. However, the converse is generally not true, see Problem 3.8.

3.5 The Gaussian Distribution

In this section, we introduce the Gaussian, or normal, distribution. This probability density function is very widely used in practice for three reasons: first, it provides a good stochastic model for many natural phenomena (see the central limit theorem in Proposition 3.1). Second, it leads to a computationally tractable formalism because the statistical properties of a normal variable are described by its first two moments alone. Third, and most importantly, normality is preserved through linear transformations, both static and dynamic.

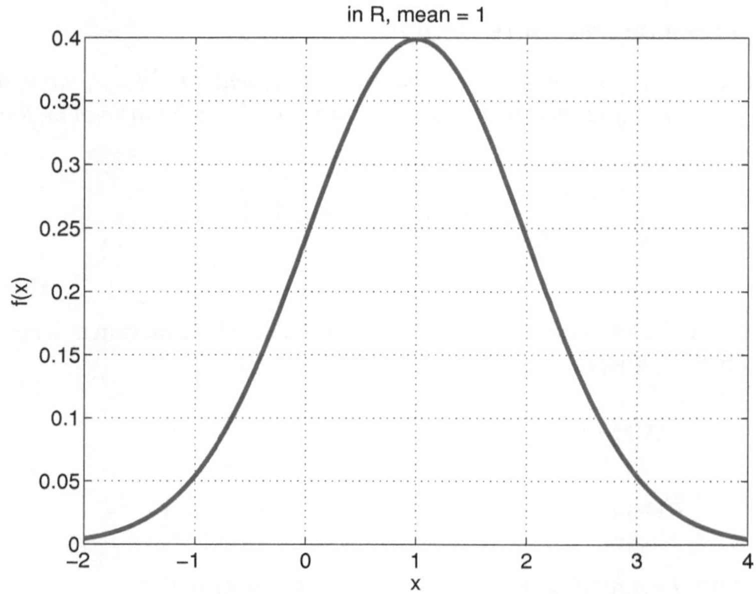

Figure 3.6. Probability density of a Gaussian random variable.

DEFINITION 3.17 *A random vector* $x \in \mathbb{R}^n$ *is* ***Gaussian*** *distributed or* ***normal*** *if its characteristic function has the form*

$$\phi_x(s) = \exp\left(j\bar{x}^T s - \frac{1}{2}s^T P s\right), \tag{3.44}$$

where

$$s \in \mathbb{C}^n,\ \bar{x} = E(x),\ P = [(x-\bar{x})(x-\bar{x})^T]. \tag{3.45}$$

For such a random vector, we use the notation

$$x = \mathcal{N}(\bar{x}, P). \tag{3.46}$$

Two vectors x and y are ***jointly Gaussian distributed*** *if* $(x^T, y^T)^T$ *is Gaussian.*

When a random vector x is Gaussian and its covariance matrix P_x is nonsingular, its probability density function can be evaluated as

$$f(x) = \frac{1}{\sqrt{(2\pi)^n \det(P_x)}} \exp\left(-\frac{1}{2}(x-\bar{x})^T P_x^{-1}(x-\bar{x})\right). \tag{3.47}$$

However, $f(x)$ cannot be properly evaluated if the covariance matrix is singular, even though the random vector still has a probability density function in this case (see Problem 3.14). This is the reason we define the Gaussian distribution indirectly through its characteristic function instead of directly through its probability density function.

For $n = 1$, the probability density function of a normally distributed variable is the familiar "bell curve" of Figure 3.6.

For a Gaussian random vector x, the "level curves" $f(x) = $ constant are

$$(x-\bar{x})^T P_x^{-1}(x-\bar{x}) = \text{constant}. \tag{3.48}$$

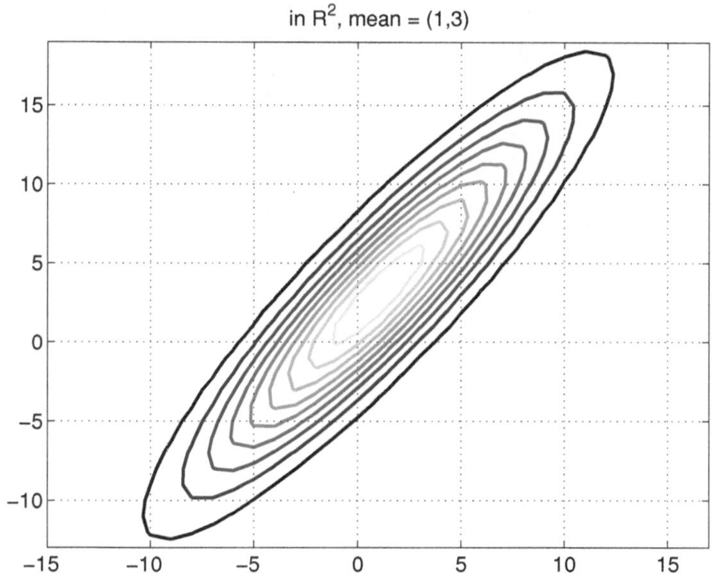

Figure 3.7. Level curves of probability density function of a second-order Gaussian vector.

In $\mathbb{R}^n$, these represent a bundle of concentric, coaxial, similar ellipsoids. Specifically, these ellipsoids are all centered at $\bar{x}$, have for principal directions the eigenvectors of P, and have axes that are proportional to the square roots of the eigenvalues of P. For instance, in $\mathbb{R}^2$ we have the ellipses shown in Figure 3.7.

The following level sets are of particular interest. For a given probability p, the level set determined by the constant c such that

$$P\left((x-\bar{x})^T P_x^{-1}(x-\bar{x}) \leq c\right) = p \tag{3.49}$$

is called the p probability ellipsoid. It can be shown [57] that p and c are related by

$$p = \int_0^c \frac{\xi^{\frac{n}{2}-1} e^{-\frac{\xi}{2}}}{2^{\frac{n}{2}} \Gamma\left(\frac{n}{2}\right)} \, \mathrm{d}\xi, \tag{3.50}$$

where the Gamma function is defined by

$$\Gamma(z) = \int_0^\infty t^{z-1} e^{-t} \, \mathrm{d}t. \tag{3.51}$$

Hence, the p probability ellipsoid is an ellipsoid whose probability is exactly p.

We now give two basic results on Gaussian distributions that justify their widespread use. According to the first result, if a random variable is the superposition of a large number of "elementary" independent, identically distributed random variables, then it is normally distributed in the limit. The remarkable fact is that this holds regardless of the statistical properties of the elementary random variables. In practice, this means that when a macroscopic phenomenon results from the superposition of many microscopic phenomena, its statistical properties can often be approximated by those of a normal distribution.

PROPOSITION 3.1 (Central Limit Theorem) *Let $x^i, 1 \leq i \leq r$, be independent, identically distributed random n-vectors with finite means $\bar{x}$ and covariances P. Let*

$$y^r = \sum_{i=1}^{r} x^i \text{ and } z^r = (P^r)^{-1}(y^r - \bar{y}^r), \tag{3.52}$$

where

$$\bar{y}^r = \sum_{i=1}^{r} \bar{x}^i.$$

Then,

$$\lim_{r\to\infty} f(z^r) = \frac{1}{\sqrt{(2\pi)^n}} \exp\left(-\frac{1}{2} z^T z\right). \tag{3.53}$$

The second basic result foreshadows the principle that Gaussian distributions are closed under linear transformation.

PROPOSITION 3.2 *If x is Gaussian with mean $\bar{x}$ and covariance P_x and $y = Ax + b$, then y is Gaussian with mean*

$$\bar{y} = A\bar{x} + b \tag{3.54}$$

and covariance

$$P_y = AP_xA^T. \tag{3.55}$$

Recall that our purpose in this chapter is to understand how linear systems behave under stochastic inputs. Now, (3.54) and (3.55) precisely quantify how the statistical properties of a random vector are transformed by a linear static operator. The generalization to a linear dynamic operator driven by a random function is studied in Section 3.8.

3.6 Random Processes

In this section we introduce time variance in random phenomena, which leads to the notion of random process. The motivation for doing so is clearly that, in navigation and guidance, the random phenomena of interest are often time varying.

DEFINITION 3.18 *A **stochastic** or **random process** is a family of random vectors $\{x(t), t \in I\}$ indexed by time. Such a family of random vectors is also called an **ensemble**.*

An equivalent definition of random process is that of a function of two variables, one of which is time, the other being a random vector. Formally, a random process is then a function $\{x(t, \omega), t \in I, \omega \in \Omega\}$, where $x(t, .)$ is a random vector (for t fixed) and $x(., \omega)$ is a function of time (for ω fixed). In the latter case, such a function is often called a **realization** or **sample function** of the process.

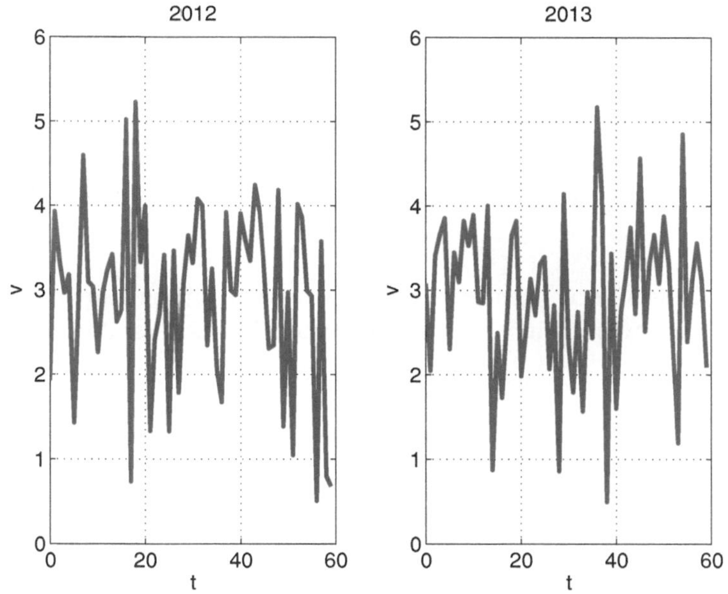

Figure 3.8. Sample functions of wind velocity at given location, in given direction, on June 5 from 12 to 1 P.M.

EXAMPLE 3.10 *Assume that we measure the wind velocity at a given location in a given direction on June 5 from noon to 1 P.M. We may then obtain plots such as those in Figure 3.8. Each one of those plots is a realization of the random process. Also, for a given time (say, 12:08 P.M), the wind velocity is a random variable.*

EXAMPLE 3.11 *Consider the free response of the double integrator system of Example 2.5, that is,*

$$\begin{bmatrix} x_1(t) \\ x_2(t) \end{bmatrix} = \begin{bmatrix} x_{10} + tx_{20} \\ x_{20} \end{bmatrix}, \quad 0 \le t \le t_f, \tag{3.56}$$

where x_{10} and x_{20} are the initial position and velocity, respectively, and t_f is the final time. Assume that the initial conditions are uniformly distributed on the rectangle $D = [-r, r] \times [-v, v]$, where r and v represent the maximum magnitude of the initial position and velocity, respectively. Hence,

$$f_{x_{10}x_{20}}(x_{10}, x_{20}) = \begin{cases} \frac{1}{4rv} & \text{if } (x_{10}, x_{20}) \in [-r, r] \times [-v, v] \\ 0 & \text{otherwise.} \end{cases} \tag{3.57}$$

Then the free response (3.56), (3.57) is a random process. At every time, $x_2(t)$ is uniformly distributed on $[-v, v]$. However, $x_1(t)$ is generally not uniformly distributed.

DEFINITION 3.19 *The **mean value function** of a random process $\{x(t), t \in I\}$ is*

$$\bar{x}(t) = E[x(t)], \quad t \in I. \tag{3.58}$$

*The **covariance kernel** of a random process $\{x(t), t \in I\}$ is*

$$P(t, \tau) = E[(x(t) - \bar{x}(t))(x(t) - \bar{x}(t))^T], \quad t \in I. \tag{3.59}$$

*The **cross-covariance kernel** of two random processes $\{x(t), t \in I\}$ and $\{y(t), t \in I\}$ is*

$$P_{xy}(t,\tau) = E[(x(t) - \bar{x}(t))(y(\tau) - \bar{y}(\tau))^T], \quad t \in I, \tau \in I. \tag{3.60}$$

*The **covariance matrix** of a random process $\{x(t), t \in I\}$ is $P(t,t)$.*
*The **cross-covariance matrix** of two random processes $\{x(t), t \in I\}$ and $\{y(t), t \in I\}$ is $P_{xy}(t,t)$.*

Note that covariance and cross-covariance matrices of random processes satisfy the following identities, inherited from their counterparts in random variables:

$$\begin{aligned} P(t) &= E[x(t)x^T(t)] - \bar{x}(t)\bar{x}^T(t) \\ P_{xy}(t) &= E[x(t)y^T(t)] - \bar{x}(t)\bar{y}^T(t). \end{aligned} \tag{3.61}$$

EXAMPLE 3.12 *Consider again the same random process as in Example 3.11, that is,*

$$\begin{bmatrix} x_1(t) \\ x_2(t) \end{bmatrix} = \begin{bmatrix} x_{10} + tx_{20} \\ x_{20} \end{bmatrix}, \quad 0 \le t \le t_f, \tag{3.62}$$

where x_{10} and x_{20} are uniformly distributed on $D = [-r, r] \times [-v, v]$. For this random process, the mean value function is

$$\begin{aligned} \bar{x}(t) &= E\begin{bmatrix} x_{10} + tx_{20} \\ x_{20} \end{bmatrix} \\ &= \begin{bmatrix} \bar{x}_{10} + t\bar{x}_{20} \\ \bar{x}_{20} \end{bmatrix} \\ &= \begin{bmatrix} 0 \\ 0 \end{bmatrix}. \end{aligned} \tag{3.63}$$

The covariance kernel is

$$\begin{aligned} P(t,\tau) &= E\left[x(t)x^T(\tau)\right] \\ &= \begin{bmatrix} \frac{r^2}{3} + (t+\tau)\frac{rv}{4} + t\tau\frac{v^2}{3} & \frac{rv}{4} + t\frac{v^2}{3} \\ \frac{rv}{4} + \tau\frac{v^2}{3} & \frac{v^2}{3} \end{bmatrix}. \end{aligned} \tag{3.64}$$

Therefore, the covariance matrix is

$$P(t) = \begin{bmatrix} \frac{r^2}{3} + \frac{trv}{2} + \frac{t^2v^2}{3} & \frac{rv}{4} + \frac{tv^2}{3} \\ \frac{rv}{4} + \frac{tv^2}{3} & \frac{v^2}{3} \end{bmatrix}. \tag{3.65}$$

For random processes, the notions of independence, correlation, and their relation are similar to their counterparts for random vectors. The key to the definitions is: given the random process $\{x(t),\ t \in I\}$, consider $m \in \mathbb{N}$ arbitrary, and $t_1, t_2, \ldots, t_m \in I$ arbitrary. Then, we apply the definition to the random vector with components $x(t_1), x(t_2), \ldots, x(t_m)$.

DEFINITION 3.20 *A random process* $\{x(t), t \in I\}$ *is* ***independent*** *if* $\forall m \in \mathbb{N}, \forall t_1, t_2, \dots, t_m \in I$, *the random vectors* $x(t_1), x(t_2), \dots, x(t_m)$ *are independent of each other. Two random processes* $\{x(t), t \in I\}$ *and* $\{y(t), t \in I\}$ *are independent of each other if* $\forall m \in \mathbb{N}, \forall t_1, t_2, \dots, t_m \in I$, *the two random vectors*

$$\left(x^T(t_1), x^T(t_2), \dots, x^T(t_m)\right)^T \text{ and } \left(y^T(t_1), y^T(t_2), \dots, y^T(t_m)\right)^T$$

are independent of each other.

DEFINITION 3.21 *A random process* $\{x(t), t \in I\}$ *is* ***uncorrelated*** *if* $\forall t_1, t_2 \in I, t_1 \neq t_2$, *the random vectors* $x(t_1)$ *and* $x(t_2)$ *are uncorrelated. Two random processes* $\{x(t), t \in I\}$ *and* $\{y(t), t \in I\}$ *are uncorrelated if* $\forall t_1, t_2 \in I$ *not necessarily distinct, the random vectors* $x(t_1)$ *and* $y(t_2)$ *are uncorrelated.*

We should note that if the random process $\{x(t), t \in I\}$ is independent, then it is uncorrelated. However, the converse is not true in general. This property is similar to that of random vectors. Also, for a zero-mean independent process, the covariance matrix satisfies

$$P(t, \tau) = 0, \quad t \neq \tau. \tag{3.66}$$

EXAMPLE 3.13 *Consider again the same random process as in Example 3.11, that is,*

$$\begin{bmatrix} x_1(t) \\ x_2(t) \end{bmatrix} = \begin{bmatrix} x_{10} + t x_{20} \\ x_{20} \end{bmatrix}, \quad 0 \le t \le t_f, \tag{3.67}$$

where x_{10} *and* x_{20} *are uniformly distributed on* $D = [-r, r] \times [-v, v]$. *This random process is not independent because knowledge of* $x(0)$ *determines* $x(t)$ *at all times. Indeed,*

$$f\left(x(t) \mid x(0)\right) = \delta\left(\left\| \begin{bmatrix} x_1(t) \\ x_2(t) \end{bmatrix} - \begin{bmatrix} x_1(0) + t x_2(0) \\ x_2(0) \end{bmatrix} \right\|\right), \tag{3.68}$$

which implies that the marginal distribution of $x_2(t)$, *conditioned on* $x(0)$, *is no longer uniform.*

DEFINITION 3.22 *A random process* $\{x(t),\ t \in I\}$ *is* ***stationary*** *if*

$$\begin{aligned} &\forall m \in \mathbb{N}, \forall t_1, t_2, \dots, t_m, \forall \tau, f\left(x(t_1), x(t_2), \dots, x(t_m)\right) \\ &= f\left(x(t_1 + \tau), x(t_2 + \tau), \dots, x(t_m + \tau)\right), \end{aligned}$$

that is, its statistical properties are invariant under time shift.

Note that if the random process $\{x(t),\ t \in I\}$ is stationary, then its mean value function is constant, its covariance matrix is constant, and its covariance kernel satisfies $P(t, \tau) = P(t - \tau, 0)$. This property of the covariance kernel of a stationary process is very reminiscent of the state-transition matrix of a linear time invariant system.

DEFINITION 3.23 *A random process $\{x(t),\ t \in I\}$ is* **wide-sense stationary** *if its mean value function and covariance kernel satisfy $\bar{x}$ = constant and*

$$\forall t, \tau, \sigma, \quad P(t, \tau) = P(t + \sigma, \tau + \sigma),$$

respectively.

In other words, a wide-sense stationary process is such that its first and second moments are those of a stationary process. Note that a stationary process is wide-sense stationary but that the converse is not true.

REMARK 3.4 *From Definition 3.23, an effective way of checking that a differentiable covariance kernel is that of a wide-sense stationary process is to ascertain that*

$$\frac{\partial P(t, \tau)}{\partial t} + \frac{\partial P(t, \tau)}{\partial \tau} \equiv 0. \tag{3.69}$$

EXAMPLE 3.14 *Consider again the same random process as in Example 3.11, that is,*

$$\begin{bmatrix} x_1(t) \\ x_2(t) \end{bmatrix} = \begin{bmatrix} x_{10} + t x_{20} \\ x_{20} \end{bmatrix}, \quad 0 \le t \le t_f, \tag{3.70}$$

where x_{10} and x_{20} are uniformly distributed on $D = [-r, r] \times [-v, v]$. The mean value function, given in (3.63)*, is constant; therefore, the process is "mean stationary." However, for the covariance kernel, given in* (3.64)*, we evaluate*

$$\frac{\partial P(t, \tau)}{\partial t} + \frac{\partial P(t, \tau)}{\partial \tau} = \begin{bmatrix} \frac{rv}{2} + (t + \tau)\frac{v^2}{3} & \frac{v^2}{3} \\ \frac{v^2}{3} & 0 \end{bmatrix}. \tag{3.71}$$

Because the right-hand side of (3.71) *is not identically zero, the random process is not "covariance stationary" and, therefore, not wide-sense stationary.*

DEFINITION 3.24 *Consider a stationary random process, $\{x(t),\ t \in I\}$, with covariance kernel $P(t, \tau) = P(t - \tau, 0) = P(t - \tau)$, where this last matrix is called the* ***autocorrelation function****. The* ***power spectral density of x*** *is*

$$\Psi_x(\omega) = F\{P(t)\} = \int_{-\infty}^{\infty} e^{-j\omega t} P(t)\, dt, \tag{3.72}$$

that is, the Fourier transform of the autocorrelation function.

Note that the autocorrelation matrices and power spectral densities are even matrix functions of time and frequency respectively, that is, if $P(t)$ is an autocorrelation matrix and $\Psi_x(\omega)$ is a power spectral density, then they satisfy

$$P(t) = P(-t)$$
$$\Psi_x(\omega) = \Psi_x(-\omega). \tag{3.73}$$

Also, the power spectral density $\Psi_x(\omega)$ indicates the (ensemble) average power contained at frequency ω in the random signal $x(t)$.

3.7 Gauss–Markov Processes

In this section, we combine the notions of Gaussian distribution and of random process. The result is a class of random processes that is especially useful in practice and, in particular, in navigation and guidance.

DEFINITION 3.25 *A random process* $\{x(t), t \in I\}$ *is* ***Gaussian*** *or* ***normal*** *if* $\forall m \in \mathbb{N}$, $\forall t_1, t_2, \ldots, t_m \in I$, *the vectors* $x(t_1), x(t_2), \ldots, x(t_m)$ *are jointly Gaussian.*

DEFINITION 3.26 *A Gaussian random process* $\{x(t), t \in I\}$ *is* ***white*** *if* $\forall m \in \mathbb{N}$, $\forall t_1, t_2, \ldots, t_m \in I$, *the vectors* $x(t_1), x(t_2), \ldots, x(t_m)$ *are independent; otherwise it is called* ***colored.***

REMARK 3.5 *For a Gaussian white process, the covariance kernel clearly satisfies* $P(t, \tau) = 0$ *when* $t \neq \tau$. *It can be shown that, for a continuous-time white process, we have*

$$P(t, \tau) = Q(t)\delta(t - \tau), \tag{3.74}$$

which is sometimes used as definition of white process.

REMARK 3.6 *To understand why we use the word "white" in Definition 3.26, realize that, for a stationary white process,* (3.74) *and* (3.72) *imply that the power spectral density satisfies*

$$\Psi_x(\omega) = Q = constant, \tag{3.75}$$

that is, the power spectral density is constant, which is a characteristic of white light.

DEFINITION 3.27 *A random process* $\{x(t), t \in I\}$ *is* ***Markovian*** *or* ***Markov*** *if* $\forall m \in \mathbb{N}$, $m \geq 2$, $\forall t_1 < t_2 < \cdots < t_m \in I$, *we have* $f(x(t_m)|x(t_{m-1}), x(t_{m-2}), \ldots, x(t_1)) = f(x(t_m)|x(t_{m-1}))$.

To understand the meaning of Definition 3.27, assume that t_{m-1} is the present time, that $t_{m-2}, \ldots, t_1$ are past times, and that t_m is in the future. Then we obtain the following interpretation: for a Markov process, the probability law describing the future is entirely specified by the present and does not require knowledge of the past. With this interpretation, the Markov property is clearly reminiscent of deterministic systems. Indeed, for a nonlinear vector differential equation $\dot{x} = f(x, t)$, where $x(t_{m-1})$ is given, we know that for $t \geq t_{m-1}$, we can compute $x(t)$ as long as it exists. In other words, we only need to know $x(t_{m-1})$, and we do not need the values of $x(t)$ for $t < t_{m-1}$.

REMARK 3.7 *For a Markov process* $\{x(t), t \in I\}$, $\forall m \in \mathbb{N}$, $\forall t_1 < t_2 < \cdots < t_m \in I$, *the joint probability density function* $f(x(t_m), x(t_{m-1}), x(t_{m-2}), \ldots, x(t_1))$ *is uniquely specified once we know* $f(x(t_1))$ *and* $f(x(t)|x(\tau))$, $\forall t, \tau \in I$. *Also, if* $\{x(t), t \in I\}$ *is Markov and* $J \subseteq I$, *then* $\{x(t), t \in J\}$ *is Markov too. Finally, a Markov process remains Markov after time reversal, that is, if* $t_m < t_{m+1} < t_{m+2} < \cdots < t_{m+k}$, *then* $f(x(t_m)|x(t_{m+1}), x(t_{m+2}), \ldots, x(t_{m+k})) = f(x(t_m)|x(t_{m+1}))$.

DEFINITION 3.28 *A random process $\{x(t), t \in I\}$ is **Gauss–Markov** if it is both Gaussian and Markov.*

EXAMPLE 3.15 *Consider again the free response of the double integrator system of Example 2.5, that is,*

$$\begin{bmatrix} x_1(t) \\ x_2(t) \end{bmatrix} = \begin{bmatrix} x_{10} + tx_{20} \\ x_{20} \end{bmatrix}, \quad 0 \le t \le t_f, \tag{3.76}$$

where x_{10} and x_{20} are the initial position and velocity, respectively, and t_f is the final time. Let us now assume that the initial conditions are jointly Gaussian, that is,

$$\begin{bmatrix} x_{10} \\ x_{20} \end{bmatrix} = \mathcal{N}\left(\begin{bmatrix} 0 \\ 0 \end{bmatrix}, P_{x_0}\right), \tag{3.77}$$

where P_{x_0} is the 2×2 covariance matrix of the initial conditions. Then the random process (3.76), (3.77) is Gauss–Markov.

Indeed, because at every time the process is a linear transformation of the initial conditions, which are Gaussian, we have $\forall m \in \mathbb{N}, \forall t_1, t_2, \ldots t_m \in [0, t_f]$, $x(t_1), \ldots, x(t_m)$ are jointly Gaussian. Therefore, the process is Gaussian. Moreover, recall that state transition is governed by the state transition matrix (2.43) so that

$$\forall t_1, t_2 \in [0, t_f], \quad x(t_2) = \begin{bmatrix} 1 & t_2 - t_1 \\ 0 & 1 \end{bmatrix} x(t_1). \tag{3.78}$$

Hence, if $0 \le t_1 < t_2 < \cdots < t_m \le t_f$, we have

$$\begin{aligned} f\left(x(t_m) \mid x(t_{m-1}), \ldots, x(t_1)\right) &= f\left(x(t_m) \mid x(t_{m-1})\right) \\ &= \delta\left(\left\| x(t_m) - \begin{bmatrix} 1 & t_m - t_{m-1} \\ 0 & 1 \end{bmatrix} x(t_{m-1}) \right\|\right). \end{aligned} \tag{3.79}$$

Therefore, the process is also Markov.

Since the process (3.76), (3.77) is Gauss–Markov, per Remark 3.7, all its statistical properties are specified by $f(x(0))$ and $f(x(t) \mid x(\tau)), t \ge \tau$. Here we have

$$f(x(0)) = \frac{1}{2\pi\sqrt{\det P_{x_0}}} \exp\left(-\frac{1}{2} x^T(0) P_{x_0}^{-1} x(0)\right) \tag{3.80}$$

$$f(x(t) \mid x(\tau)) = \delta\left(\left\| x(t) - \begin{bmatrix} 1 & t - \tau \\ 0 & 1 \end{bmatrix} x(\tau) \right\|\right). \tag{3.81}$$

Also, the mean value function is

$$\bar{x}(t) = 0, \tag{3.82}$$

the covariance matrix is

$$P(t) = \begin{bmatrix} 1 & t \\ 0 & 1 \end{bmatrix} P_{x_0} \begin{bmatrix} 1 & 0 \\ t & 1 \end{bmatrix}, \tag{3.83}$$

and the probability density function of f is

$$f(x, t) = \frac{1}{2\pi\sqrt{\det P(t)}} \exp\left(-\frac{1}{2} x^T P^{-1}(t) x\right). \tag{3.84}$$

3.8 Linear Gauss–Markov Models

In this section, we introduce the framework of linear dynamic systems that are driven by Gaussian stochastic inputs. This framework is particularly useful in our study of navigation and guidance. Indeed, in Chapter 2, we have justified our use of the linearity assumption. In Sections 3.1 through 3.7, we have justified our use of stochastic models, both time invariant and time varying, with Gaussian assumptions.

The **standard model** used henceforth is

$$\begin{aligned}\dot{x}(t) &= A(t)x(t) + B(t)u(t) + w(t), \quad t \geq t_0\\ y(t) &= C(t)x(t) + v(t),\end{aligned} \tag{3.85}$$

where $x \in \mathbb{R}^n$ contains the excursions of positions and velocities with respect to their nominal values, together with possibly other states such as those of noise models, $u \in \mathbb{R}^m$ contains deterministic controls (force, thrust, torque etc.) or measured accelerations, $y \in \mathbb{R}^p$ contains the readings of position and velocity measurement devices, w contains the appropriate state disturbance, and v contains the measurement noise.

EXAMPLE 3.16 *To illustrate the formalism of* (3.85), *consider a generalization of the double integrator introduced and motivated in Example 2.3. This generalization is used here to model the navigation of a point mass along a straight line, assuming that we measure the acceleration with an error* $\epsilon_a(t)$ *and the position with an error* $\epsilon_p(t)$. *Let* u *be the reading of the accelerometer. Then, the equations of motion have the form*

$$\begin{aligned}\dot{x} &= \begin{bmatrix} 0 & 1 \\ 0 & 0 \end{bmatrix}\begin{bmatrix} x_1 \\ x_2 \end{bmatrix} + \begin{bmatrix} 0 \\ 1 \end{bmatrix} u + \begin{bmatrix} 0 \\ \epsilon_a \end{bmatrix}\\ y &= \begin{bmatrix} 1 & 0 \end{bmatrix}\begin{bmatrix} x_1 \\ x_2 \end{bmatrix} + \epsilon_p,\end{aligned} \tag{3.86}$$

where x_1 *and* x_2 *are the true position and velocity, respectively.*

The **standard assumptions** on the model given in (3.85) are as follows:

1. The initial condition $x(t_0)$ is Gaussian:

$$x(t_0) = \mathcal{N}\left(\bar{x}(t_0), P(t_0)\right). \tag{3.87}$$

2. The disturbance $w(t)$ is a zero-mean, Gaussian, white process that is independent of $x(t_0)$:

$$\begin{aligned}E[w(t)] &= 0,\\ E[w(t)w^T(\tau)] &= R_w(t)\delta(t-\tau),\\ E[w(t)(x(t_0) - \bar{x}(t_0))^T] &= 0.\end{aligned} \tag{3.88}$$

3. The measurement noise $v(t)$ is a zero-mean, Gaussian, white process that is independent of $x(t_0)$:

$$\begin{aligned}E[v(t)] &= 0,\\ E[v(t)v^T(\tau)] &= R_v(t)\delta(t-\tau),\\ E[v(t)(x(t_0) - \bar{x}(t_0))^T] &= 0.\end{aligned} \tag{3.89}$$

4. The processes $v(t)$ and $w(t)$ are uncorrelated:

$$E[w(t)v^T(\tau)] = 0. \tag{3.90}$$

We can now state the fundamental properties of the model (3.85) under assumptions (3.87) through (3.90).

PROPOSITION 3.3 *Consider model* (3.85) *under assumptions* (3.87) *through* (3.90). *Then, the process* $\{x(t), t \in I\}$ *is Markov.*

PROOF. First, recall the variation of constants formal expression:

$$x(t) = \Phi(t, t_0)x(t_0) + \int_{t_0}^{t} \Phi(t, \tau)B(\tau)u(\tau)\,\mathrm{d}\tau + \int_{t_0}^{t} \Phi(t, \tau)w(\tau)\,\mathrm{d}\tau. \tag{3.91}$$

Accordingly, we see that

$$x(t_m) = \Phi(t_m, t_{m-1})x(t_{m-1}) + \int_{t_{m-1}}^{t} \Phi(t_m, \tau)\left(B(\tau)u(\tau) + w(\tau)\right)\mathrm{d}\tau \tag{3.92}$$

does not depend on $x(\tau), \quad \tau < t_{m-1}$.

PROPOSITION 3.4 *Consider model* (3.85) *under assumptions* (3.87) *through* (3.90). *Then, the process* $\{x(t), t \in I\}$ *is Gaussian.*

PROOF. Notice in (3.91) that $x(t_0)$ is Gaussian and $\int_{t_0}^{t} \Phi(t, \tau)w(\tau)\,\mathrm{d}\tau$ is also Gaussian. From Propositions 3.3 and 3.4, the statistical properties of the process $\{x(t), t \geq t_0\}$ are completely defined by its first and second moments; that is, we need only specify $E[x(t)] = \bar{x}(t)$ and $E[(x(t) - \bar{x}(t))(x(t) - \bar{x}(t))^T] = P(t)$. We have the following.

PROPOSITION 3.5 *Consider model* (3.85) *under assumptions* (3.87) *through* (3.90). *Then, the mean value function satisfies*

$$\bar{x}(t) = \Phi(t, t_0)\bar{x}(t_0) + \int_{t_0}^{t} \Phi(t, \tau)B(\tau)u(\tau)\,\mathrm{d}\tau. \tag{3.93}$$

PROOF. Evaluate the expected value of (3.91).

PROPOSITION 3.6 *Consider model* (3.85) *under assumptions* (3.87) *through* (3.90). *Then, the covariance matrix satisfies the* ***Lyapunov equation****:*

$$\dot{P}(t) = A(t)P(t) + P(t)A^T(t) + R_w(t). \tag{3.94}$$

PROOF. From (3.91) and (3.93), we have

$$x(t) - \bar{x}(t) = \Phi(t, t_0)\left(x(t_0) - \bar{x}(t_0)\right) + \int_{t_0}^{t} \Phi(t, \tau)w(\tau)\,\mathrm{d}\tau. \tag{3.95}$$

Therefore,

$$
\begin{aligned}
P(t) &= \Phi(t,t_0)P(t_0)\Phi^T(t,t_0) + 0 + 0 + \int_{t_0}^{t}\int_{t_0}^{t} \Phi(t,\tau)R_w(\tau)\delta(t-\sigma)\Phi^T(t,\sigma)\,\mathrm{d}\tau\,\mathrm{d}\sigma \\
&= \Phi(t,t_0)P(t_0)\Phi^T(t,t_0) + \int_{t_0}^{t} \Phi(t,\tau)R_w(\tau)\Phi^T(t,\tau)\,\mathrm{d}\tau. \qquad (3.96)
\end{aligned}
$$

Using Leibniz's rule (see Reminder 2.1), we obtain

$$
\dot{P}(t) = A(t)P(t) + P(t)A^T(t) + R_w(t). \qquad (3.97)
$$

Propositions 3.5 and 3.6 are the fundamental results that allow us to compute $\bar{x}(t)$ and $P(t)$, $t \geq t_0$. From these, the probability density function of $x(t)$ is known. Also note that, using the same line of argument, the covariance kernel can be shown to satisfy

$$
P_x(t,\tau) = \Phi(t,t_0)P(t_0)\Phi^T(\tau,t_0) + \int_{t_0}^{t} \Phi(t,\sigma)R_w(\sigma)\Phi^T(\tau,\sigma)\,\mathrm{d}\sigma. \qquad (3.98)
$$

For the statistical properties of the output, we have $y(t) = C(t)x(t) + v(t)$. Therefore,

$$
\bar{y}(t) = C(t)\bar{x}(t) \qquad (3.99)
$$

and

$$
\begin{aligned}
P_y(t,\tau) &= E[(y(t)-\bar{y}(t))(y(t)-\bar{y}(t))^T] \\
&= C(t)P_x(t,\tau)C^T(\tau) + R_v(t)\delta(t-\tau), \qquad (3.100)
\end{aligned}
$$

because $v(t)$ and $x(t)$ are uncorrelated. Also, the covariance kernel of the output satisfies

$$
\begin{aligned}
P_y(t,\tau) = {} & C(t)\Phi(t,t_0)P(t_0)\Phi^T(\tau,t_0)C(\tau) + \int_{t_0}^{t} C(t)\Phi(t,\sigma)R_w(\sigma)\Phi^T(\tau,\sigma)C(\tau)\,\mathrm{d}\sigma \\
& + R_v(t)\delta(t-\tau). \qquad (3.101)
\end{aligned}
$$

We now give the version of Propositions 3.5 and 3.6 that is relevant to linear, time invariant dynamic systems under stationary random inputs. First, recall that for a linear, time invariant deterministic system, the Laplace transform of the output is obtained by multiplying the Laplace transform of the input by the transfer function. The stochastic version of this result is as follows.

PROPOSITION 3.7 *Consider a BIBO stable, single-input, single-output linear time invariant system with transfer function $G(s)$, driven by a stationary Gaussian random input w. Then, at steady-state, the output y is a stationary Gaussian random process, with mean and power spectral density*

$$
\begin{aligned}
\bar{y} &= G(0)\bar{w} \\
\Psi_y(\omega) &= \Psi_w(\omega)|G(j\omega)|^2, \qquad (3.102)
\end{aligned}
$$

respectively.

REMARK 3.8 *In practice, Proposition 3.7 is often used as follows. The power spectral density $\Psi_y(\omega)$ of the output of a system can be measured relatively easily. We can then consider the problem of, given $\Psi_y(\omega)$, finding a stable transfer function $G(s)$ such that, if it were driven by a white noise process with $\Psi_w(\omega) = 1$, the output would have the statistical properties that were measured for $\Psi_y(\omega)$. This is called the* ***stochastic realization problem****, and its solution can be used for modeling a colored noise process as the output of a system that is driven by white noise.*

3.9 Summary of Key Results

The key results in Chapter 3 are as follows:

1. Proposition 3.2, which quantifies how the statistical properties of a Gaussian vector are changed under a linear affine transformation
2. Propositions 3.5 and 3.6, which quantify the statistical properties of a random process obtained by driving a linear dynamic system with Gaussian white noise
3. Proposition 3.7, which quantifies the statistical properties of the output of a stable linear time invariant system under stationary Gaussian input

3.10 Bibliographic Notes for Further Reading

The material in Chapter 3 is standard and is well covered in many texts, including [52], [71], [32], and [45].

3.11 Homework Problems

PROBLEM 3.1 *Let F be a σ-algebra on the sample space Ω.*

1. *Show that if F is not empty, then $\Omega \in F$ and $\emptyset \in F$.*
2. *Show that if $A \in F$ and $B \in F$, then $A \cap B \in F$.*

PROBLEM 3.2 *Let (Ω, F, P) be a probability space, and $B \in \Omega$ be an event of nonzero probability. Show that the conditional probability given B is a probability function in its own right, that is, that it satisfies*

$$\begin{aligned} \forall A,\ P(A|B) &\geq 0, \\ P(\Omega|B) &= 1, \\ \forall A_1, A_2, \ldots \in F, A_i \cap A_j = \emptyset \text{ for } i \neq j \Rightarrow P\left(\bigcup_{i=1}^{\infty} A_i|B\right) &= \sum_{i=1}^{\infty} P(A_i|B). \end{aligned}$$

PROBLEM 3.3 *A random vector is called uniformly distributed on a domain $D \subset \mathbb{R}^n$ if its probability density function is constant on D.*

1. *Obtain the probability density function of a random vector of $\mathbb{R}^2$, uniformly distributed over $D = [a_1, b_1] \times [a_2, b_2]$.*
2. *Generalize this result to a random vector of $\mathbb{R}^n$, uniformly distributed over $D = [a_1, b_1] \times [a_2, b_2] \times \cdots \times [a_n, b_n]$.*

PROBLEM 3.4 *Consider the random variable x with probability density function*

$$f(x) = \begin{cases} 2x & \text{for } x \in [0, 1] \\ 0 & \text{otherwise.} \end{cases}$$

Compute $\bar{x}$ and P_x.

PROBLEM 3.5 *Consider a random vector $x \in \mathbb{R}^2$, uniformly distributed over $D = [a_1, b_1] \times [a_2, b_2]$. Compute $\bar{x}$ and P_x. Assume that x_1 and x_2 are the Cartesian coordinates of a vehicle. Evaluate the mean square error:*

$$M = E[(x_1 - \bar{x}_1)^2 + (x_2 - \bar{x}_2)^2].$$

PROBLEM 3.6 *A vehicle navigates in a fixed plane and measures its polar coordinates (r, θ) with respect to the origin. These coordinates are related to the Cartesian coordinates (x_1, x_2) by*

$$x_1 = r \cos \theta$$

$$x_2 = r \sin \theta.$$

Assume that (r, θ) is random, uniformly distributed on $D = [r_1, r_2] \times [\theta_1, \theta_2]$. Compute $\bar{x}_1$, $\bar{x}_2$, and $M = E[(x_1 - \bar{x}_1)^2 + (x_2 - \bar{x}_2)^2]$.

PROBLEM 3.7 *Let x and y be two zero-mean random variables, with correlation coefficient*

$$\rho = \frac{E[xy]}{\sqrt{E[x^2]E[y^2]}}.$$

Show that $|\rho| \leq 1$.

Hint: *Use the fact that, for all scalars c, $E[(cx - y)^2] \geq 0$.*

PROBLEM 3.8 *Treat both of the following questions:*

1. *Show that if the two random vectors x and y are independent, then they are uncorrelated.*
2. *Consider a random variable x possessing a symmetric probability density function, that is, with $f_x(-x) = f_x(x)$, and define $y = x^2$. Check that x and y are uncorrelated, even though they are obviously not independent.*

PROBLEM 3.9 *Show that if $x = \mathcal{N}(\bar{x}, P)$ and if $y = Ax$, then $y = \mathcal{N}(A\bar{x}, APA^T)$.*

Hint: *Compute the characteristic function of y:*

$$\phi_y(r) = E[e^{jy^T r}].$$

PROBLEM 3.10 *Show that if the vectors x and y are jointly Gaussian and uncorrelated, then they are independent.*

Hint: *Assume that $P_{xy} = 0$. Then show that $\phi_{xy}(s, r) = \phi_x(s)\phi_y(r)$, that is, the joint characteristic function factors into the product of the two marginal characteristic functions.*

PROBLEM 3.11 *A gun shoots a cannonball from the origin* $(x, y) = (0, 0)$ *at time* $t = 0$. *Assuming flat Earth, and neglecting aerodynamic forces, the equations of motion are*

$$\ddot{x} = 0$$

$$\ddot{y} = -g.$$

The initial velocities are $\dot{x}(0) = v_1$ *and* $\dot{y}(0) = v_2$. *The* **range** *R is the value of x when* $y = 0,\ t > 0$. *Assume that* (v_1, v_2) *is uniformly distributed on* $D = [a_1, b_1] \times [a_2, b_2]$. *Compute the mean and variance of the range.*

PROBLEM 3.12 *Let x and y be two scalar random variables uniformly distributed on the disk* $x^2 + y^2 \leq 1$. *Are they independent? What is* $f(x|y)$ *for* $y = 1$?

PROBLEM 3.13 *Let* x_1 *and* x_2 *be two scalar random variables with probability density function*

$$f(x_1, x_2) = \begin{cases} 4x_1x_2e^{-(x_1^2+x_2^2)} & \text{if } x_1, x_2 \geq 0 \\ 0 & \text{otherwise.} \end{cases}$$

Are they independent?

PROBLEM 3.14 *Show that if x is a Gaussian vector, then x has the form* $x = M\xi$, *where M is a real rectangular matrix and* ξ *is a Gaussian vector with uncorrelated components. Is every subvector of x also Gaussian?*

PROBLEM 3.15 *Let* $x \in \mathbb{R}^n$ *be a random vector with known probability density function* $f_x(x)$. *Let* $g : \mathbb{R}^n \to \mathbb{R}^m : x \mapsto y = g(x)$ *be a function of x. Show that the probability density function of y is given formally by*

$$f_y(y) = \frac{1}{(2\pi)^m} \int_{\mathbb{R}^{m+n}} \exp(-js^T(y - g(x)))f_x(x)\,\mathrm{d}x\,\mathrm{d}s.$$

Hint: *Compute the characteristic function of y and use the inversion formula* (3.37).

PROBLEM 3.16 *Let* $x \in \mathbb{R}^n$ *be a random vector with mean value* $\bar{x}$ *and covariance matrix P. Let* $Q \in \mathbb{R}^{n\times n}$ *be a symmetric matrix. Show that*

$$E[(x - \bar{x})^T Q(x - \bar{x})] = \text{tr}[PQ],$$

where tr$[A]$ *denotes the* **trace** *of the square matrix A, that is, the sum of its diagonal elements.*

Hint: *Use the fact that* $\text{tr}[AB] = \text{tr}[BA]$ *whenever AB is square.*

PROBLEM 3.17 *Let* $\{x(t), t \in I\}$ *be a Markov random process. Show that,* $\forall m \in \mathbb{N}$, $\forall t_1 < t_2 < \cdots < t_m \in I$, *the joint probability density function* $f(x(t_m), x(t_{m-1}), \ldots, x(t_1))$ *is uniquely specified once we know* $f(x(t_1))$ *and* $\forall t, \tau \in I$, $f(x(t)|x(\tau))$.

Hint: *Use Bayes's rule:* $f_{x_1x_2}(x_1, x_2) = f_{x_1}(x_1|x_2)f_{x_2}(x_2)$.

PROBLEM 3.18 *Consider*

$$\dot{x}(t) = A(t)x(t) + w(t), \tag{3.103}$$

where $w(t) \in \mathbb{R}^n$ is white Gaussian with covariance $R_w(t)\delta(t-\tau)$. We know that the covariance matrix of x satisfies

$$P(t) = \Phi(t, t_0)P(t_0)\Phi^T(t, t_0) + \int_{t_0}^{t} \Phi(t, \tau)R_w(\tau)\Phi^T(t, \tau)\,\mathrm{d}\tau,$$

where $P(t_0)$ is constant and $\Phi(t, t_0)$ is the state transition matrix.

1. *Use Leibniz's rule to check that $P(t)$ satisfies*

$$\dot{P}(t) = A(t)P(t) + P(t)A^T(t) + R_w(t).$$

2. *Assume that A and R_w are constant and that the system (3.103) is stable. Give a formula to compute the steady-state covariance of x.*

PROBLEM 3.19 *Consider a BIBO stable linear time invariant system with impulse response $g(t)$ and transfer function $G(s) = \int_{-\infty}^{+\infty} e^{-st}g(t)\,\mathrm{d}t$. Let the input $w(t)$ be a stationary random process with mean $\bar{w}$ and autocorrelation function $P_w(t)$. Let $P_y(t)$ and $P_{yw}(t)$ be the autocorrelation function of the output and the cross-correlation function of the input and output, respectively, at steady state. Let $*$ denote convolution, that is,*

$$(a * b)(t) = \int_{-\infty}^{+\infty} a(\tau)b(t-\tau)\,\mathrm{d}\tau.$$

1. *Show that $\bar{y} = \bar{w}G(0)$ where $G(0)$ is sometimes called the DC gain.*
2. *Show that $P_{yw}(t) = g(t) * P_w(t)$.*
3. *Show that $P_y(t) = P_{yw}(t) * g(-t)$.*
4. *Show that $\Psi_y(\omega) = |G(j\omega)|^2\Psi_w(\omega)$, where Ψ_y and Ψ_w are the power spectral densities of the output and input, respectively.*

Hint: *Use the input–output relation $y(t) = (g * u)(t)$ and the relation between convolution in the time domain and product in the frequency domain.*

PROBLEM 3.20 *Compute the power spectral density of the output of the following linear time invariant BIBO stable systems driven by stationary Gaussian white noise of covariance $P(t) = \delta(t)$:*

1. $G(s) = \frac{k}{sT+1}$
2. $G(s) = \frac{k\omega_n^2}{s^2+2\zeta\omega_n s+\omega_n^2}, \ \zeta > 0, \ \omega_n > 0$

PROBLEM 3.21 *Consider a vehicle navigating along a straight line, with equation of motion $\ddot{x} = u_t$, where x is the position and u_t is the true acceleration.*

We measure the position and acceleration with respective errors $v(t)$ and $\epsilon_a(t)$. The error $v(t)$ is modeled as a zero-mean white Gaussian process with covariance

$R_v\delta(t)$. The error $\epsilon_a(t)$ is modeled as a zero-mean Gaussian **colored** *process with power spectral density*

$$\frac{\sigma^2}{1+\omega^2 T^2}.$$

The measurement errors are mutually independent and uncorrelated with the initial condition. Write down a linear Gauss–Markov model for the motion of this vehicle.

Hint: *Let $\epsilon_a(t)$ be generated by a filter driven by white noise. Then, append the dynamics of this filter to those of the system.*

PROBLEM 3.22 *Consider a vehicle navigating in a plane, using Cartesian coordinates. The equations of motion are $\ddot{x}_1 = u_{1t}$, $\ddot{x}_2 = u_{2t}$, where (x_1, x_2) are the Cartesian coordinates of the vehicle and (u_{1t}, u_{2t}) are the components of the true acceleration of the vehicle. We measure the coordinates and the acceleration with respective errors $v(t) = [v_1(t), v_2(t)]^T$ and $\epsilon_a(t) = [\epsilon_{a1}(t), \epsilon_{a2}(t)]^T$. The error $v(t)$ is modeled as a zero-mean white Gaussian process with covariance*

$$\begin{bmatrix} R_{v11} & R_{v12} \\ R_{v21} & R_{v22} \end{bmatrix} \delta(t).$$

The error $\epsilon_a(t)$ is modeled as a zero-mean white Gaussian process with covariance

$$\begin{bmatrix} R_{\epsilon 11} & R_{\epsilon 12} \\ R_{\epsilon 21} & R_{\epsilon 22} \end{bmatrix} \delta(t).$$

The errors are mutually independent and uncorrelated with the initial conditions. Write down a linear Gauss–Markov model for the motion of this vehicle. Specify the form of R_w.

PROBLEM 3.23 *Consider a vehicle navigating in a plane, using polar coordinates (r, θ). The equations of motion are*

$$\dot{r} = v_r,$$
$$\dot{\theta} = \frac{v_\theta}{r},$$
$$\dot{v}_r = \frac{v_\theta^2}{r} - \frac{k}{r^2} + u_{rt},$$
$$\dot{v}_\theta = -\frac{v_r v_\theta}{r} + u_{\theta t},$$

where u_{rt} and $u_{\theta t}$ are the true radial and tangential components of acceleration, respectively. Assume that we have a nominal trajectory $r^0(t)$, $\theta^0(t)$, $v_r^0(t)$, $v_\theta^0(t)$, $u_{rt}^0(t)$, $u_{\theta t}^0(t)$, and let δr, $\delta\theta$, δv_r, δv_θ, δu_{rt}, $\delta u_{\theta t}$ be the corresponding deviations.

We measure $(\delta r, \delta\theta)^T$ and $(\delta u_{rt}, \delta u_{\theta t})^T$ with respective errors $v(t) = [v_1(t), v_2(t)]^T$ and $\epsilon_a(t) = [\epsilon_{ar}(t), \epsilon_{a\theta}(t)]^T$. The errors are modeled as independent zero-mean white Gaussian processes with respective covariances,

$$\begin{bmatrix} R_{v11} & R_{v12} \\ R_{v21} & R_{v22} \end{bmatrix} \delta(t)$$

and

$$\begin{bmatrix} R_{\epsilon 11} & R_{\epsilon 12} \\ R_{\epsilon 21} & R_{\epsilon 22} \end{bmatrix} \delta(t),$$

and are uncorrelated with the initial conditions. Write down a linearized Gauss–Markov model for the motion of this vehicle. Specify the form of R_w.

4 Navigation

In this chapter, we present the theory that is used in the analysis and design of navigation systems for aerospace vehicles, with an emphasis on the fundamentals rather than on their hardware implementation. The purpose of navigation is twofold: to estimate the position and velocity of a vehicle based on the output of imperfect sensors and to assess the accuracy of these estimates. Mathematically, this corresponds to computing the first and second moments (expected value and covariance) of a particular random variable. Note that one should not neglect the importance of the accuracy assessment. Indeed, for a vehicle traveling in the vicinity of Earth, the question "Where is the vehicle?" can always be answered by "On the Sun, to within one light-hour." (Recall that the Earth is 8 light-minutes away from the Sun.) Such an answer, although it is correct, is totally useless for guiding the vehicle in the vicinity of the Earth. It is therefore very important to quantify the navigation error.

We start, in Section 4.1, by considering the navigation problem under the most restrictive assumptions: perfect sensors, nonredundant measurements, static estimation, nonrecursive processing, and perfect clock. Then, in subsequent sections, we remove these assumptions sequentially to build up the theory. Specifically, Section 4.2 considers position fixing with imperfect, nonredundant measurements, and Section 4.3 treats the case of imperfect redundant measurements. Section 4.5 considers inertial navigation, and Section 4.6 introduces recursive navigation, which is optimized in Section 4.7 as the Kalman filter for linear navigation and extended in Section 4.8 to the case of nonlinear navigation. Section 4.9 treats the case of imperfect clocks. Along the way, Section 4.4 presents several navigation aid systems that facilitate position fixing, and Section 4.10 discusses hardware used for navigation. Sections 4.11, 4.12, and 4.13 present a summary of the key results in the chapter, bibliographic notes for further reading, and homework problems, respectively.

4.1 Position Fixing: The Ideal Case

Let us start by considering the simplest possible situation that one might face in navigation: **position fixing**, which consists of computing the position of a vehicle based on the output of imperfect position sensors. Let $x \in \mathbb{R}^n$ contain the coordinates of the vehicle. These could, for instance, be Cartesian coordinates, longitude–latitude coordinates, and so on, and n could range from 1 to 3. Let $y \in \mathbb{R}^p$ contain the

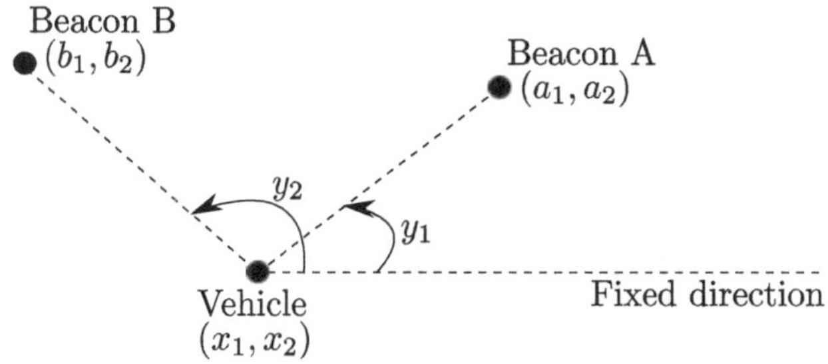

Figure 4.1. Vehicle navigation using bearing fixes to known locations.

measurements. Clearly, for the navigation problem to be solvable, we need $p \geq n$, otherwise we do not have enough information from the sensors. For the sake of simplicity, let us for now assume that $p = n$, that is, there is no redundant measurement. Let us also assume for now that our sensors are perfect, that is, they give readings that are not corrupted by noise.

In general, depending on the configuration, we can use geometry, trigonometry, astronomy, and so on, to derive a relationship of the form

$$y = g(x, t), \tag{4.1}$$

or, if we use nominal quantities x^0, y^0 with excursions $\delta x = x - x^0$, $\delta y = y - y^0$, we have

$$\delta y = \left(\frac{\partial g}{\partial x}\right)_{x^0}^T \delta x. \tag{4.2}$$

For fixed t (assuming that we have a perfect clock), (4.1) and (4.2) represent n equations with n unknowns. A simplistic view of the navigation problem is that it consists of solving these equations for x or δx.

EXAMPLE 4.1 *Consider the planar navigation of a vehicle using bearing fixes to known locations, as shown in Figure 4.1. Let* (x_1, x_2) *be the Cartesian coordinates of the vehicle,* (a_1, a_2) *and* (b_1, b_2) *be the Cartesian coordinates of two known locations, and* (y_1, y_2) *be the angles between the lines of sight to the two locations and a given, fixed direction. Then, we have*

$$\begin{aligned} y_1 &= \arctan[(x_2 - a_2)/(x_1 - a_1)] \\ y_2 &= \arctan[(x_2 - b_2)/(x_1 - b_1)]. \end{aligned} \tag{4.3}$$

Assuming that we measure (y_1, y_2) *and that* $(a_1(t), a_2(t))$ *and* $(b_1(t), b_2(t))$ *are known as functions of time, the navigation problem is simply to solve* (4.3) *for* (x_1, x_2). *This can be done by rewriting* (4.3) *as*

$$\begin{bmatrix} \tan y_1 & -1 \\ \tan y_2 & -1 \end{bmatrix} \begin{bmatrix} x_1 \\ x_2 \end{bmatrix} = \begin{bmatrix} a_1 \tan y_1 - a_2 \\ b_1 \tan y_2 - b_2 \end{bmatrix}, \tag{4.4}$$

which is now linear with respect to the unknown (x_1, x_2) *and can be solved as*

$$\begin{bmatrix} x_1 \\ x_2 \end{bmatrix} = \begin{bmatrix} \tan y_1 & -1 \\ \tan y_2 & -1 \end{bmatrix}^{-1} \begin{bmatrix} a_1 \tan y_1 - a_2 \\ b_1 \tan y_2 - b_2 \end{bmatrix}. \tag{4.5}$$

It is important to notice the generality of Example 4.1. Indeed, the vehicle could be a boat on a lake, measuring angles between the lines of sight to known landmarks and the north direction, or it could be a spacecraft in the ecliptic plane, measuring bearing angles between known planets and the Sun.

REMARK 4.1 *In* (4.1), y_i *is called a* ***fix****. The nature of the surface* $y_i =$ *constant determines the name of the fix. Accordingly, we may have linear fixes, conic fixes, circular fixes, hyperbolic fixes, and so on. In particular, Example 4.1 deals with linear fixes.*

According to the preceding simplistic view, navigation by position fixing consists of, given y and t, solving the nonlinear system of equations (4.1) for the unknown vector x. One way to do this is to use **Newton's method**, which is presented in Appendix A.4 and yields the iteration (A.34), repeated here for convenience:

$$x^{k+1} = x^k + \left(\frac{\partial g}{\partial x}\right)_{x^k}^{-T} \left(y - g(x^k, t)\right). \tag{4.6}$$

REMARK 4.2 *When a vehicle is moving, x changes as a function of time. If the updates* (4.6) *are performed "fast enough," the iteration* (4.6) *remains "locked" on the current coordinates of the vehicle.*

REMARK 4.3 *If* $g(x, t)$ *is linear in the unknown x (such as in* (4.2)*), then the iteration* (4.6) *converges in one step!*

REMARK 4.4 *When the Jacobian matrix,* $(\frac{\partial g}{\partial x})$*, is almost singular, we can expect numerical difficulties in applying Newton's method. Remark 4.6 warns that, in this case, we can also expect large navigation errors due to imperfect sensors.*

It is worthwhile at this point to recall all the assumptions underlying (4.1) and (4.2). We assume that we are only interested in the position of the vehicle and not in the velocity. We assume that we have a number of sensors equal to, and not greater than, the number of unknowns. We also assume that the sensors are perfect. Finally, we assume that we have a perfect clock. In the remainder of this chapter, we remove these assumptions one after the other to develop a theory of navigation.

4.2 Position Fixing: Error Analysis

In this section, we introduce imperfection in the sensors and analyze the subsequent navigation error. Let us first discuss to what extent we may be able to account for uncertainties in the readings of the sensors. In general, the measurements given by imperfect sensors have the form

$$y = g(x, t) + v, \tag{4.7}$$

where the vector v represents the measurement errors. Because, for given y, v, t we should be able to compute x, we expect that there should exist a function $x = h(y - v, t)$ that expresses the unknown coordinates in terms of perfect sensor readings. Then, assuming that we know the probability density function $f_v(v)$ of the measurement errors, we should be able to evaluate $f_x(x|y)$, the conditional probability density function of the coordinates given the measurements. Such a procedure, although it is theoretically feasible, is exceedingly cumbersome.

More practically, we assume that the measurement errors are additive, small, and Gaussian, and we analyze their effect. Assume that the measurement vector v in y causes an error ϵ_x in x. Because, in the case of perfect measurements, we had $y = g(x, t)$, we now have

$$y = g(x + \epsilon_x, t) + v. \tag{4.8}$$

Under the assumption that v is small, a Taylor series expansion of (4.8) yields

$$y = g(x, t) + v + \left(\frac{\partial g}{\partial x}\right)_x^T \epsilon_x + \text{H.O.T.} \tag{4.9}$$

Assuming that ϵ_x is also small, and neglecting the higher-order terms, (4.9) becomes

$$v + \left(\frac{\partial g}{\partial x}\right)_x^T \epsilon_x = 0. \tag{4.10}$$

Assuming that the Jacobian matrix in (4.10) is nonsingular, we have

$$\epsilon_x = -\left(\frac{\partial g}{\partial x}\right)_x^{-T} v. \tag{4.11}$$

Now, a typical assumption about measurement errors is that they are Gaussian with zero-mean and known covariance, that is,

$$v = \mathcal{N}(0, R_v). \tag{4.12}$$

Then, recalling Proposition 3.2, we obtain that the coordinate error ϵ_x is also Gaussian with known covariance, specifically,

$$\epsilon_x = \mathcal{N}\left(0, \left(\frac{\partial g}{\partial x}\right)_x^{-T} R_v \left(\frac{\partial g}{\partial x}\right)_x^{-1}\right). \tag{4.13}$$

Let us therefore define

$$R_{\epsilon_x} = \left(\frac{\partial g}{\partial x}\right)_x^{-T} R_v \left(\frac{\partial g}{\partial x}\right)_x^{-1} \tag{4.14}$$

as the covariance of the coordinate error. We can now use this expression to quantify the navigation error as follows. In general, depending on the particular system of coordinates used, the differential element of length is related to the differential increments in coordinates by

$$(\mathrm{d}l)^2 = \mathrm{d}x^T Q(x)\, \mathrm{d}x, \tag{4.15}$$

where the matrix $Q(x)$ is symmetric, positive definite, and represents the local metric of space. Let M be the mean square error, that is, the expected value of the square of the navigation error. Then, we have

$$M = E[\epsilon_x^T Q \epsilon_x] = E\left[||\epsilon_x||_Q^2\right], \tag{4.16}$$

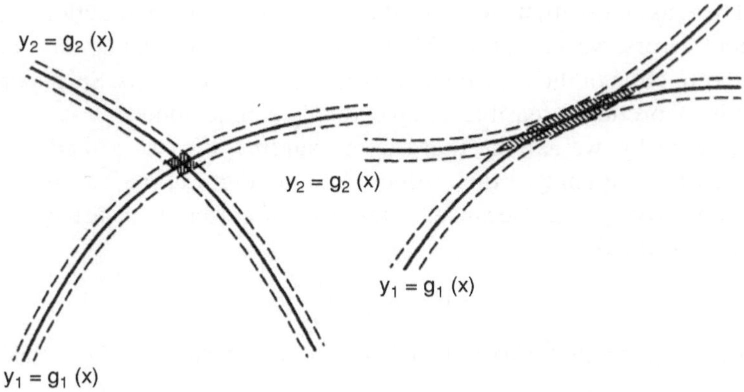

Figure 4.2. Accuracy analysis.

leading to (see Problem 3.16)

$$
\begin{aligned}
M &= \operatorname{tr}(QR_{\epsilon_x}) \\
&= \operatorname{tr}\left[\left(\frac{\partial g}{\partial x}\right)_x^{-T} R_v \left(\frac{\partial g}{\partial x}\right)_x^{-1} Q\right].
\end{aligned}
\tag{4.17}
$$

This last formula allows the navigator to quantify the accuracy of navigation given the accuracy of the sensors. It can also be used in mission planning to decide which combination of fixes should be used at what times during the mission. Finally, it can also be used to decide where to place land-based navigation aids.

REMARK 4.5 *The error analysis presumes knowledge of the mean (zero) and of the covariance matrix, R_v, of the measurement error. In practice, this knowledge arises from* ***calibration****, which is the process of determining the statistical properties of the error in the reading of a sensor. Hence, we assume that the sensors are calibrated before use, which is standard practice.*

REMARK 4.6 *Equation* (4.11) *has a geometric interpretation. Its ith row gives $v_i = \left(\frac{\partial g_i}{\partial x}\right)^T \epsilon_x$, that is, $v_i = (\nabla g_i) \cdot \epsilon_x$, where the gradient ∇g_i is determined by the type of fix g_i and is orthogonal to the fix surface. We see that in* (4.17)*, when the Jacobian matrix $\left(\frac{\partial g}{\partial x}\right)^T$ is almost singular, the mean square of the navigation error is large. Because the rows of $\left(\frac{\partial g}{\partial x}\right)^T$ contain the gradients of the fix functions, this matrix is almost singular when the fix functions are almost linearly dependent. This informal argument leads to the following* ***rule of thumb*** *for position fixing:* ***have the fix surfaces be as mutually orthogonal as possible****. Figure 4.2 illustrates the benefit of this rule of thumb in planar navigation. For perfect sensors, solving* (4.1) *is equivalent to computing the intersection of two fix lines. However, when the sensors are noisy, the navigation locates the vehicle in the shaded region, which is relatively large when the nominal fix lines are tangent, that is, when the Jacobian of the fix functions is singular [42].*

REMARK 4.7 *In* (4.12)*, R_v need not be diagonal, that is, the measurement errors may be correlated.*

EXAMPLE 4.2 *First, let us illustrate the derivation of the metric $Q(x)$ of (4.15). In planar Cartesian coordinates (x_1, x_2), let (ϵ_1, ϵ_2) be differential increments of the coordinates. Then, Pythagoras's theorem yields that the differential element of length satisfies*

$$(\mathrm{d}l)^2 = (\epsilon_1)^2 + (\epsilon_2)^2. \tag{4.18}$$

Comparing with (4.15), *we obtain*

$$Q(x_1, x_2) = \begin{bmatrix} 1 & 0 \\ 0 & 1 \end{bmatrix} \tag{4.19}$$

for Cartesian coordinates. In planar polar coordinates (r, θ), let $(\epsilon_r, \epsilon_\theta)$ be differential increments of the coordinates. Then, the differential element of length satisfies

$$(\mathrm{d}l)^2 = (\epsilon_r)^2 + r^2(\epsilon_\theta)^2. \tag{4.20}$$

Comparing with (4.15), *we obtain*

$$Q(r, \theta) = \begin{bmatrix} 1 & 0 \\ 0 & r^2 \end{bmatrix} \tag{4.21}$$

for polar coordinates.

EXAMPLE 4.3 *Let us continue Example 4.1 with an error analysis. Recall that*

$$\frac{\mathrm{d}}{\mathrm{d}u} \arctan(u) = \frac{1}{1+u^2}. \tag{4.22}$$

Therefore,

$$\begin{aligned}
\frac{\partial y_1}{\partial x_1} &= \frac{-(x_2 - a_2)}{(x_1 - a_1)^2 + (x_2 - a_2)^2} \\
\frac{\partial y_1}{\partial x_2} &= \frac{x_1 - a_1}{(x_1 - a_1)^2 + (x_2 - a_2)^2} \\
\frac{\partial y_2}{\partial x_1} &= \frac{-(x_2 - b_2)}{(x_1 - b_1)^2 + (x_2 - b_2)^2} \\
\frac{\partial y_2}{\partial x_2} &= \frac{x_1 - b_1}{(x_1 - b_1)^2 + (x_2 - b_2)^2}.
\end{aligned} \tag{4.23}$$

Hence, the Jacobian of the fix function is

$$\left(\frac{\partial g}{\partial x}\right)^T = \begin{bmatrix} \frac{\partial g_1}{\partial x_1} & \frac{\partial g_1}{\partial x_2} \\ \frac{\partial g_2}{\partial x_1} & \frac{\partial g_2}{\partial x_2} \end{bmatrix} = \begin{bmatrix} \nabla g_1^T \\ \nabla g_2^T \end{bmatrix}. \tag{4.24}$$

If the measurements are corrupted by a small measurement error vector v, the resulting error in Cartesian coordinates is

$$\epsilon_x = -\left(\frac{\partial g}{\partial x}\right)^{-T} v. \tag{4.25}$$

If, in addition, we also assume that v is Gaussian with zero mean and covariance R_v, then the coordinate error is also Gaussian, with zero mean and covariance

$$R_{\epsilon_x} = \left(\frac{\partial g}{\partial x}\right)^{-T} R_v \left(\frac{\partial g}{\partial x}\right)^{-1}. \tag{4.26}$$

Finally, because for Cartesian coordinates the metric $Q(x)$ is the identity matrix (see Example 4.2), the trace of the matrix in (4.26) *is the mean square of the navigation error. Assume that we are given the following specific data:* $(a_1, a_2) = (1, 1)$, $(b_1, b_2) = (-1, 2)$, *and* $(y_1, y_2) = (1.107, 1.893)$. *We are also given that the two goniometers provide uncorrelated measurements, with standard deviation one degree. From this, we have*

$$R_v = \begin{bmatrix} \sigma^2 & 0 \\ 0 & \sigma^2 \end{bmatrix}$$

$$\sigma^2 = (1^\circ)^2 = 3.047 \times 10^{-4}. \tag{4.27}$$

Equation (4.5) *yields* $(x_1, x_2) = (0, -1)$ *as the position of the vehicle. For the error analysis,* (4.23) *and* (4.24) *yield*

$$\left(\frac{\partial g}{\partial x}\right)^T = \begin{bmatrix} \frac{2}{5} & \frac{-1}{5} \\ \frac{3}{10} & \frac{1}{10} \end{bmatrix}$$

$$\left(\frac{\partial g}{\partial x}\right)^{-T} = \begin{bmatrix} 1 & 2 \\ -3 & 4 \end{bmatrix}. \tag{4.28}$$

Then, (4.17) *implies that the mean square error is* $M = \sigma^2 = 9.14 \times 10^{-3}$, *yielding the standard deviation* $\sqrt{M} = 9.56 \times 10^{-2}$.

4.3 Position Fixing: Redundant Measurements

In this section, we remove the assumption of nonredundant sensors. In other words, we assume that we have more sensors than coordinates to be determined, that is, $p > n$. In that case, owing to the imperfection of the sensors, the equation $y = g(x, t)$ is in general infeasible with respect to the unknown x. Our approach is to attempt to find the "best" x in some sense, taking the statistical properties of the measurement errors into account.

We proceed as follows. Recall that our model of errors has the form $y = g(x, t) + v$, where v is the measurement error vector. Assume that we know $f_v(v)$, the probability density function of v. Also recall, from (3.13), that $f_v(a)\,dv = P[a \le v \le a + dv]$. Therefore, the values of v where $f_v(v)$ is large are, in some sense, more "likely" than those where $f_v(v)$ is small. We are therefore led to formulate the optimization problem:

$$\max_x \bar{L}(x) = f_v(y - g(x, t)). \tag{4.29}$$

In (4.29), the function $\bar{L}(x)$ is called the **likelihood function**, hence the name **maximum likelihood method** for this procedure.

In practice, we assume that the errors are small and Gaussian with zero-mean and known covariance, that is,

$$v = \mathcal{N}(0, R_v). \tag{4.30}$$

Then, the probability density function of v, from (3.47), is

$$f_v(v) = \frac{1}{\sqrt{(2\pi)^p \det R_v}} \exp\left(-\frac{1}{2} v^T R_v^{-1} v\right), \tag{4.31}$$

and the optimization problem (4.29) becomes

$$\min_x L(x) = \frac{1}{2}(y - g(x,t))^T R_v^{-1}(y - g(x,t)). \tag{4.32}$$

The first-order necessary conditions for optimality are

$$\frac{\partial L}{\partial x}(x) = 0, \tag{4.33}$$

which provide n nonlinear equations in n unknowns in the vector x. The second-order conditions are

$$\frac{\partial^2 L}{\partial x^2}(x) \geq 0. \tag{4.34}$$

The nonlinear equations (4.33) are in general difficult to solve, and one may resort to Newton's method as presented in Appendix A.4. For linearized navigation in the neighborhood of a nominal position, they do have an explicit closed-form solution. Indeed, we have

$$\delta y = \left(\frac{\partial g}{\partial x}\right)^T_{x^0} \delta x + v = C\delta x + v, \tag{4.35}$$

where the definition of C is obvious. The optimization problem (4.32) becomes

$$\min_{\delta x} L(\delta x) = \frac{1}{2}(\delta y - C\delta x)^T R_v^{-1}(\delta y - C\delta x). \tag{4.36}$$

Equation (4.33) is then linear, with explicit solution

$$\delta x = [C^T R_v^{-1} C]^{-1} C^T R_v^{-1} \delta y. \tag{4.37}$$

Moreover, the second order condition (4.34) is clearly satisfied because

$$\frac{\partial^2 L}{\partial (\delta x)^2} = C^T R_v^{-1} C. \tag{4.38}$$

EXAMPLE 4.4 *Here we illustrate the use of the maximum likelihood method and convince ourselves that it leads to very reasonable results in a simple case. Consider the one-dimensional navigation of a particle along a straight line. Assume that we measure the range x with two sensors such that*

$$\begin{aligned} y_1 &= x + v_1 \\ y_2 &= x + v_2. \end{aligned} \tag{4.39}$$

Let us further assume that

$$v = \begin{bmatrix} v_1 \\ v_2 \end{bmatrix} = \mathcal{N}\left(\begin{bmatrix} 0 \\ 0 \end{bmatrix}, \begin{bmatrix} 1 & 0 \\ 0 & 3 \end{bmatrix}\right), \tag{4.40}$$

that is, the measurement errors are uncorrelated, and the reading of the second range sensor, y_2, is three times less reliable than that of the first range sensor, y_1. The maximum *likelihood method then requires solving the optimization problem*

$$\min_x L(x) = \min_x \frac{1}{2} \begin{bmatrix} y_1 - x \\ y_2 - x \end{bmatrix}^T \begin{bmatrix} 1 & 0 \\ 0 & 3 \end{bmatrix}^{-1} \begin{bmatrix} y_1 - x \\ y_2 - x \end{bmatrix}$$
$$= \min_x \frac{1}{2} \left((y_1 - x)^2 + \frac{1}{3}(y_2 - x)^2 \right). \tag{4.41}$$

The first-order necessary condition is

$$\frac{\mathrm{d}L(x)}{\mathrm{d}x} = -(y_1 - x) - \frac{1}{3}(y_2 - x) = 0, \tag{4.42}$$

which yields

$$x = \frac{3y_1 + y_2}{4}. \tag{4.43}$$

The second-order condition (4.34) *is satisfied because* $\frac{\mathrm{d}^2 L}{\mathrm{d}x^2} = \frac{4}{3}$. *Remarkably, the maximum likelihood estimate* (4.43) *is* ***a convex weighted sum of the readings of the sensors, in which the relative weights are proportional to the confidence we have in the sensors!***

When we use redundant measurements and the maximum likelihood method, we can perform the error analysis as follows. Assume that, with perfect measurements, we have a nominal solution (x^0, y^0) to the overdetermined navigation problem; that is, we have $y^0 = g(x^0, t)$. Also assume that a small zero-mean Gaussian measurement vector v corrupts the reading of the sensor, causing a small perturbation (ϵ_x, ϵ_y) to (x^0, y^0). Then, we have

$$\epsilon_y = C\epsilon_x + v, \tag{4.44}$$

where

$$C = \left(\frac{\partial g}{\partial x} \right)^T_{x^0}. \tag{4.45}$$

We can use the maximum likelihood method to find the most likely value of ϵ_x in (4.44). This requires minimizing the quadratic form $v^T R_v^{-1} v$ with respect to ϵ_x, subject to the linear constraint (4.44). The solution of this minimization problem is

$$\epsilon_x = [C^T R_v^{-1} C]^{-1} C^T R_v^{-1} \epsilon_y. \tag{4.46}$$

Hence, the covariance of ϵ_x is

$$R_{\epsilon_x} = [C^T R_v^{-1} C]^{-1} C^T R_v^{-1} R_v R_v^{-1} C [C^T R_v^{-1} C]^{-1}$$
$$= [C^T R_v^{-1} C]^{-1}. \tag{4.47}$$

EXAMPLE 4.5 *Let us perform an error analysis for the situation in Example 4.4. Here we have*

$$R_v = \begin{bmatrix} 1 & 0 \\ 0 & 3 \end{bmatrix}, \quad C = \begin{bmatrix} 1 \\ 1 \end{bmatrix}. \tag{4.48}$$

Therefore, from (4.47), *we have*

$$R_{\epsilon_x} = \left[\begin{bmatrix} 1 & 1 \end{bmatrix} \begin{bmatrix} 1 & 0 \\ 0 & 3 \end{bmatrix}^{-1} \begin{bmatrix} 1 \\ 1 \end{bmatrix} \right]^{-1}$$

$$= \left(1 + \frac{1}{3}\right)^{-1}$$

$$= \frac{3}{4} = 0.75. \tag{4.49}$$

The meaning of this result is as follows. If we use only the reading of the second sensor, y_2, *we can expect a navigation error of variance 3. If we use only the reading of the first sensor,* y_1, *we can expect a navigation error of variance 1. However, if we optimally combine the readings of the two sensors according to* (4.43), *we can expect a navigation error of variance 0.75, which is better than when we use either sensor alone.*

EXAMPLE 4.6 *Assume that we add a third measurement in the situation of Example 4.3. Assume that we use a third location,* $(c_1, c_2) = (2, -2)$, *yielding a third reading* $y_3 = -0.4643$. *Assume that this measurement is uncorrelated with the others and has identical statistical properties. We still have* $(x_1, x_2) = (0, -1)$ *as the location of the vehicle. However, now we have*

$$C = \begin{bmatrix} 2/5 & -1/5 \\ 3/10 & 1/10 \\ -1/5 & -2/5 \end{bmatrix}, \tag{4.50}$$

which differs from (4.28) *by the addition of a third row. We also have*

$$R_v = \begin{bmatrix} \sigma^2 & 0 & 0 \\ 0 & \sigma^2 & 0 \\ 0 & 0 & \sigma^2 \end{bmatrix}. \tag{4.51}$$

Then, (4.47) *and* (4.17) *yield a mean square error* $M = 8.3\sigma^2$, *that is, a reduction of more than 60% over the two-measurement case.*

Figure 4.3 shows the 0.99 probability ellipses for the two and three measurements cases. For the figure, σ *is chosen to be 10 degrees instead of 1 degree so that the probability ellipses are more visible.*

REMARK 4.8 *Based on* (4.47), *it can be shown that one always gains by adding a new measurement in terms of navigation error, and this no matter how bad the additional measurement is. Of course, the amount of confidence one gains by adding a measurement depends on the quality of the measurement.*

4.4 Examples of Fixes

In this section, we give examples of position fixing systems that are used in land, sea, atmospheric, and space navigation. We also give some of the acronyms associated with these systems. Here again, the emphasis is on the fundamentals as opposed to the hardware implementation. Many of the electronic navigation aids to which we refer in this section were developed during and after World War II. Also many of

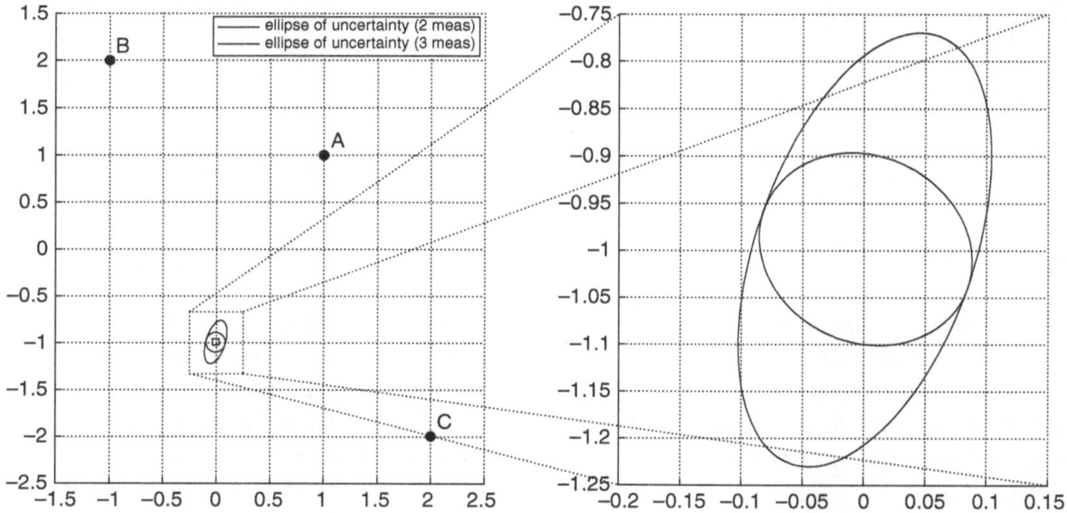

Figure 4.3. Some 0.99-probability ellipses for the two and three measurement cases with $\sigma = 10$ degrees.

them have greatly benefited from the advent of the digital computer in the 1960s. The performance of all these position-fixing systems can be analyzed using the methods introduced in Sections 4.2 and 4.3.

In planar navigation, bearing measurements (i.e., measurements of the angles between lines of sight to a known location and a given direction) yield a linear fix (see, e.g., Example 4.1). This type of measurements is used for instance in the International Civil Aviation Organization (ICAO) system and the VHF Omni-directional Ranging (VOR) system (see Figure 4.4). In three-dimensional navigation, bearing measurements yield a conic fix (see Figure 4.5): by measuring the angle y between

Figure 4.4. Narco VHT-2 Superhomer VOR receiver (mid-1950s era). Image courtesy of Smithsonian Institution, National Air and Space Museum.

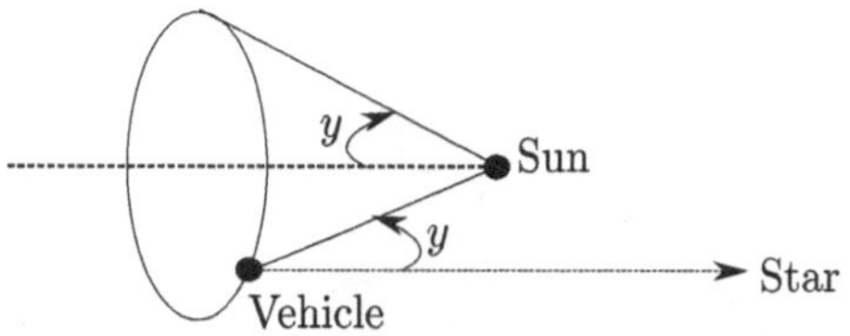

Figure 4.5. Three-dimensional bearing measurements: conic fix. The cone has aperture y and axis directed toward the star.

the line of sight to the Sun and the line of sight to a star, the navigator knows that the spacecraft is located on a circular cone with apex at the Sun, with aperture y, and with axis directed to the star.

In planar navigation, range measurements yield circular fixes. This type of measurement is used, for instance, in the Distance Measuring Equipment (DME) (see Figure 4.6), Tactical Air Navigation (TACAN) (see Figure 4.7), and VHF Omnidirectional Ranging Tactical Air Navigation (VORTAC) systems (see Figure 4.8). In three-dimensional navigation, range measurements yield spherical fixes. This is used, for instance, in space navigation, where the navigator measures the apparent diameter of a planet (in degrees) and then, from knowledge of the exact planet diameter, infers the range to the planet.

In planar navigation, measurements of the angle between the lines of sight to two known locations yield a circular fix (see Figure 4.9 and Problem 4.1). This is used mostly in astronautics but could also be used to navigate on a lake. In three-dimensional navigation, the same type of measurement yields for a fix surface a (spindle) torus that is obtained by revolution of a circle about a chord.

In planar navigation, measurement of the difference of distances to two known locations yields a hyperbolic fix, where the two locations are foci of the hyperbola (see Figure 4.10). This type of measurement is used, for instance, in the Long Range

Figure 4.6. Distance Measuring Equipment (DME). Image courtesy of Smithsonian Institution, National Air and Space Museum.

Figure 4.7. TACAN indicator type CA-1477/2. Image courtesy of Smithsonian Institution, National Air and Space Museum.

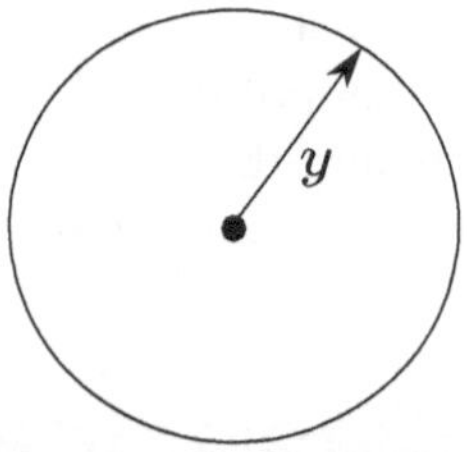

Figure 4.8. Two-dimensional range measurement.

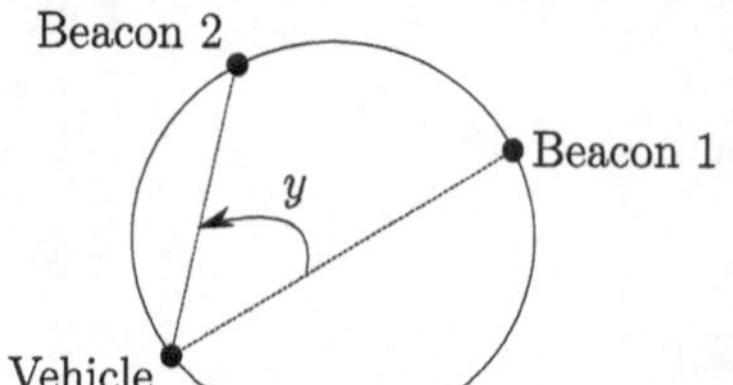

Figure 4.9. Angle between the lines of sight to two known locations.

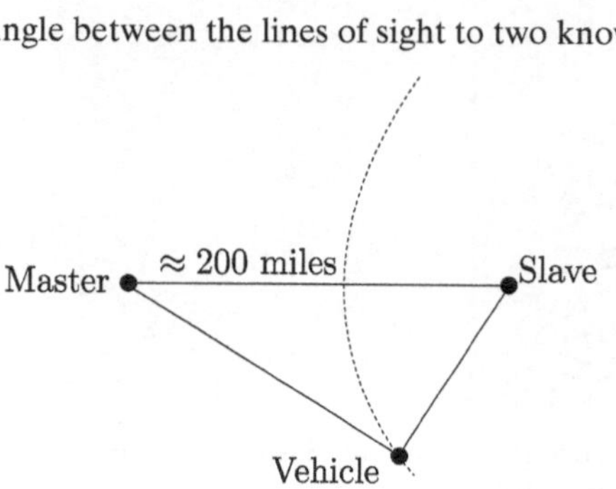

Figure 4.10. Two-dimensional measurement of difference of distance to two known locations.

Figure 4.11. Texas Instruments TI 9100 LORAN-C receiver (certification set for the first-general aviation LORAN-C receiver). Image courtesy of Smithsonian Institution, National Air and Space Museum.

Navigation (LORAN) (see Figure 4.11) and DECCA systems. (The DECCA system was shut down in spring 2000.) In three-dimensional navigation, the same type of measurement yields for a fix surface a hyperboloid of revolution.

On a spherical Earth, measurement of the apparent elevation of a star above the horizon yields a circular fix (see Figure 4.12). More precisely, the fix line is a small circle, which is the intersection of Earth's surface with a plane, where the orthogonal projection of the star on Earth is the center of the circle. Indeed, measuring the elevation of a star is equivalent to measuring the complementary angle, that is, the angle between the lines of sight to the star and to the center of the Earth. This measurement yields a fix surface that is a circular cone, with apex at the center of the Earth and axis directed toward the star. The intersection of this cone with the surface of Earth is clearly a small circle, as claimed earlier. Before the advent of the electronic age, it was customary to aid sea navigation by tabulating this fix function in **ephemerid tables** for the brightest stars in the sky and the standard longitude latitude coordinates.

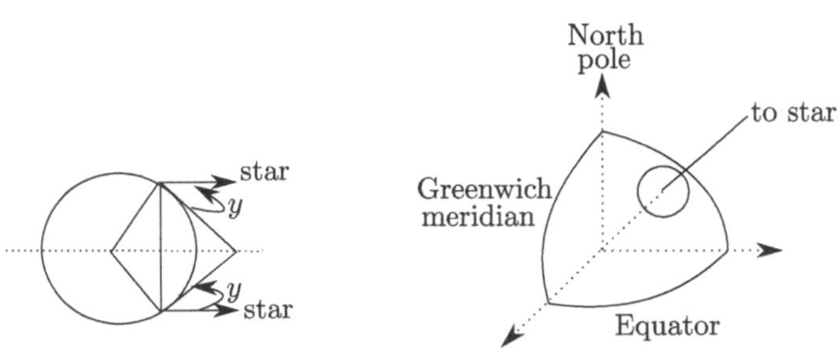

Figure 4.12. Measurement of the apparent elevation of a star.

Figure 4.13. Artist's rendition of GPS block III satellite. Image courtesy of Smithsonian Institution, National Air and Space Museum.

The state of the art in navigational aids is to use satellites for global positioning (see Figure 4.13). Several such systems are currently operational, such as the NAVSTAR Global Positioning System (GPS) and the Global Navigation Satellite System (GLONASS). Typically, each global positioning system consists of three subsystems: a set of ground stations, a constellation of satellites (24 for both GPS and GLONASS), and a passive receiver carried by a user. The ground stations track the satellites, determine their positions accurately through radar telemetry, and transmit these positions to the respective satellites. The constellation is designed so that every point on Earth is in the coverage of at least four satellites at every time. The satellites are equipped with accurately synchronized atomic clocks and continually broadcast their coordinates together with the time when they are located at these coordinates. The receiver records the time at which it receives a broadcast from any satellite using its own clock, which is not synchronized to the satellites' clocks. On receiving signals from at least four satellites, the receiver can, in principle, determine simultaneously its location and the time as follows.

Let $x \in \mathbb{R}^3$ contain the unknown Cartesian coordinates of the receiver, and let $b \in \mathbb{R}$ be the unknown bias of the receiver clock with respect to the satellites' clocks. Also, let $x_i \in \mathbb{R}^3,\ 1 \le i \le 4$ be the known Cartesian coordinates of four satellites, let $t_i \in \mathbb{R},\ 1 \le i \le 4$ be the known times when the satellites are at these locations, and let $\tau_i \in \mathbb{R},\ 1 \le i \le 4$ be the known times, on the receiver clock, when it receives the signals from the satellites. Then, we have

$$||x - x_i|| - c(b + \Delta t_i) = 0, \quad 1 \le i \le 4, \tag{4.52}$$

where c is the speed of light and $\Delta t_i = \tau_i - t_i$ is known and represents the apparent travel time of the ith satellite's broadcast. Equations (4.52) are a system of four equations for the four unknowns $(x^T, b)^T$, which can be solved, for example, using Newton's method, as described in Appendix A.4.

In global positioning systems, the errors are mainly due to atmospheric deformations of electromagnetic wave fronts. For improved accuracy, these effects are mitigated in differential global positioning systems (DGPS) by adding a fourth subsystem: a set of ground stations at perfectly known locations. Each of these stations receives signals from the satellites, computes the current positioning error at its location, and broadcasts that positioning error so that receivers in its vicinity can recalibrate.

4.5 Inertial Navigation

In this section, we remove the assumption that we use only the instantaneous readings of the sensors (as opposed to their time histories). The simplest situation one might face in this context is that of **inertial navigation**, which consists of measuring the acceleration of the vehicle with respect to an inertial frame and integrating twice to obtain the position. This is similar to **dead-reckoning** (see Problem 2.15), which relies on measurements of velocity. To understand the difficulty of inertial navigation, we should realize that it suffers from three separate basic limitations: the equivalence principle, the coupling of attitude and position errors, and the accumulation of drifts and errors. In this section, we study these limitations to the extent that they degrade the performance of inertial navigation systems (INS).

We start by stating the following result, which is the basic axiom of the theory of general relativity.

PROPOSITION 4.1 (Einstein's Equivalence Principle [54]) *The effect of gravitational forces cannot be distinguished from that of inertial forces.*

EXAMPLE 4.7 *To illustrate the equivalence principle, Einstein proposes the following thought experiment. Consider a windowless elevator cage in free space. Assume that a rope is attached to the top of the cage, and that an agent has the capability to pull the cage in such a way that the cage has an acceleration that is orthogonal to the floor and is equal in magnitude to* 9.81 *m/s*^{2}. *Assume that a passenger in the cage releases an object. Because the motion of this object is rectilinear uniform, and the motion of the cage is accelerated, the passenger sees the object fall to the floor of the cage with an acceleration equal to* 1 *g. Moreover, the passenger is unable to determine whether this fall is due to an acceleration of the cage or to the presence of a massic planet below the cage floor.*

As a consequence of the equivalence principle, we have the following. If the vector $\ddot{\vec{r}}$ represents the acceleration of a vehicle with respect to an inertial frame, and if $\vec{g}(\vec{r}, t)$ represents the local acceleration of gravity, then the reading of an accelerometer mounted on the vehicle is $\vec{a}_T = \ddot{\vec{r}} - \vec{g}(\vec{r}, t)$, which is called the **thrust acceleration**. It is then clear that inaccuracies in readings of the accelerometers are combined with inaccuracies in computing the local gravity to degrade the performance of the navigation.

Figure 4.14 represents the basic block diagram of all inertial navigation systems. It features a double integrator with feedback through a gravity computer. For a vehicle navigating in the vicinity of Earth, the computation of gravity is called **geodesy**.

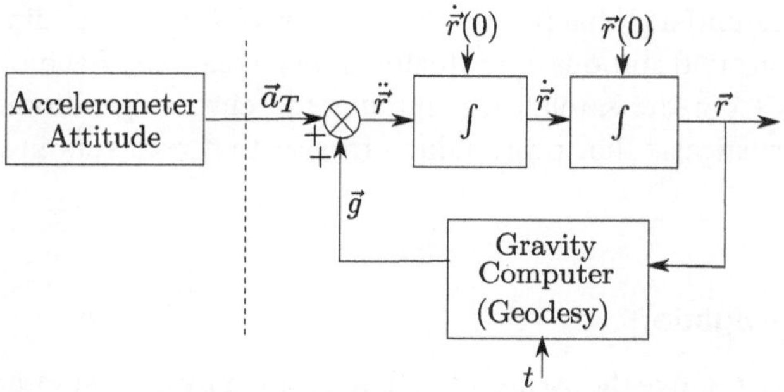

Figure 4.14. Basic organization of all inertial navigation systems.

The basic sensors in any inertial navigation system are accelerometers and gyroscopes. Accelerometers measure components of thrust acceleration, and gyroscopes measure components of angular velocity. These sensors are typically arranged in one of two standard configurations that determine the left side of Figure 4.14: **inertially stabilized** or **strapped down**.

4.5.1 Inertially Stabilized Inertial Navigation Systems

The idea of inertially stabilized INS is very simple: measure the components of thrust acceleration *in inertial space*, and then integrate them twice, as shown in Figure 4.14. To that effect, the platform P of Figure 4.15 carries three mutually orthogonal accelerometers, and three mutually orthogonal gyroscopes (see Figure 4.16).

The gimbals A, B, and C carry motors that can change the orientation of the platform with respect to the vehicle, whereas the axis (AA') is fixed in the vehicle. The principle of operation of this platform is very simple; the components of the angular velocity of P are sensed by gyroscopes and fed back to the motors A, B, and C so that the orientation of P remains constant in inertial space. As a consequence, the accelerometers sense components of thrust acceleration $\vec{a}_T$ in inertial space.

REMARK 4.9 *The three degrees of freedom platform shown in Figure 4.15 has a singular "locked" position. In practice, a fourth gimbal is used to prevent locking.*

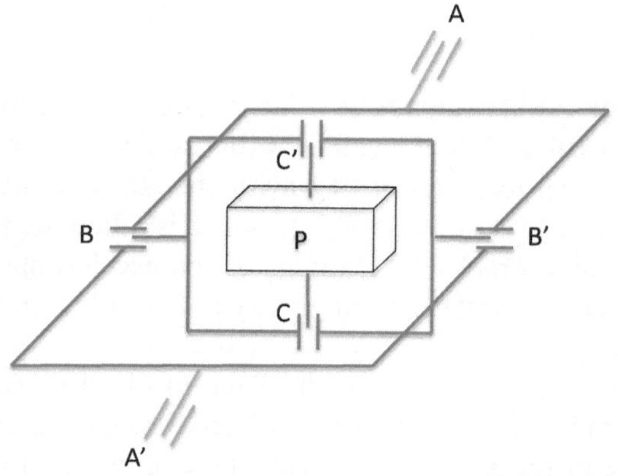

Figure 4.15. Inertially stabilized inertial navigation system schematic.

Figure 4.16. Model of ship's inertial navigation stable platform made by Charles Stark Draper Laboratories. Image courtesy of Smithsonian Institution, National Air and Space Museum.

REMARK 4.10 *Instead of motors at A, B, and C, one can use momentum wheels mounted directly on the platform. The potential difficulty, however, is that they need to be "de-spun" from time to time.*

REMARK 4.11 *The question of how to feed back angular rates sensed by gyroscopes to stabilize a platform was a major motivation for the development of classical control theory in the twentieth century.*

4.5.2 Strapped Down Inertial Navigation Systems

In contrast to inertially stabilized INS, strapped-down INS use measurements of the thrust acceleration $\vec{a}_T$ *in the vehicle frame*. Typically, the vehicle carries three accelerometers and three gyroscopes that are fixed with respect to the vehicle (hence the phrase "strapped down"). Because the components of thrust acceleration $\vec{a}_T$ are measured in the vehicle basis, we are led to formulate two questions: How are the components of $\vec{a}_T$ in the vehicle basis and the inertial basis related? And how is the angular orientation of a moving basis related to its components of angular velocity? The answers to these questions are a coordinate transformation and Poisson's equation, respectively. Specifically, we have the following standard results in kinematics of rotations.

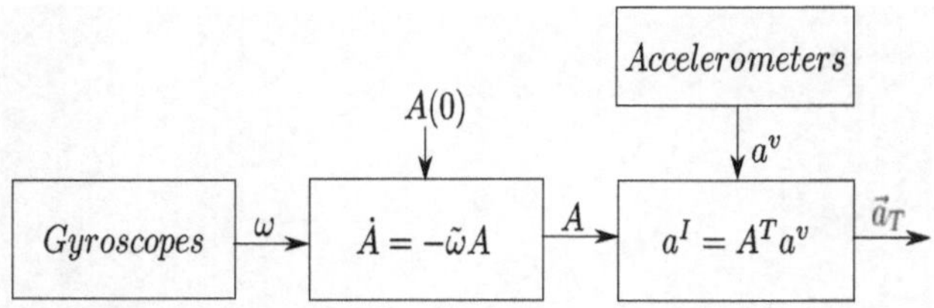

Figure 4.17. Strapped-down inertial navigation system schematic.

PROPOSITION 4.2 *If $A \in \mathbb{R}^{3\times3}$ is the direction cosine matrix of a vehicle with respect to an inertial basis, $a^V \in \mathbb{R}^3$ is the matrix of components of a vector $\vec{a}$ in the vehicle basis, and $a^I \in \mathbb{R}^3$ is the matrix of components of the same vector $\vec{a}$ in the inertial basis, then*

$$a^I = A^T a^V. \tag{4.53}$$

PROPOSITION 4.3 *If $\omega \in \mathbb{R}^{3\times1}$ is the matrix of components of the angular velocity of the vehicle basis, and $A \in \mathbb{R}^{3\times3}$ is the direction cosine matrix of the vehicle with respect to an inertial basis, then they satisfy* ***Poisson's differential equation****:*

$$\dot{A} = \begin{bmatrix} 0 & \omega_3 & -\omega_2 \\ -\omega_3 & 0 & \omega_1 \\ \omega_2 & -\omega_1 & 0 \end{bmatrix} A = -\tilde{\omega} A. \tag{4.54}$$

Let the readings of the strapped down gyroscope be ω, and that of the accelerometers be a^V. From Propositions 4.2 and 4.3, the components of thrust acceleration in the inertial basis can be obtained using the block diagram in Figure 4.17.

REMARK 4.12 *In* (4.54)*, the matrix $\tilde{\omega}$ is skew-symmetric while the matrix A is orthogonal. It is indeed a remarkable property of Poisson's equation that for any orthogonal initial condition, its solution is orthogonal at all times (see Problem 2.5).*

REMARK 4.13 *For best accuracy in integrating Poisson's equation* (4.54)*, one often uses Euler parameters, also known as* ***quaternions****.*

Also note that in the block diagrams of Figures 4.14 and 4.17, disturbances and errors intervene everywhere and tend to accumulate. As a consequence, inertial navigation used alone is practical only for short trips, for example, the flight of a ballistic missile (see, for instance, Problem 2.15 for a similar analysis applied to dead-reckoning).

EXAMPLE 4.8 *Consider the case of inertial navigation in a horizontal plane (i.e., with $\vec{g} = \vec{0}$,) using strapped-down sensors. Let $\{\vec{I}_1, \vec{I}_2\}$ be an inertial basis, let $\{\vec{z}_1, \vec{z}_2\}$ be a basis that is strapped down to the vehicle, and let*

$$\vec{r} = \vec{I}_1 x_1 + \vec{I}_2 x_2 \tag{4.55}$$

be the position vector of the vehicle, where (x_1, x_2) are its Cartesian coordinates. Let

$$\begin{aligned} \vec{a} &= a_1^I \vec{I}_1 + a_2^I \vec{I}_2 \\ &= a_1^V \vec{z}_1 + a_2^V \vec{z}_2 \end{aligned} \tag{4.56}$$

be the acceleration of the vehicle, where (a_1^I, a_2^I) and (a_1^V, a_2^V) are the Cartesian components of acceleration in the inertial and vehicle frames, respectively. The strapped-down

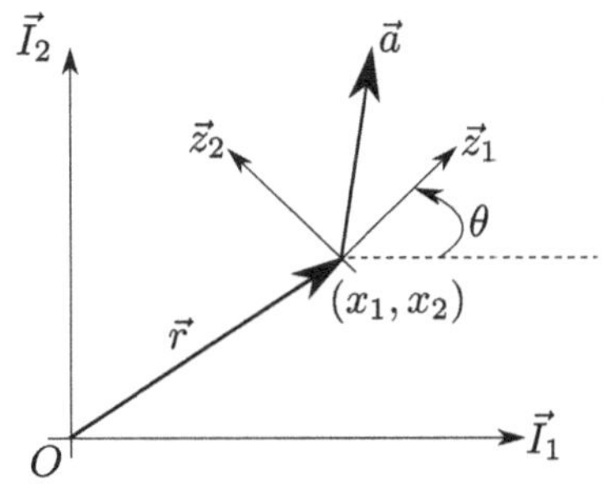

Figure 4.18. Strapped-down inertial navigation in a horizontal plane.

accelerometers measure (a_1^V, a_2^V)*, whereas a gyroscope measures the angular velocity* $\dot{\theta} = \omega$*. The problem of strapped-down inertial navigation is then to reconstruct* (x_1, x_2) *based on these measurements.*

From the geometry of Figure 4.18, we have

$$\begin{aligned}\begin{bmatrix}\vec{z}_1\\ \vec{z}_2\end{bmatrix} &= \begin{bmatrix}\cos\theta & \sin\theta\\ -\sin\theta & \cos\theta\end{bmatrix}\begin{bmatrix}\vec{I}_1\\ \vec{I}_2\end{bmatrix}\\ &= A\begin{bmatrix}\vec{I}_1\\ \vec{I}_2\end{bmatrix}.\end{aligned} \tag{4.57}$$

Therefore, from Proposition 4.2, (a_1^I, a_2^I) *and* (a_1^V, a_2^V) *are related by*

$$\begin{aligned}\begin{bmatrix}a_1^I\\ a_2^I\end{bmatrix} &= \begin{bmatrix}\cos\theta & -\sin\theta\\ \sin\theta & \cos\theta\end{bmatrix}\begin{bmatrix}\vec{I}_1\\ \vec{I}_2\end{bmatrix}\\ &= A^T\begin{bmatrix}a_1^V\\ a_2^V\end{bmatrix}.\end{aligned} \tag{4.58}$$

Moreover, the matrix A satisfies the differential equation

$$\begin{aligned}\dot{A} &= \begin{bmatrix}-\omega\sin\theta & \omega\cos\theta\\ -\omega\cos\theta & -\omega\sin\theta\end{bmatrix}\\ &= \begin{bmatrix}0 & \omega\\ -\omega & 0\end{bmatrix}\begin{bmatrix}\cos\theta & \sin\theta\\ -\sin\theta & \cos\theta\end{bmatrix}\\ &= \begin{bmatrix}0 & \omega\\ -\omega & 0\end{bmatrix}A,\end{aligned} \tag{4.59}$$

which is the two-dimensional version of Poisson's equation (4.54)*. Then, the onboard navigation computer should integrate the differential equations*

$$\begin{aligned}\dot{a}_{11}(t) &= \omega(t)a_{21}(t)\\ \dot{a}_{21}(t) &= -\omega(t)a_{11}(t)\\ \dot{a}_{12}(t) &= \omega(t)a_{22}(t)\\ \dot{a}_{22}(t) &= -\omega(t)a_{12}(t)\\ \ddot{x}_1(t) &= a_{11}(t)a_1^V(t) + a_{21}(t)a_2^V(t)\\ \ddot{x}_2(t) &= a_{12}(t)a_1^V(t) + a_{22}(t)a_2^V(t),\end{aligned} \tag{4.60}$$

that is, a total of eight first-order differential equations. Note that the first four equations in (4.60) *generate* $A(t)$ *and require* $A(0)$, *whereas the last two equations in* (4.60) *generate* $\vec{r}(t)$ *and require* $\vec{r}(0)$ *and* $\dot{\vec{r}}(0)$.

Also note that $A(t)$ *can be simply generated by*

$$\dot{\theta}(t) = \omega(t)$$
$$\begin{bmatrix} a_{11}(t) & a_{12}(t) \\ a_{21}(t) & a_{22}(t) \end{bmatrix} = \begin{bmatrix} \cos\theta(t) & \sin\theta(t) \\ -\sin\theta(t) & \cos\theta(t) \end{bmatrix}, \tag{4.61}$$

which requires $\theta(0)$. *Then, for navigation, it is sufficient to integrate* (4.61) *together with the last two equations of* (4.60), *that is, a total of five first-order differential equations.*

4.6 Asymptotic Observers

In this section, we introduce **recursive navigation**, whereby the navigator uses the entire time history of past measurements to estimate the position and velocity of the vehicle. This is in contrast to position fixing, where the navigator uses only present measurements. Recursive navigation uses knowledge of the vehicle kinematics and current measurements to update the estimate of position and velocity. To understand how this can be done, we first treat the case of recursive navigation with perfect sensors.

Consider again a standard linear time varying dynamic system,

$$\begin{aligned} \dot{x}(t) &= A(t)x(t) + B(t)u(t) \\ y(t) &= C(t)x(t), \end{aligned} \tag{4.62}$$

and assume that we want to recursively reconstruct the state vector $x(t)$ from knowledge of the past history of the output $y(t)$ and the input $u(t)$. The idea is to use a dynamic system of the form

$$\dot{\hat{x}}(t) = A(t)\hat{x}(t) + B(t)u(t) + G(t)\left(C(t)\hat{x}(t) - y(t)\right), \tag{4.63}$$

which can be justified as follows. The first two terms on the right-hand side of (4.63) replicate the dynamics of (4.62). With only these two terms, if the initial condition $\hat{x}(0)$ coincided with $x(0)$, then the state vector of (4.63) would perfectly follow that of (4.62). However, if the two initial conditions do not coincide, this may show up by the expected output $C(t)\hat{x}(t)$ being different from the actual output $y(t)$. Hence the idea of correcting the dynamics of (4.63) by feeding back the difference $C(t)\hat{x}(t) - y(t)$ through a gain $G(t)$.

To analyze the performance of (4.63) as an observer for (4.62), define the observation error as

$$\tilde{x}(t) = x(t) - \hat{x}(t). \tag{4.64}$$

Subtracting (4.63) from the first line of (4.62) then yields

$$\begin{aligned} \dot{\tilde{x}}(t) &= A(t)\tilde{x}(t) + G(t)C(t)\tilde{x}(t) \\ &= [A(t) + G(t)C(t)]\tilde{x}(t), \end{aligned} \tag{4.65}$$

which is a homogeneous linear differential equation. Therefore, if we can find $G(t)$ in (4.63) such that (4.65) is asymptotically stable, then we guarantee that

$$\lim_{t\to\infty} \tilde{x}(t) = 0, \tag{4.66}$$

hence the name **asymptotic observer** for (4.63).

Finding such a stabilizing time varying gain can be a difficult task in general. There are cases, however, where this can be done relatively easily. In particular, for linear time invariant systems with time invariant observers, we have the following standard result.

PROPOSITION 4.4 *The time invariant pair (A, C) is observable if and only if, for every self-conjugate set of eigenvalues, there exists a real constant matrix G such that $A + GC$ has the given eigenvalues.*

REMARK 4.14 *The problem addressed by Proposition 4.4 is called the **pole assignment problem** and has received substantial attention in the literature – so much so that in practice, control design toolboxes perform the required computations with a single command.*

EXAMPLE 4.9 *Consider an instance of the double integrator, which was introduced and motivated in Example 2.3. Here it is used to model the one-dimensional navigation of a vehicle along a straight line. Assume that we use a perfect accelerometer and a perfect range sensor. The kinematics of the vehicle are*

$$\begin{bmatrix}\dot{x}_1\\ \dot{x}_2\end{bmatrix} = \begin{bmatrix}0 & 1\\ 0 & 0\end{bmatrix}\begin{bmatrix}x_1\\ x_2\end{bmatrix} + \begin{bmatrix}0\\ 1\end{bmatrix}u$$

$$y = \begin{bmatrix}1 & 0\end{bmatrix}\begin{bmatrix}x_1\\ x_2\end{bmatrix}, \tag{4.67}$$

where x_1 and x_2 are the range and velocity, respectively, u is the reading of the accelerometer, and y is the reading of the range sensor. Following (4.64), *the navigation computer integrates online the following equation:*

$$\dot{\hat{x}} = \begin{bmatrix}0 & 1\\ 0 & 0\end{bmatrix}\hat{x} + \begin{bmatrix}0\\ 1\end{bmatrix}u + \begin{bmatrix}g_1\\ g_2\end{bmatrix}(\hat{x}_1 - y), \tag{4.68}$$

where the coefficients g_1 and g_2 have to be determined for stability of the observation error. According to (4.65), *this error system is*

$$\dot{\tilde{x}} = \begin{bmatrix}0 & 1\\ 0 & 0\end{bmatrix}\tilde{x} + \begin{bmatrix}g_1\\ g_2\end{bmatrix}\begin{bmatrix}1 & 0\end{bmatrix}\tilde{x}$$

$$= \begin{bmatrix}g_1 & 1\\ g_2 & 0\end{bmatrix}\tilde{x} = A_c\tilde{x}. \tag{4.69}$$

The characteristic polynomial of the state matrix in (4.69) *is*

$$\det(sI - A_c) = s^2 - g_1 s - g_2. \tag{4.70}$$

For stability, the Routh–Hurwitz criterion requires

$$g_1 < 0, \; g_2 < 0. \tag{4.71}$$

Assume that, for the sake of definiteness, we want observer eigenvalues to be

$$s_1 = -1 - j, \; s_2 = -1 + j. \tag{4.72}$$

Then, the desired characteristic polynomial of (4.69) *is*

$$(s - s_1)(s - s_2) = s^2 + 2s + 2, \tag{4.73}$$

which, by identification with (4.70), *yields*

$$g_1 = -2, \; g_2 = -2. \tag{4.74}$$

For navigation, the onboard computer therefore integrates

$$\dot{\hat{x}} = \begin{bmatrix} g_1 & 1 \\ g_2 & 0 \end{bmatrix} \hat{x} + \begin{bmatrix} 0 \\ 1 \end{bmatrix} u - \begin{bmatrix} g_1 \\ g_2 \end{bmatrix} y, \tag{4.75}$$

that is,

$$\dot{\hat{x}} = \begin{bmatrix} -2 & 1 \\ -2 & 0 \end{bmatrix} \hat{x} + \begin{bmatrix} 0 \\ 1 \end{bmatrix} u - \begin{bmatrix} -2 \\ -2 \end{bmatrix} y. \tag{4.76}$$

REMARK 4.15 *As Example 4.9 shows, in inertial navigation, one typically treats the reading of accelerometers as* ***inputs*** *to the kinematic equations of the vehicle, instead of treating them as outputs. Conversely, the readings of position and velocity sensors are generally treated as outputs.*

REMARK 4.16 *If the pair* (A, C) *is observable, the eigenvalues of* $A + GC$ *can be made arbitrarily fast by choice of* G*. While it may seem desirable to cause the observation error to decay as fast as possible, we must remember that so far we have been assuming perfect measurements. We remove this assumption in the following section.*

4.7 The Kalman Filter

In this section, we treat recursive navigation under the assumption of imperfect sensors. The basic ingredient of this treatment is the Gauss–Markov model introduced in Section 3.8. The idea is, for such a model, to design an observer following the method of Section 4.6, but where the gain is chosen to minimize the expected value of the square of the magnitude of the estimation error.

Consider again the linear Gauss–Markov model for navigation:

$$\begin{aligned} \dot{x}(t) &= A(t)x(t) + B(t)u(t) + w(t), \quad t \geq t_0 \\ y(t) &= C(t)x(t) + v(t), \end{aligned} \tag{4.77}$$

where we make the standard assumptions of Section 3.8, namely,

$$\begin{aligned}
x(t_0) &= \mathcal{N}\left(\bar{x}(t_0), P(t_0)\right), \\
E[w(t)] &= 0, \\
E[w(t)w^T(\tau)] &= R_w(t)\delta(t-\tau), \\
E[w(t)(x(t_0)-\bar{x}(t_0))^T] &= 0, \\
E[v(t)] &= 0, \\
E[v(t)v^T(\tau)] &= R_v(t)\delta(t-\tau), \\
E[v(t)(x(t_0)-\bar{x}(t_0))^T] &= 0, \\
E[w(t)v^T(\tau)] &= 0.
\end{aligned} \tag{4.78}$$

For the system given in (4.77), we consider the same observer as in (4.63), that is,

$$\dot{\hat{x}}(t) = A(t)\hat{x}(t) + B(t)u(t) + G(t)\left(C(t)\hat{x}(t) - y(t)\right), \tag{4.79}$$

where $\hat{x}$ denotes the estimate of the state vector. Here also, defining

$$\tilde{x}(t) = x(t) - \hat{x}(t) \tag{4.80}$$

as the estimation error and subtracting (4.79) from the first line of (4.77) yields

$$\dot{\tilde{x}}(t) = [A(t) + G(t)C(t)]\tilde{x}(t) + G(t)v(t) + w(t), \tag{4.81}$$

which, unlike (4.65), is no longer a homogeneous differential equation, but is driven by the input

$$n(t) = G(t)v(t) + w(t). \tag{4.82}$$

Because of this input, we generally can no longer have

$$\lim_{t\to\infty} \tilde{x}(t) = 0, \tag{4.83}$$

hence the idea of minimizing the covariance of the estimation error. To do this, notice that $n(t)$ is Gaussian, zero-mean, and white, with covariance

$$E[n(t)n^T(\tau)] = [G(t)R_v(t)G^T(t) + R_w(t)]\delta(t-\tau). \tag{4.84}$$

Define

$$P(t) = E[(\tilde{x} - \bar{\tilde{x}})(\tilde{x} - \bar{\tilde{x}})^T] \tag{4.85}$$

as the covariance of the estimation error. In Proposition 3.6, we have seen that $P(t)$ satisfies

$$\begin{aligned}
\dot{P}(t) &= [A(t) + G(t)C(t)]P(t) + P(t)[A(t) + G(t)C(t)]^T + G(t)R_v(t)G^T(t) + R_w(t) \\
P(t_0) &= P_0,
\end{aligned} \tag{4.86}$$

where P_0 denotes the initial error covariance. The idea is now simply to choose the time history of the gain $G(t)$ in (4.81) to minimize the trace of $P(t)$ in (4.86). The following result is proved in Chapter 9 using optimal control.

PROPOSITION 4.5 *Let $t_f > t_0$ be arbitrary, and consider the performance index*

$$J(G(.)) = \text{tr}(P(t_f)) = \sum_{i=1}^{n} P_{ii}(t_f). \tag{4.87}$$

Then, a gain that minimizes (4.87) *subject to* (4.86) *must have the form*

$$G(t) = -P(t)C^T(t)R_v^{-1}(t), \tag{4.88}$$

where the optimal covariance satisfies the ***differential Ricatti equation:***

$$\begin{aligned} \dot{P}(t) &= A(t)P(t) + P(t)A^T(t) - P(t)C^T(t)R_v^{-1}(t)C(t)P(t) + R_w(t) \\ P(t_0) &= P_0. \end{aligned} \tag{4.89}$$

REMARK 4.17 *The development* (4.77)–(4.89) *is a very informal derivation of the Kalman filter as an optimal observer. It can be shown that, in fact, the Kalman filter is optimal even among the more general class of nonlinear causal time varying estimators.*

REMARK 4.18 *As* (4.88) *and* (4.89) *make clear, the time history of the optimal gain $G(t)$ can be computed* ***before*** *flight. Therefore, the online computational burden of the Kalman filter is rather small.*

We now interpret (4.88) and (4.89) to show that they end up treating the measurements in reasonable ways. To that effect, assume that all the quantities in (4.88) and (4.89) are scalars. Recall that P represents the uncertainty in our estimate, R_v represents the uncertainty in our measurements, and R_w represents the magnitude of "state disturbances."

1. First consider (4.88). As P decreases in magnitude, G decreases in magnitude. This means that when we are very confident in our estimate, we tend to disregard measurements. This is quite reasonable because, if we know very well where we are, we should not listen to noisy sensors.
2. Consider (4.88) again. When R_v increases in magnitude, G decreases in magnitude. This means that when some measurements are unreliable, we give them little weight in feedback, that is, we listen to them less. Again, this is quite reasonable.
3. Now consider (4.89). When R_w increases in magnitude, $\dot{P}$ increases also. This means that when state disturbances are large, this tends to increase the rate at which the uncertainty in our estimate grows. Again, this is quite reasonable.
4. When $G = 0$, that is, when we completely disregard the sensors, (4.86) becomes $\dot{P} = AP + PA^T + R_w$. Comparing this equation with (4.89), we realize that the term

$$-PC^T R_V^{-1} CP = D \tag{4.90}$$

in (4.89) represents the decrease in growth rate of the uncertainty of the estimate due to the use of the measurements.

5. Consider (4.90). As P increases in magnitude, $-D$ increases also. This means that the greater the uncertainty of the estimates, the more we gain by using measurements. Again, this is quite reasonable.
6. Consider (4.90) again. When R_v decreases in magnitude, $-D$ increases. This means that the smaller the uncertainty of the measurements, the more we gain by using them. Again, this is quite reasonable.
7. Finally, and most remarkably, the term $-D$ in (4.90) can be viewed as a "signal-to-noise ratio" in the output (4.77) (line 2), comparing the intensity of the signal $C(t)x(t)$ to that of the noise $v(t)$. Hence, the better the signal-to-noise ratio of our measurements is, the more we gain by using them. This is also quite reasonable.

REMARK 4.19 *Equations* (4.88) *and* (4.89) *require the output noise covariance matrix* R_v *to be nonsingular. When this is not the case, there is a theoretical solution based on differentiating the components of* y *until white noise appears. In practice, one should remember that* R_v *and* R_w *are parameters in the design, and assigning them different values solves this problem.*

REMARK 4.20 *Even though the presence of a driving term* $n(t) = G(t)v(t) + w(t)$ *in* (4.81) *prevents*

$$\lim_{t\to\infty} \tilde{x}(t) = 0,$$

we hope that for every initial condition of the estimator, the estimation error satisfies

$$\lim_{t\to\infty} \bar{\tilde{x}}(t) = 0,$$

that is, if the noise input were turned off in (4.81), *the estimation error in* (4.81) *would go to zero as* $t \to \infty$. *Under the conditions of "stochastic observability" of* (A, C) *and boundedness of* R_v *and* R_w, *this is guaranteed.*

REMARK 4.21 *The same theory can be developed for discrete-time systems.*

REMARK 4.22 *If the matrices* A, B, C, R_v, *and* R_w *are time invariant, it is convenient to use the steady state filter obtained by setting*

$$\dot{P} = 0 = AP + PA^T - PC^T R_v^{-1} CP + R_w. \tag{4.91}$$

This last matrix equation is known as the ***algebraic Ricatti equation (ARE)***.

EXAMPLE 4.10 *Consider a generalization of the double integrator that was introduced and motivated in Example 2.3. Here this generalization is used to model the one-dimensional navigation of a vehicle with noisy acceleration and range measurements. The equations of motion are*

$$\begin{aligned} \dot{x} &= \begin{bmatrix} 0 & 1 \\ 0 & 0 \end{bmatrix} x + \begin{bmatrix} 0 \\ 1 \end{bmatrix} u + \begin{bmatrix} 0 \\ \epsilon_a \end{bmatrix} \\ y &= \begin{bmatrix} 1 & 0 \end{bmatrix} x + v(t), \end{aligned} \tag{4.92}$$

where $\epsilon_a(t)$ *and* $v(t)$ *represent noise in the readings of the accelerometer and range sensor, respectively. We assume that they are Gaussian, zero-mean, uncorrelated, and stationary with covariances* σ_w^2 *and* σ_v^2*, respectively. Equation* (4.92) *is then a standard linear time invariant Gauss–Markov model as in* (4.77) *with assumptions* (4.78)*, with*

$$R_v = \sigma_v^2, \; R_w = \begin{bmatrix} 0 & 0 \\ 0 & \sigma_w^2 \end{bmatrix}. \tag{4.93}$$

Let

$$P = \begin{bmatrix} p_1 & p_2 \\ p_2 & p_3 \end{bmatrix} \tag{4.94}$$

be the steady-state covariance matrix of the estimation error. We know that this is a symmetric positive semidefinite matrix. The algebraic Riccati equation (4.91) *reads*

$$\begin{bmatrix} 0 & 1 \\ 0 & 0 \end{bmatrix}\begin{bmatrix} p_1 & p_2 \\ p_2 & p_3 \end{bmatrix} + \begin{bmatrix} p_1 & p_2 \\ p_2 & p_3 \end{bmatrix}\begin{bmatrix} 0 & 0 \\ 1 & 0 \end{bmatrix}$$
$$- \begin{bmatrix} p_1 & p_2 \\ p_2 & p_3 \end{bmatrix}\begin{bmatrix} 1 \\ 0 \end{bmatrix}\frac{1}{\sigma_v^2}\begin{bmatrix} 1 & 0 \end{bmatrix}\begin{bmatrix} p_1 & p_2 \\ p_2 & p_3 \end{bmatrix} + \begin{bmatrix} 0 & 0 \\ 0 & \sigma_w^2 \end{bmatrix} = 0, \tag{4.95}$$

yielding for the three independent entries of the matrix P *of* (4.94) *the equations*

$$\begin{aligned} 2p_2 - \frac{p_1^2}{\sigma_v^2} &= 0, \\ p_3 - \frac{p_1 p_2}{\sigma_v^2} &= 0, \\ -\frac{p_2^2}{\sigma_v^2} + \sigma_w^2 &= 0, \end{aligned} \tag{4.96}$$

whose solution is

$$\begin{aligned} p_2 &= \sigma_v \sigma_w, \\ p_1 &= \sigma_v \sqrt{2\sigma_v \sigma_w}, \\ p_3 &= \sigma_w \sqrt{2\sigma_v \sigma_w}, \end{aligned} \tag{4.97}$$

where the "+" sign has been chosen in the right-hand side of (4.97) *(line 1) to ensure that* $P > 0$*. Hence, the solution of the algebraic Riccati equation is*

$$P = \begin{bmatrix} \sigma_v \sqrt{2\sigma_v \sigma_w} & \sigma_v \sigma_w \\ \sigma_v \sigma_w & \sigma_w \sqrt{2\sigma_v \sigma_w} \end{bmatrix}, \tag{4.98}$$

and the optimal gain is

$$G = -PC^T R_v^{-1} = \begin{bmatrix} \sqrt{\frac{2\sigma_w}{\sigma_v}} \\ \frac{\sigma_w}{\sigma_v} \end{bmatrix}. \tag{4.99}$$

Finally, the optimal navigator is

$$
\begin{aligned}
\dot{\hat{x}}_1 &= \hat{x}_2 - \sqrt{\frac{2\sigma_w}{\sigma_v}}(\hat{x}_1 - y) \\
\dot{\hat{x}}_2 &= u - \frac{\sigma_w}{\sigma_v}(\hat{x}_1 - y),
\end{aligned} \tag{4.100}
$$

which, remarkably, does not depend on the individual properties of the sensors but only on the ratio σ_w/σ_v. Equations (4.100) *are to be integrated online by the onboard navigation computer.*

REMARK 4.23 *As Example 4.10 illustrates, the Kalman filter does not preserve the structural properties of the physical system in general. Indeed, in* (4.92)*, the velocity x_2 is exactly the time derivative of the range x_1. However, according to* (4.100)*, the optimal estimate of velocity is* ***not*** *the time derivative of the optimal estimate of range. In other words, $\hat{x}_2 \neq \dot{\hat{x}}_1$.*

A few practical considerations are in order regarding the Kalman filter. First, the solution of the nonlinear differential Riccati equation (4.89) is equivalent to solving a linear differential equation of order $2n$, the Hamiltonian system. Second, (4.89) suggests that if the covariance of the state disturbance R_w is too small, the error covariance matrix $P(t)$ tends to zero. Then, (4.88) indicates that the navigator turns off the measurement gain, that is, the filter "goes to sleep." A simple solution to this problem is to "wake it up" by artificially boosting R_w.

REMARK 4.24 *So far we have assumed that continuous measurements are available for navigation. If the measurements are discrete (such as in the case of hourly or daily fixes), it can be shown that the optimal strategy for navigation is to propagate $\hat{x}$ and P open-loop between samples, then, at samples, to use an optimal discrete update strategy based on maximum likelihood. Hence, $\hat{x}$ and P have discontinuities at the sampling instants. It can be shown that by using such a method and letting the sampling period go to zero, one recovers the continuous-time Kalman filter.*

4.8 The Extended Kalman Filter

In this section, we remove the assumption of linearity underlying the navigation model of Section 4.7. Hence, we consider the nonlinear Markov model for the motion of a vehicle in some coordinate system:

$$
\begin{aligned}
\dot{x}(t) &= f(x, u, t) + w(t) \\
y &= g(x, t) + v(t),
\end{aligned} \tag{4.101}
$$

where the state vector x contains position variables, velocity variables, and possibly some coloring filter states (see Problem 3.21), u contains components of thrust acceleration or the readings of accelerometers, y contains the readings of position and velocity sensors, and v and w are Gaussian white noise processes representing disturbances or accelerometer errors and position/velocity sensor errors, respectively.

REMARK 4.25 *Notice that* (4.101) *is linear with respect to* $w(t)$ *and* $v(t)$*. The linearity with respect to* $v(t)$ *is due to the assumption that sensor errors are* ***additive****, whereas the linearity with respect to* $w(t)$ *is due to the fact that, in general, the equations of motion are linear with respect to the accelerations.*

To design a navigator for (4.101), we use the same idea as for the standard Kalman filter; that is, we postulate a dynamic observer of the form

$$\dot{\hat{x}} = f(\hat{x}, u, t) + G(t)[g(\hat{x}, t) - y(t)]. \tag{4.102}$$

Assume that the observation error,

$$\tilde{x} = x - \hat{x}, \tag{4.103}$$

is small, and develop (4.102) around $\hat{x}$, neglecting nonlinear terms to obtain

$$\dot{\tilde{x}} = \left(\frac{\partial f}{\partial x}\right)^T_{\hat{x}} \tilde{x} + w + G\left(\frac{\partial g}{\partial x}\right)^T_{\hat{x}} \tilde{x} + Gv. \tag{4.104}$$

If we define

$$\begin{aligned} A(\hat{x}, u, t) &= \left(\frac{\partial f(x, u, t)}{\partial x}\right)^T_{\hat{x}} \\ C(\hat{x}, t) &= \left(\frac{\partial g(x, t)}{\partial x}\right)^T_{\hat{x}}, \end{aligned} \tag{4.105}$$

then (4.104) has the form

$$\dot{\tilde{x}}(t) = [A(\hat{x}, u, t) + G(t)C(\hat{x}, t)]\tilde{x}(t) + G(t)v(t) + w(t), \tag{4.106}$$

which resembles the error equation (4.81) obtained earlier, except that now the matrices A and C are functions of the state estimate $\hat{x}$. However, we apply the same idea as before, to optimize the covariance of $\tilde{x}$, which leads to the **extended Kalman filter (EKF)**, consisting of (4.102) and (4.105), together with

$$\begin{aligned} G &= -PC^T R_v^{-1} \\ \dot{P} &= AP + PA^T - PC^T R_v^{-1} CP + R_w. \end{aligned} \tag{4.107}$$

REMARK 4.26 *Whereas in the standard Kalman filter, the optimal gain can be computed offline and stored in memory before flight, now the optimal gain must be computed online during flight, because it depends on the estimate of the state, which is unknown before flight.*

REMARK 4.27 *If the measurements are not continuous but intermittent, the optimal navigation strategy is as before: open-loop propagation of* $\hat{x}$ *and* P *between sampling times, and optimal discrete updates of* $\hat{x}$ *and* P *at the samples.*

REMARK 4.28 *The optimality properties of the extended Kalman filter are quite difficult to prove theoretically; this is done in nonlinear estimation. However, this method works quite well in practice.*

4.9 Clock Corrections

So far in this chapter, we have assumed that the navigator uses a perfect clock. In this section, we remove this assumption and consider the problem of estimating simultaneously the position and velocity of a vehicle, together with the local time. In view of the theory developed in Sections 4.1 through 4.8, this generalization is very simple: we consider t as an additional state variable; that is, starting from

$$\begin{aligned} \dot{x}(t) &= f(x(t), u(t), t) + w(t), \quad x(t_0) = x_0 \\ y(t) &= g(x(t), t) + v(t), \end{aligned} \tag{4.108}$$

we write

$$\begin{aligned} \dot{x}(t) &= f(x(t), u(t), \tau(t)) + w(t), \quad x(t_0) = x_0 \\ \dot{\tau}(t) &= 1 + w_\tau(t), \quad \tau(t_0) = t_0 \\ y(t) &= g(x(t), \tau(t)) + v(t) \end{aligned} \tag{4.109}$$

where w_τ represents the rate at which the onboard clock drifts. The new state vector is now

$$\begin{bmatrix} x \\ \tau \end{bmatrix}, \tag{4.110}$$

and we can use all of the methods developed earlier to estimate this augmented state vector based on the outputs of the sensors.

REMARK 4.29 *If* (4.108) *is time invariant, that is, if* $\partial f/\partial t = 0$ *and* $\partial g/\partial t = 0$, *then it can easily be shown that the time variable* τ *is unobservable in* (4.109). *This is quite expected. For instance, we are used to observing the motion of heavenly bodies to know what time it is. If these heavenly bodies were stationary, observing them would not tell us what time it is.*

EXAMPLE 4.11 *Consider the navigation along a straight line of a vehicle that uses two perfect sensors. Assume that the first sensor measures the range with respect to an object that is itself moving with constant known velocity. The second sensor measures the absolute velocity. Let* x_1 *and* x_2 *represent the range and velocity, whereas* y_1 *and* y_2 *are the readings of the sensors. We have*

$$y_1 = x_1 - vt, \quad y_2 = x_2. \tag{4.111}$$

Hence, the equations of motion are

$$\begin{aligned} \dot{x} &= \begin{bmatrix} 0 & 1 \\ 0 & 0 \end{bmatrix} x + \begin{bmatrix} 0 \\ 1 \end{bmatrix} u \\ y &= \begin{bmatrix} 1 & 0 \\ 0 & 1 \end{bmatrix} x - \begin{bmatrix} vt \\ 0 \end{bmatrix}. \end{aligned} \tag{4.112}$$

These equations can be augmented into

$$\begin{bmatrix} \dot{x}_1 \\ \dot{x}_2 \\ \dot{\tau} \end{bmatrix} = \begin{bmatrix} 0 & 1 & 0 \\ 0 & 0 & 0 \\ 0 & 0 & 0 \end{bmatrix} \begin{bmatrix} x_1 \\ x_2 \\ \tau \end{bmatrix} + \begin{bmatrix} 0 \\ 1 \\ 0 \end{bmatrix} u + \begin{bmatrix} 0 \\ 0 \\ 1 \end{bmatrix}$$

$$y = \begin{bmatrix} 1 & 0 & -v \\ 0 & 1 & 0 \end{bmatrix} \begin{bmatrix} x_1 \\ x_2 \\ \tau \end{bmatrix}, \tag{4.113}$$

which are now in a form suitable for navigation. An asymptotic observer has the form

$$\begin{bmatrix} \dot{\hat{x}}_1 \\ \dot{\hat{x}}_2 \\ \dot{\hat{\tau}} \end{bmatrix} = \begin{bmatrix} 0 & 1 & 0 \\ 0 & 0 & 0 \\ 0 & 0 & 0 \end{bmatrix} \begin{bmatrix} \hat{x}_1 \\ \hat{x}_2 \\ \hat{\tau} \end{bmatrix} + \begin{bmatrix} 0 \\ 1 \\ 0 \end{bmatrix} u + \begin{bmatrix} 0 \\ 0 \\ 1 \end{bmatrix}$$

$$+ \begin{bmatrix} g_{11} & g_{12} \\ g_{21} & g_{22} \\ g_{31} & g_{32} \end{bmatrix} \begin{bmatrix} \hat{x}_1 - v\hat{\tau} - y_1 \\ \hat{x}_2 - y_2 \end{bmatrix}, \tag{4.114}$$

where the matrix G is chosen so that the observer in (4.114) *is stable. Here the third variable of the observer is an estimate of time.*

4.10 Navigation Hardware

Although the emphasis in this chapter is on the principles governing the use of data for navigation, we ought to be aware of the types of hardware used to collect those data, together with their principles of operation. In this section, we present several families of measurement devices that are used in navigation.

Clocks are used to measure time. The general principle of operation of modern clocks is as follows: given an oscillator whose resonant frequency is known and invariant under disturbances, one counts the number of oscillation cycles between two events to determine how much time has elapsed. The oscillator itself may be an inertia-spring mechanical system, a quartz crystal, or an atomic resonator.

Goniometers are used to measure angles. There are optical goniometers, such as the sextant or star tracker. There are also electronic goniometers, such as the radio goniometer and the radar. Goniometers are generally operated by measuring directly the angle between two "lines of sight." In optical goniometry, the line of sight is optically determined by a human operator. In radio goniometry, the line of sight is determined using two antennas and measuring the phase delay between the signals received by the two antennas. The goniometer is rotated until the delay is zero, which indicates that the line of the two antennas is orthogonal to the line of sight. Another possibility is to rotate a directional antenna and use the intensity of the received signal as a direction indicator (see Figures 4.19 and 4.20).

Telemeters are used to measure distances. There are optical telemeters and electronic telemeters, such as the radio telemeter, the radar, the sonar, and the laser telemeter (see Figure 4.21). Optical telemeters are operated by measuring the angle between the lines of sight of two cameras observing the same object. Electronic telemeters are operated by measuring the round-trip time of a signal emitted by the telemeter and bouncing off the observed object.

Figure 4.19. Periscopic aircraft sextant, D-1, Kollsman Instrument Corporation. Image courtesy of Smithonian Institution, National Air and Space Museum.

Velocimeters are used to measure relative velocities. These typically are electronic devices such as the Doppler velocimeter and the radar. Doppler velocimeters are operated by measuring the frequency shift of a signal caused by the Doppler effect.

Accelerometers are used to measure the components of thrust acceleration (see Section 4.5 and Figure 4.22). These are typically mechanical devices, such as the translational accelerometer, the pendulum, and the gyroscopic accelerometer. Translational accelerometers are operated by measuring the displacement of a proof-mass that is recalled by a spring and suspended in a viscous fluid. See Problems 4.6 through 4.8 for the principles of operation of gyroscopic accelerometers.

Figure 4.20. Amelia Earhart's Lockheed Model 10 Electra with the circular Radio Direction Finder (goniometer) aerial visible above the cockpit. Image courtesy of U.S. Air Force.

Figure 4.21. Telemeter: E-6 fire control range meter. Image courtesy of Smithsonian Institution, National Air and Space Museum.

Figure 4.22. V2 gyroscopic accelerometer. Image courtesy of Smithsonian Institution, National Air and Space Museum.

Gyrsocopes are used to measure components of angular velocity. There are mechanical gyroscopes, which use the gyroscopic rigidity of fast-spinning rotors (see Problems 4.6 through 4.8), and there are laser gyroscopes, which use the **Sagnac effect**. This effect works as follows. Consider a monochromatic laser beam that is split into two beams. Assume that the two beams are injected in the two opposite directions of a nonrotating, closed, fiber optic contour. Then, the two beams interfere, forming a pattern of interference that can be observed. If now the fiber optic contour starts rotating, one can show that the pattern of interference is *fixed with respect to inertial space*. Hence, given the wavelength of the beam, it is possible to determine the angular rate of the contour by counting the interference peaks that pass through a given point in the contour.

Digital computers together with very large scale integration are perhaps the most important hardware achievements that allow modern navigation.

4.11 Summary of Key Results

The key results in Chapter 4 are as follows:

1. Equation (4.17), which quantifies the navigation error in position fixing with nonredundant measurements
2. Equations (4.32)–(4.34), which provide the most likely estimate in position fixing with redundant measurements
3. Equations (4.45) and (4.47), which provide the covariance of the estimate in position fixing with redundant measurements
4. Proposition 4.1 and Figure 4.14, which bring to light the fundamental limitations of inertial navigation
5. Propositions 4.2 and 4.3, which allow strapped-down inertial navigation
6. Equations (4.63)–(4.65), which model recursive navigation
7. Proposition 4.5, which provides the Kalman filter for optimal navigation
8. Equations (4.108)–(4.109), which show how to model a navigation system for clock corrections

4.12 Bibliographic Notes for Further Reading

Navigation by position fixing is presented in [9], where the front cover features a beautiful picture of conic fixes used in interplanetary navigation. Newton's method for finding the roots of a system of nonlinear algebraic equations is covered in [4]. Metric tensors, used to compute the lengths related to differential increments in coordinates, are presented in [21]. The process of calibrating measurement devices is discussed in [68]. The method of maximum likelihood for estimation is covered in [74]. Reference [58] surveys several position-fixing systems. Global positioning and inertial navigation systems are discussed in [31] and [23]. Asymptotic observers are presented in [17] and [35]. The Kalman filter and its extension are discussed in [5].

As the examples of Section 4.4 show, position fixing, as presented in Sections 4.1 through 4.3, has been used extensively in atmospheric and space flight ([9], [8], [58]). Inertial navigation using inertially stabilized platforms is used in atmospheric flight, space flight, and submarine navigation [77]. Inertial navigation using strapped-down

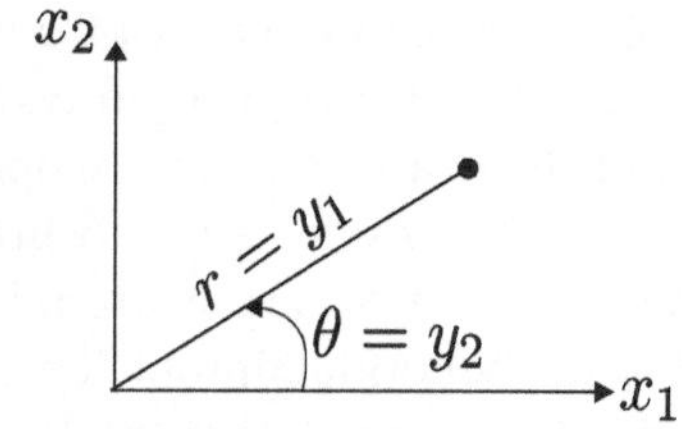

Figure 4.23. Layout for Problem 4.2.

sensors is used in ballistic missiles [77]. One of the earliest aerospace applications of recursive navigation, especially Kalman filtering, was in navigating the Apollo missions to the Moon ([9], [8]).

4.13 Homework Problems

PROBLEM 4.1 *In the plane, show that the locus of points that see two given points, A and B, under a given angle y consists of arcs of circle through A and B. (Show how to construct this locus.) This can be done in a number of ways.*

Hint: *Treating the problem geometrically rather than algebraically saves tedious computations.*

In Problems 4.2, 4.3, 4.4, and 4.5, perform the same analysis as in the text for some position fixing systems. This implies the following:

1. Find the expression for the fix function, $y = g(x)$.
2. Compute the fix Jacobian
$$C = \left(\frac{\partial g}{\partial x}\right)^T.$$
3. Outline an iterative method for solving the navigation problem.
4. Give an expression for the error covariance R_{ϵ_x}, assuming that y is measured with an error v of covariance R_v.

PROBLEM 4.2 (x_1, x_2) *are Cartesian coordinates of the vehicle, and* (y_1, y_2) *are polar coordinates as shown in Figure 4.23.*

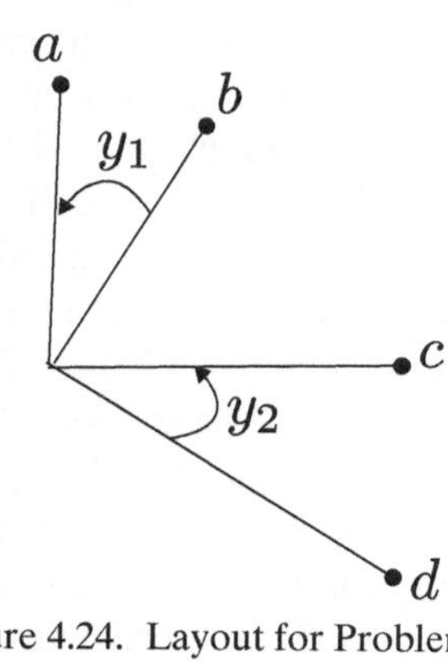

Figure 4.24. Layout for Problem 4.3.

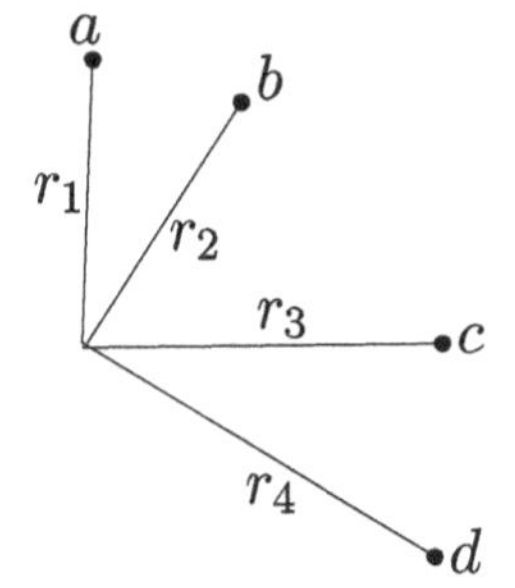

Figure 4.25. Layout for Problem 4.4.

PROBLEM 4.3 *(x_1, x_2) are Cartesian coordinates of the vehicle, y_1 is the angle under which beacons 1 and 2 are seen, y_2 is the angle under which beacons 3 and 4 are seen, and (a_1, a_2), (b_1, b_2), (c_1, c_2), (d_1, d_2) are the Cartesian coordinates of the beacons, as shown in Figure 4.24.*

PROBLEM 4.4 *(x_1, x_2) are Cartesian coordinates of the vehicle. Let r_1, r_2, r_3, r_4 be the distances from the vehicle to the four beacons as shown in Figure 4.25. We define $y_1 = r_1 - r_2$ and $y_2 = r_3 - r_4$.*

PROBLEM 4.5 *(x_1, x_2, x_3) are Cartesian coordinates of the vehicle, and (y_1, y_2, y_3) are spherical coordinates, as shown in Figure 4.26.*

PROBLEM 4.6 *A symmetric rotor has mass m and axial moment of inertia I. The rotor is spun with a high constant angular velocity Ω around its axis in a gravity field. Point O of the axis is kept fixed. Assume for simplicity that Ω is so large that the angular momentum of the rotor around O is always aligned with the axis. Show that gravity causes the rotor to* ***precess*** *with a rate $\omega_p = mgl/I\Omega$, that is, the axis of the rotor describes a circular cone around $\hat{1}_3$ with angular velocity ω_p. The rotor is shown in Figure 4.27.*

Hint: *Apply the conservation of angular momentum law $\dot{H}^0 = L^0$, where H^0 is the angular momentum around O and L^0 is the moment of external forces about O. Let $u \in \mathbb{R}^3$ be the unit vector along the axis. The preceding law yields a differential equation for u.*

REMARK 4.30 *This simplified model of the spinning top helps us understand the phenomenon of* ***gyroscopic rigidity****, which is the basis of operation of mechanical gyros.*

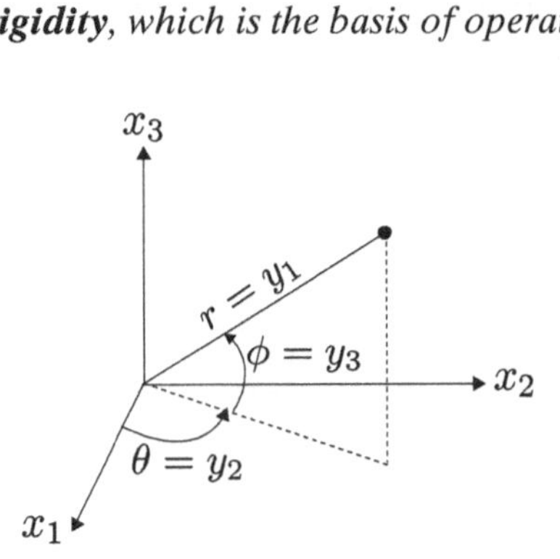

Figure 4.26. Layout for Problem 4.5.

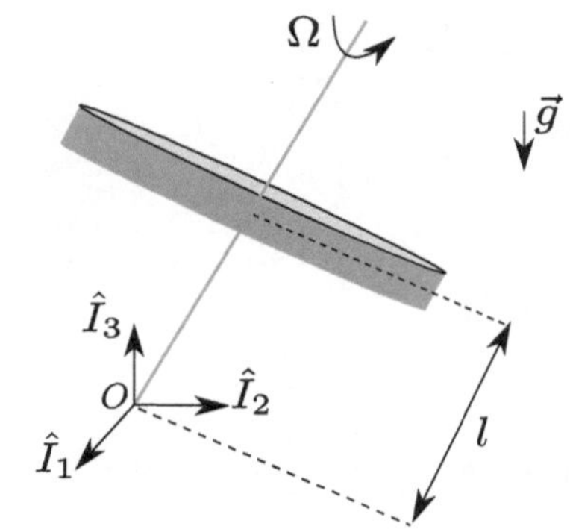

Figure 4.27. Rotor schematic for Problem 4.6.

Because of this rigidity, when subject to the gravity torque, the top does not tumble like an inverted pendulum, but it reacts by precessing around $\hat{I}_3$ and, on the average, it keeps pointing toward the vertical.

REMARK 4.31 *As you have found, the precession rate and gravity are related by*

$$g = \frac{I\Omega\omega_p}{ml}.$$

Therefore, by measuring the precession rate, it is possible to compute gravity. Now, the equivalence principle tells us that the effects of gravitational and inertial forces cannot be distinguished. Hence, this device allows us to measure the vertical acceleration of O. This is the principle of operation of **gyroscopic accelerometers**.

PROBLEM 4.7 *A symmetric rotor has axial moment of inertia I and is spun with a high constant angular velocity Ω. For simplicity, we assume that Ω is so large that the angular momentum of the rotor around its center of mass c is always aligned with its axis. (The unit vectors $\hat{x}_1, \hat{x}_2, \hat{x}_3$ are fixed in the frame $O_1O_2O_3O_4$, that is, the rotor axis is always along $\hat{x}_3$.) Let x be the roll angle, that is, the rotation angle of the frame $O_1O_2O_3O_4$ along $\hat{x}_1$. Because of the spinning rotor, the frame reacts to a roll of angle x by pitching with an angle β measured at the bearings BB'. Assume that the bearings BB' provide a restoring torque $L = -(k\beta + d\dot{\beta})\hat{x}_2$ along the pitch axis $\hat{x}_2$. The setup is shown in Figure 4.28.*

1. *Use the law of conservation of angular momentum to find the equation relating x and β.*

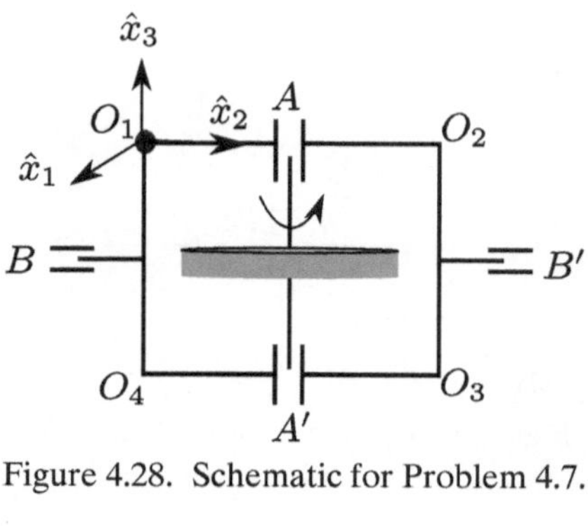

Figure 4.28. Schematic for Problem 4.7.

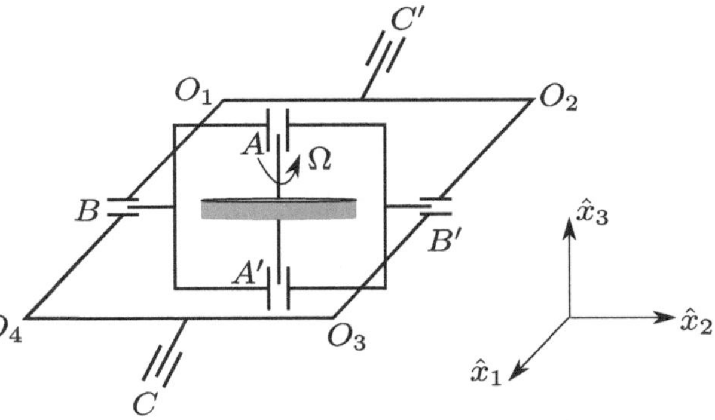

Figure 4.29. Schematic for Problem 4.8.

2. *When $k/d \gg 1$, this device is commonly called a **rate gyro**, and when $k/d \ll 1$, it is called an **integrating gyro**. Can you explain these names based on your result in the previous subquestion?*

REMARK 4.32 *This simplified model helps us understand the operation of the **single degree of freedom gyroscope**. The input axis is $\hat{x}_1$ and the output axis is $\hat{x}_2$, that is, the gyro responds to rotations around $\hat{x}_1$ by rotating around $\hat{x}_2$. In practice, the bearings AA' carry a motor to keep the spin rate Ω constant, and the restoring torque L at the bearings BB' is provided by electronic feedback amplifiers.*

PROBLEM 4.8 *Consider the same symmetric rotor as in Problem 4.6, with the same simplifying assumption about the angular momentum, but with an additional gimbal at C, C', as shown in Figure 4.29. Let β and γ be the deflections at the bearings BB' and CC', respectively. These bearings provide restoring torques $L_\beta = -(k_1\beta + d_1\dot{\beta})\hat{x}_2$ along the axis BB' and $L_\gamma = -(k_2\gamma + d_2\dot{\gamma})\hat{x}_1$ along the axis CC'.*

Assume that $k_1 \gg d_1$, so that the angles β and γ remain small. Let $\dot{\phi}$ represent the pitch rate, that is, the angular velocity of the frame $O_1O_2O_3O_4$ around $\hat{x}_2$. Let $\dot{\psi}$ represent the roll rate, that is, the angular velocity of the frame $O_1O_2O_3O_4$ around $\hat{x}_1$.

Use the law of conservation of angular momentum to find a linear equation relating $\dot{\phi}$ and $\dot{\psi}$ with γ and β.

REMARK 4.33 *This simplified model helps us understand the operation of the **two-degree of freedom rate gyro**. From the equations you obtain, it should be clear that a pitch angular velocity causes a roll deflection, and similarly a roll velocity causes a pitch deflection. This device is therefore able to measure two components of angular velocity.*

PROBLEM 4.9 *Consider again the bank-to-turn aircraft of Problem 2.14. The triple integrator equations are*

$$\dot{\psi} = k_1\phi, \; k_1 \neq 0$$
$$\ddot{\phi} = k_2 u, \; k_2 \neq 0,$$

where ψ is the heading angle, ϕ is the roll angle, and u is the aileron deflection. Assume that the output y is the heading angle, so that the system is observable. Design

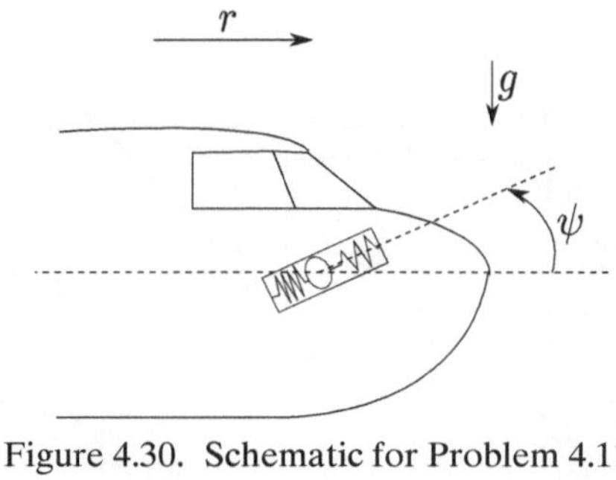

Figure 4.30. Schematic for Problem 4.11.

an asymptotic observer for this system, and specify the ranges of gains that ensure stability of the error.

PROBLEM 4.10 *Consider the linear time invariant system*

$$\begin{bmatrix}\dot{x}_1\\ \dot{x}_2\end{bmatrix} = \begin{bmatrix}A_{11} & A_{12}\\ 0 & A_{22}\end{bmatrix}\begin{bmatrix}x_1\\ x_2\end{bmatrix} + \begin{bmatrix}B_1\\ B_2\end{bmatrix}u$$

$$y = \begin{bmatrix}0 & C_2\end{bmatrix}\begin{bmatrix}x_1\\ x_2\end{bmatrix}. \tag{4.115}$$

We have seen in Problem 2.9 that x_1 is unobservable. Show that if the pair (A_{22}, C_2) is completely observable, and A_{11} is asymptotically stable, then an asymptotic observer can be designed to reconstruct the state from the outputs, despite the unobservability.

REMARK 4.34 *When a linear time invariant system is not completely observable, it turns out that it is always possible to find a coordinate transformation to write it in the form of* (4.115)*, where the pair (A_{22}, C_2) is completely observable. Then, the system is called* ***detectable*** *if A_{11} is stable. Hence, if a system is not completely observable but is detectable, its state can still be reconstructed by an asymptotic observer.*

PROBLEM 4.11 *As we have seen, inertially stabilized platforms are subject to angular drifts. This causes some components of gravity to be mistaken for inertial acceleration, entailing navigation errors. Consider the estimation of horizontal range of a vehicle where misalignment of the inertial platform is taken into account. Assume that the Earth is flat and that gravity is constant. Let r be the horizontal range, v be the horizontal velocity, ψ be the small misalignment angle of the inertial platform, and ρ be the drift rate of the platform. The situation is show in Figure 4.30. Then we have*

$$\dot{r} = v$$

$$\dot{v} = a,$$

where a is the true acceleration. The reading of the accelerometer is

$$y_a = a + g\psi + w_a,$$

where w_a is white accelerometer noise. In addition, the misalignment and drift rate obey the double integrator dynamics

$$\dot{\psi} = \rho$$

$$\dot{\rho} = w_\rho,$$

where w_ρ is white noise.

1. *Develop navigation equations to obtain estimates $(\hat{r}, \hat{v}, \hat{\psi}, \hat{\rho})^T$ based on measurements of y_a alone. What are the estimation error equations? Is the estimation stable?*
2. *Assume that we use additional range measurements, $y_r = r + w_r$, where w_r is a range sensor white noise. Develop navigation equations to obtain estimates $(\hat{r}, \hat{v}, \hat{\psi}, \hat{\rho})^T$ based on measurements of y_a and y_r together. What are the estimation error equations? How should the gains be selected so that the estimation is stable?*

PROBLEM 4.12 *The equations of motion for a vehicle are given by*

$$\dot{x}(t) = f(x(t), u(t), t)$$
$$y(t) = g(x(t), t),$$

where, as usual, x contains positions and velocities, u contains the readings of the accelerometers, and y contains position measurements. Show that if $\partial f/\partial t = 0$ and $\partial g/\partial t = 0$, it is not possible to correct the onboard clock of the vehicle using recursive navigation.

Hint: *Let τ be an additional state. Linearize the equations around a nominal trajectory, and show that $\delta\tau$ is unobservable, with dynamics that are not asymptotically stable.*

PROBLEM 4.13 *For the global positioning equation* (4.52),

1. *Outline an iterative algorithm for the computing of the unknown $(x^T, b)^T$.*
2. *Perform an error analysis, assuming that the vector of travel times Δt_i is corrupted by a measurement error with zero mean and known covariance matrix $R_{\Delta t_i} \in \mathbb{R}^{4\times 4}$.*

PROBLEM 4.14 *For the global positioning equation* (4.52),

1. *Outline a procedure for computing the unknown $(x^T, b)^T$ in the presence of redundant information, that is, when the number of satellites is strictly greater than four.*
2. *Show how to perform the error analysis in that case.*

5 Homing Guidance

In this chapter, we present the fundamentals used in the analysis and design of terminal homing guidance systems. The purpose of **terminal guidance** is to cause a pursuer (typically a missile) to hit, or come close to, a preselected target. Moreover, we require that this interception take place despite unpredictable maneuvers of the target, disturbances, and navigation uncertainties. Here we make a rough distinction between three types of terminal guidance: homing, ballistic, and midcourse. This distinction is based on the amount of control authority that is applied during most of the flight and does not lead to categories with sharply defined boundaries. Ballistic and midcourse guidance are studied in Chapters 6 and 7, respectively.

In **homing guidance**, we make the following assumptions:

1. The pursuer is actively controlled during the entirety of the engagement.
2. The pursuer has a velocity with constant norm (this is typical of the terminal phase of an engagement in atmospheric flight).
3. The pursuer is equipped with a passive seeker (e.g., a heat seeker), or a semipassive seeker (such as in laser-guided, smart bombs).

Under the preceding assumptions, we study solutions to the terminal homing guidance problem. We also allow the possibility of limiting the radius of curvature of the pursuer's trajectory.

Section 5.1 presents the fundamentals of planar homing guidance and identifies the main candidate strategies for homing. These are studied in some detail in the subsequent sections: Section 5.2 is devoted to pursuit guidance, Section 5.3 to fixed-lead guidance, Section 5.4 to constant bearing guidance, Section 5.5 to proportional guidance (also known as proportional navigation), Section 5.6 to various aspects of linearized proportional guidance, and Section 5.7 to beam rider guidance. Sections 5.8, 5.9, and 5.10 present a summary of the key results in the chapter, bibliographic notes for further reading, and homework problems, respectively.

5.1 Fundamentals of Homing

Figure 5.1 depicts the geometry of planar homing. In this figure, M represents the missile, T represents the target, *LOS* is the line of sight, that is, the line from missile to target, β is the line-of-sight angle, that is, the angle between the LOS and a given,

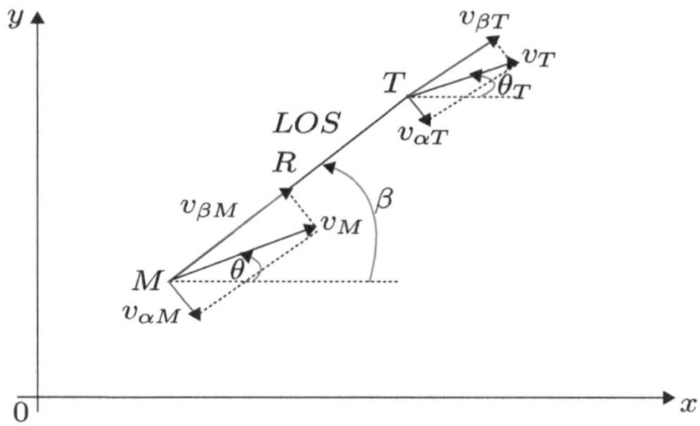

Figure 5.1. Geometry of planar homing.

fixed direction, v_M is the missile velocity, with heading angle θ and components $v_{\alpha M}$, $v_{\beta M}$ orthogonal to and along the LOS, respectively, v_T is the target velocity, with heading angle θ_T and components $v_{\alpha T}$, $v_{\beta T}$, and, finally, R is the range, that is, the distance from missile to target.

From Figure 5.1, the fundamental equations of homing, in polar form, are

$$\dot{R} = v_{\beta T} - v_{\beta M} = v_T \cos(\beta - \theta_T) - v_M \cos(\beta - \theta)$$
$$\dot{\beta} = -\frac{v_{\alpha T} - v_{\alpha M}}{R} = -\frac{v_T \sin(\beta - \theta_T) - v_M \sin(\beta - \theta)}{R}. \tag{5.1}$$

We can now formally state the Homing Problem.

Homing Problem: For the pursuer (missile) and target of Figure 5.1, find a guidance law for θ as a function of β and $\dot{\beta}$ to cause a collision or near-collision.

Note that the **Escape Problem**, which is to find θ_T to avoid interception, can be treated using results presented in Chapter 9.

REMARK 5.1 *From the preceding problem formulation, it should be clear that the homing problem is inherently a "final value" problem, that is, we care mostly about the relative position of the pursuer and target at the end of the engagement – hence the phrase "terminal guidance." However,* (5.1) *(line 2), makes it clear that near interception, because we require $R \to 0$, we may run the risk that $\dot{\beta} \to \infty$, which in practice means that the missile may encounter maneuver limitations.*

Before any further analysis of this problem, we must understand two fundamental properties of homing. The first property is that *it is advantageous for the pursuer to use a homing strategy that minimizes the requirement to turn.* There are several reasons for this. First, the time-optimal trajectory that leads to any intercept is clearly a straight line (see Problem 9.1). In other words, if the pursuer does not follow a straight line to the point of intercept, it could conceivably use another straight-line trajectory that would bring it to the point of intercept sooner, and therefore the turning trajectory is not time optimal. Second, for atmospheric flight, turning requires generating lift, which induces drag, which in turn consumes energy and slows down the pursuer.

The second property is as follows.

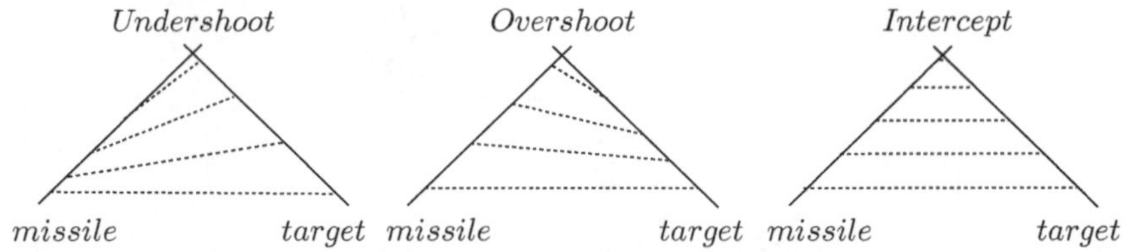

Figure 5.2. Three interception scenarios.

PROPOSITION 5.1 (Constant Bearing Principle) *In a plane, two vehicles moving with constant velocities are on a collision course if and only if their line of sight does not rotate.*

This principle is illustrated in Figure 5.2, which depicts three scenarios where a pursuer attempts to intercept a target. In all three cases, we assume that both pursuer and target have constant velocities, and we represent the time history of the line of-sight. The three cases lead to undershoot, overshoot, and intercept, respectively.

Hence, the constant bearing principle allows a navigator to anticipate an imminent collision with another vehicle. One particularly useful feature of this principle is that it is independent of winds and currents that might affect the motions. This principle is used in a variety of navigation situations. At sea, it can be used by a helmsman to decide whether the ship is on a collision course with another ship and evasive action should be taken. In air navigation, it can be used by a pilot to land an aircraft on a runway in the presence of side wind of unknown magnitude.

We can now list the following obvious candidates for homing strategies.

1. **Pursuit:** Here, the missile always heads toward the target, that is, $\theta = \beta$. This is probably the simplest homing guidance strategy one may think of. However, it typically leads to turning trajectories, which is a disadvantage, as argued earlier.
2. **Fixed Lead:** Here the missile always heads ahead of the target, that is, $\theta = \beta - \theta_0$, where θ_0 is a constant. When the target follows a straight line, there is a straight fixed-lead pursuer trajectory that yields interception.
3. **Constant Bearing:** In this strategy, the line of sight does not rotate, that is, $\dot{\beta} = 0$. This is clearly inspired by the principle presented earlier. However, its implementation requires knowledge of the target maneuvers.
4. **Proportional Guidance** (more commonly known as **Proportional Navigation**): Here the missile turns with a rate that is proportional to the turning rate of the line of sight, that is, $\dot{\theta} = \lambda\dot{\beta}$.
5. **Beam Rider:** Here the target is tracked by an electromagnetic beam from a fixed station, and the missile flies along the beam.

REMARK 5.2 *Classical control provides an illuminating interpretation of proportional navigation, as follows. First, view $\dot{\beta} = 0$, that is, constant bearing, as a reference condition. Then, it is natural to feed back any excursion from the reference into a missile maneuver, yielding $\dot{\theta} = \lambda\dot{\beta}$, where λ is the gain of a proportional controller. Hence, proportional navigation can be viewed as an attempt to achieve constant bearing through proportional feedback.*

Note that the preceding strategies are not mutually exclusive; for example, a pursuer may follow a trajectory that is both fixed-lead and constant bearing. In the remainder of this chapter, we analyze the performance of the five candidate guidance strategies.

5.2 Pursuit Guidance

In this section, we consider pursuit as a guidance strategy. For the sake of simplicity, let us assume that the target is not maneuvering; that is, referring to (5.1), assume

$$\begin{aligned} \theta_T &= 0, \\ v_T &= \text{constant}. \end{aligned} \tag{5.2}$$

Let $\theta = \beta$ (to reflect our pursuit strategy) and $\theta_T = 0$ in the fundamental equations (5.1) to obtain the equations for pursuit:

$$\begin{aligned} \dot{R} &= v_T \cos\beta - v_M, \\ \dot{\beta} &= \frac{v_T \sin\beta}{R}. \end{aligned} \tag{5.3}$$

From this last equation, we realize that $\dot{\beta} \neq 0$ unless $\beta = 0$ (i.e., the pursuer is in a tail chase, right behind the target) or $\beta = \pi$ (i.e., the pursuer is on a head-on collision course with the target). As a consequence, *in pursuit guidance, the missile always turns during the engagement, unless the engagement is a tail chase or a head-on interception.*

Combining lines 1 and 2 of (5.3) and using the implicit function theorem (as described in Appendix A.5) yields

$$\begin{aligned} \frac{\dot{R}}{\dot{\beta}} &= \frac{\mathrm{d}R}{\mathrm{d}\beta} \\ &= -\left(\frac{v_T \cos\beta - v_M}{v_T \sin\beta}\right) R \\ &= (-\operatorname{cotan}\beta + \gamma \operatorname{cosec}\beta) R, \end{aligned} \tag{5.4}$$

where

$$\gamma = \frac{v_M}{v_T} \tag{5.5}$$

is the velocity ratio, and the range R is now considered a function of the line-of-sight angle β. We integrate the differential equation (5.4) by separation of variables, as follows. First, write it as

$$\frac{\mathrm{d}R}{R} = (-\operatorname{cotan}\beta + \gamma \operatorname{cosec}\beta)\, \mathrm{d}\beta, \tag{5.6}$$

which is integrable as

$$\log R = -\log|\sin\beta| + \gamma \log|\tan(\beta/2)| + \text{constant}. \tag{5.7}$$

Assume without loss of generality that

$$0 \leq \beta \leq \pi. \tag{5.8}$$

Then, (5.7) implies

$$\log\left(\frac{R\sin\beta}{\left(\tan\frac{\beta}{2}\right)^{\gamma}}\right) = \text{constant}. \tag{5.9}$$

Hence,

$$\frac{R\sin\beta}{\left(\tan\frac{\beta}{2}\right)^{\gamma}} = \frac{R_0\sin\beta_0}{\left(\tan\frac{\beta_0}{2}\right)^{\gamma}} = K, \tag{5.10}$$

where R_0 and β_0 are the values of R and β at some arbitrary time, respectively. This last equation determines R as a function of β.

As the missile approaches the target, we have

$$R \to 0. \tag{5.11}$$

Therefore, (5.10) requires

$$\frac{\left(\tan\frac{\beta}{2}\right)^{\gamma}}{\sin\beta} \to 0. \tag{5.12}$$

Now, $\sin\beta$ is finite. Therefore, the only possibility for (5.12) to be achieved is if the numerator of the left-hand side tends to zero, which requires

$$\beta \to 0. \tag{5.13}$$

But then, the ratio in the left-hand side of (5.12) becomes

$$\frac{\left(\tan\frac{\beta}{2}\right)^{\gamma}}{\sin\beta} \to \left(\frac{\beta}{2}\right)^{\gamma}\frac{1}{\beta} \propto \beta^{\gamma-1}. \tag{5.14}$$

Hence, for (5.12) to be achieved, we must have

$$\gamma > 1, \tag{5.15}$$

that is, *the missile must be faster than the target.*

We now examine the turning rate requirements of pursuit guidance. Write (5.10) in the form

$$R = \frac{K\left(\tan\frac{\beta}{2}\right)^{\gamma}}{\sin\beta}, \tag{5.16}$$

and use this expression in (5.3) (line 2) to obtain the differential equation

$$\dot{\beta} = -\frac{v_T\sin^2\beta}{K\left(\tan\frac{\beta}{2}\right)^{\gamma}}. \tag{5.17}$$

As the missile approaches the target, (5.13) and (5.17) imply

$$\dot{\beta} \propto \beta^{2-\gamma}. \tag{5.18}$$

As a consequence, for $\gamma > 2$, the terminal turning rate is infinite, for $\gamma = 2$, the terminal turning rate is finite, and for $\gamma < 2$, the terminal turning rate is zero. In other words, we have seen that the missile is required to be faster than the target to "catch up with it." However, if the missile is much faster than the target, it ends up saturating its turning rate at the end of the engagement.

We can perform a similar analysis of the terminal turning acceleration. As the missile approaches the target, (5.18) implies that

$$\ddot{\beta} \propto \beta^{1-\gamma}\dot{\beta} \propto \beta^{3-2\gamma}. \tag{5.19}$$

Hence, for $\gamma > 1.5$, the terminal turning acceleration is infinite, for $\gamma = 1.5$, the terminal turning acceleration is finite, and for $\gamma < 1.5$, the terminal turning acceleration is zero.

In summary, we must have $1 \leq \gamma \leq 2$ for a finite turning rate, and $1 \leq \gamma \leq 1.5$ for a finite turning acceleration. This is easily generalized to higher-order derivatives (see Problem 5.1).

REMARK 5.3 *Because of the requirement in turning rate outlined earlier, pursuit guidance is used only in specific cases where the target moves very slowly ($\gamma \gg 1$) (e.g., an airborne missile engaging a ship) or when the missile attacks head-on or tail-on. In Section 5.2.2, we show that when $\gamma \gg 1$ and the pursuer has limitations in turning rate, the miss distance is negligible.*

5.2.1 Terminal Phase Analysis

In preparation for the analysis of miss due to turn rate limitations, we develop an approximate closed-form solution for the terminal portion of the trajectory. Starting from the fundamental equations of pursuit (5.3), we have seen that at the end of the engagement, we have $\beta \to 0$; that is, the engagement always ends in a tail chase. Now, for $\beta \ll 1$, we can use the small angle approximation

$$\cos\beta \approx 1, \quad \sin\beta \approx \beta, \tag{5.20}$$

which, when introduced in (5.3), yields

$$\begin{aligned} \dot{R} &\approx v_T - v_M \\ \dot{\beta} &\approx -\frac{v_T\beta}{R}. \end{aligned} \tag{5.21}$$

Integrating (5.21) (line 1) yields

$$R = (v_T - v_M)(t - t_f), \tag{5.22}$$

where t_f is the time of impact, and we have used the obvious boundary condition:

$$R(t_f) = 0. \tag{5.23}$$

Now, using (5.22) in (5.21) (line 2) yields

$$\dot{\beta} = \frac{\mathrm{d}\beta}{\mathrm{d}t} = -\frac{v_T\beta}{(v_T - v_M)(t - t_f)}, \tag{5.24}$$

which can be integrated by separation of variables:

$$\frac{\mathrm{d}\beta}{\beta} = -\frac{\mathrm{d}t}{(\gamma - 1)(t - t_f)}, \tag{5.25}$$

where γ is the velocity ratio introduced in (5.5). Hence,

$$\log \beta = \frac{1}{\gamma - 1} \log(t - t_f) + \text{constant}, \tag{5.26}$$

implying that

$$\log\left(\frac{\beta}{(t - t_f)^{\frac{1}{\gamma-1}}}\right) = \text{constant} = \log\left(\frac{\beta_0}{(t_0 - t_f)^{\frac{1}{\gamma-1}}}\right), \tag{5.27}$$

where the boundary conditions β_0 and t_0 satisfy

$$\beta(t_0) = \beta_0. \tag{5.28}$$

Equation (5.27) implies that during the terminal portion of the trajectory, when $\beta \ll 1$, we have the approximation

$$\beta \approx \beta_0 \left(\frac{t_f - t}{t_f - t_0}\right)^{\frac{1}{\gamma-1}}. \tag{5.29}$$

Note that this approximation predicts accurately the values of γ for which the final turning rate and acceleration are infinite, bounded, or zero. More specifically, (5.29) implies that, for $\gamma < 2$, $\dot{\beta} \to 0$ as $t \to t_f$; whereas, for $\gamma > 2$, $\dot{\beta} \to -\infty$ as $t \to t_f$. Indeed, from (5.29), for $\beta \ll 1$,

$$\begin{aligned}\dot{\beta} &= \frac{\beta_0}{(t_f - t_0)^{\frac{1}{\gamma-1}}} \frac{(-1)}{\gamma - 1}(t_f - t)^{\frac{1}{\gamma-1}-1} \\ &= \frac{\beta_0}{(1-\gamma)(t_f - t_0)} \left(\frac{t_f - t}{t_f - t_0}\right)^{\frac{2-\gamma}{\gamma-1}}.\end{aligned} \tag{5.30}$$

Furthermore,

$$\ddot{\beta} = \frac{\beta_0(2-\gamma)}{(\gamma-1)^2(t_f - t_0)^2} \left(\frac{t_f - t}{t_f - t_0}\right)^{\frac{3-2\gamma}{\gamma-1}}, \quad \beta \ll 1, \tag{5.31}$$

which implies that for $\gamma > 2$, $\ddot{\beta} \to -\infty$ as $t \to t_f$; for $1.5 < \gamma < 2$, $\ddot{\beta} \to +\infty$ as $t \to t_f$; and for $\gamma < 1.5$, $\ddot{\beta} \to 0$ as $t \to t_f$. We also note that for $\gamma = 2$, the approximation given by (5.29) for β is a linear function of time.

5.2.2 Approximate Miss Distance Analysis

We now consider the possibility that the radius of curvature of the pursuer's trajectory be limited. This issue is clearly motivated by the conclusions we reached earlier: the pursuer must be faster than the target, but if the pursuer is fast enough, its terminal turning rate is infinite.

Because, by assumption, the magnitude of the pursuer velocity is constant, its only component of acceleration is centripetal, of the form

$$a_c = \dot{\theta} v_M. \tag{5.32}$$

Hence, imposing a limitation on the magnitude of a_c is equivalent to imposing a limitation on the magnitude of the turn rate, of the form

$$|\dot{\theta}| \leq |\dot{\theta}|_{\max}, \tag{5.33}$$

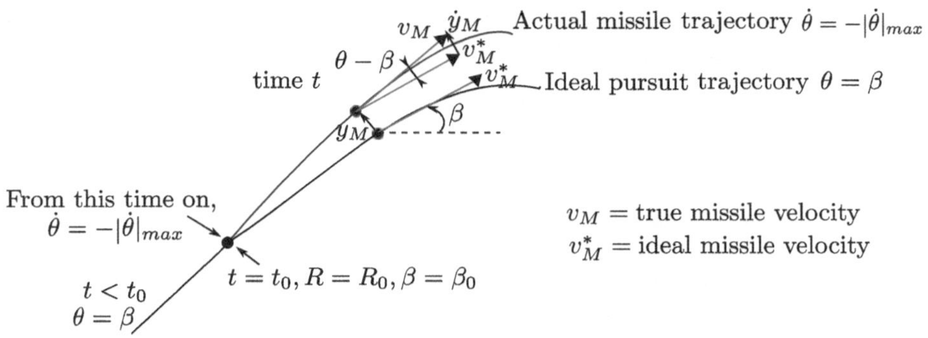

Figure 5.3. Scenario for approximate miss distance analysis.

where $|\dot\theta|_{\max}$ is a given maximal turning rate. Once the turning rate reaches its maximum feasible value, it saturates and the missile can no longer perfectly pursue the target. Figure 5.3 depicts a typical scenario.

Note that after the turning rate limitation has been reached, the actual trajectory of the pursuer is an arc of circle. For $t < t_0$, we have $|\dot\beta| < |\dot\theta|_{\max}$, the missile is able to follow a pursuit trajectory, and $\theta = \beta$. At $t = t_0$, $|\dot\beta| = |\dot\theta|_{\max}$, that is, the turning rate limitation is encountered. And for $t > t_0$, we have $|\dot\beta| > |\dot\theta|_{\max}$, and the missile follows a trajectory such that $|\dot\theta| = |\dot\theta|_{\max}$. (Note that in Figure 5.3, we have assumed that $\dot\theta < 0$.)

For $t > t_0$, let y_m be the transverse displacement of the missile from the ideal pursuit trajectory. At any time $t > t_0$, let v_M^* be the missile velocity for ideal pursuit, and let v_M be the missile velocity on the actual trajectory. We have

$$\dot y_m = v_M(\theta - \beta), \ |\theta - \beta| \ll 1. \tag{5.34}$$

Because for $t > t_0$, we have

$$\dot\theta = -|\dot\theta|_{\max} = \text{constant}, \tag{5.35}$$

we can integrate this differential equation into

$$\theta = \beta_0 - |\dot\theta|_{\max}(t - t_0), \tag{5.36}$$

where we have used for boundary condition $\theta(t_0) = \beta_0$, which represents the line-of-sight angle at which the missile encounters the turning rate limitation. Recall that near-intercept, β satisfies the approximation (5.29). Using that, together with (5.34), yields the differential equation

$$\dot y_m = v_M\left(\beta_0 - |\dot\theta|_{\max}(t - t_0) - \beta_0\left(1 - \frac{t - t_0}{t_f - t_0}\right)^{\frac{1}{\gamma-1}}\right), \tag{5.37}$$

which we integrate, using the obvious boundary condition $y_m(t_0) = 0$, to obtain

$$y_m = v_M\left(\beta_0(t - t_0) - |\dot\theta|_{\max}\frac{(t - t_0)^2}{2} + \frac{\gamma - 1}{\gamma}\beta_0(t_f - t_0)\left(1 - \frac{t - t_0}{t_f - t_0}\right)^{\frac{\gamma}{\gamma-1}} - \frac{\gamma - 1}{\gamma}\beta_0(t_f - t_0)\right). \tag{5.38}$$

Let us assume that the ideal and actual pursuit trajectories are close enough that the miss distance can be evaluated as

$$M \approx y_m(t_f). \tag{5.39}$$

Then, from (5.38), we obtain

$$M = y_m(t_f) = v_M(t_f - t_0)\left(\frac{\beta_0}{\gamma} - |\dot{\theta}|_{\max}\frac{(t_f - t_0)}{2}\right). \tag{5.40}$$

We now eliminate $(t_f - t_0)$ and $|\dot{\theta}|_{\max}$ from this expression to obtain a more useful approximation. Recall that, near intercept, the range is approximated by (5.22) and (5.23). Let R_0 be the range at which the turning rate limitation is encountered, that is, $R(t_0) = R_0$. Using this boundary condition in (5.22) yields

$$t_f - t_0 = \frac{R_0}{v_M - v_T} = \frac{\gamma R_0}{(\gamma - 1)v_M}. \tag{5.41}$$

Now, we have also seen that near intercept, β satisfies (5.30). Hence,

$$\dot{\beta}(t_0) = \dot{\beta}_0 = -|\dot{\theta}|_{\max} = \frac{\beta_0}{(1-\gamma)(t_f - t_0)}. \tag{5.42}$$

Substitute this in (5.40) to obtain

$$M = v_M\frac{\gamma R_0}{(\gamma - 1)v_M}\left(\frac{\beta_0}{\gamma} + \frac{\beta_0}{(1-\gamma)^2}\right), \tag{5.43}$$

which can be simplified as

$$M = \frac{R_0\beta_0(\gamma - 2)}{2(\gamma - 1)^2}, \quad \beta_0 \ll 1. \tag{5.44}$$

This last expression quantifies the miss distance in pursuit guidance due to the turning rate limitation. The advantage it has over (5.40) is to depend solely on R_0 and β_0, which are the range and the line-of-sight angle at which the pursuer encounters the limitation. This equation is, of course, valid only if, from the time when the limitation is encountered, the remainder of the trajectory is saturated, that is, $|\dot{\theta}| = |\dot{\theta}|_{\max}$. This, in turn, only holds when $\gamma > 2$. If $\gamma < 2$, the missile may have enough turning capability to return to the ideal pursuit trajectory. Indeed, recall that when $\gamma < 2$, the final turning rate of the ideal trajectory is zero.

Finally, from (5.44), we notice that

$$\lim_{\gamma \to \infty} M = 0. \tag{5.45}$$

This last equation implies that, despite the miss due to turn rate limitations, pursuit guidance is quite effective when $\gamma \gg 1$, that is, when the pursuer is much faster than the target.

5.2.3 Exact Miss Distance Analysis

We now consider the problem of obtaining an exact expression for the miss distance M. First, we rewrite the equations of motion in dimensionless form. Let

$$r = \frac{R}{R_0} \tag{5.46}$$

be the dimensionless range and

$$x_T = \frac{X_T}{R_0}, \quad y_T = \frac{Y_T}{R_0}, \quad x_M = \frac{X_M}{R_0}, \quad y_M = \frac{Y_M}{R_0} \tag{5.47}$$

be the dimensionless coordinates of the target and missile, respectively. We have

$$\begin{aligned}
\dot{X}_T &= v_T \cos\theta_T, \\
\dot{Y}_T &= v_T \sin\theta_T, \\
\dot{X}_M &= v_M \cos\theta, \\
\dot{Y}_M &= v_M \sin\theta, \\
X &= X_T - X_M, \\
Y &= Y_T - Y_M, \\
R &= \sqrt{X^2 + Y^2}, \\
\beta &= \arctan\left(\frac{Y}{X}\right).
\end{aligned} \tag{5.48}$$

We introduce the dimensionless time

$$\tau = \frac{v_T}{R_0} t \tag{5.49}$$

so that

$$\frac{\mathrm{d}}{\mathrm{d}t} = \frac{\mathrm{d}\tau}{\mathrm{d}t}\frac{\mathrm{d}}{\mathrm{d}\tau} = \frac{v_T}{R_0}\frac{\mathrm{d}}{\mathrm{d}\tau}. \tag{5.50}$$

Then, the equations of motion for the dimensionless coordinates and time take the form

$$\begin{aligned}
\frac{\mathrm{d}x}{\mathrm{d}\tau} &= \cos\theta_T - \gamma\cos\theta, \\
\frac{\mathrm{d}y}{\mathrm{d}\tau} &= \sin\theta_T - \gamma\sin\theta, \\
r &= \sqrt{x^2 + y^2}, \\
\beta &= \arctan\left(\frac{y}{x}\right),
\end{aligned} \tag{5.51}$$

where $x = x_T - x_M$ and $y = y_T - y_M$.

Alternatively, we have

$$\begin{aligned}
\frac{\mathrm{d}r}{\mathrm{d}\tau} &= \cos(\beta - \theta_T) - \gamma\cos(\beta - \theta) \\
\frac{\mathrm{d}\beta}{\mathrm{d}\tau} &= \frac{\sin(\beta - \theta_T) - \gamma\sin(\beta - \theta)}{r}.
\end{aligned} \tag{5.52}$$

Equations (5.51) and (5.52) are the fundamental equations of homing in dimensionless form, in Cartesian and polar coordinates, respectively. We can use them to perform an exact numerical calculation of the miss distance, as follows. Let

$$\theta_T = 0, \tag{5.53}$$

that is, assume a nonmaneuvering target. (This assumption is not necessary, however, we need to know how the target maneuvers.) Assume that, at

$$\tau = \tau_0, \quad \beta = \beta_0, \tag{5.54}$$

the missile reaches its turn-rate limitation. Hence,

$$\frac{d\theta}{d\tau} = \text{constant} = \left.\frac{d\beta}{d\tau}\right|_{\tau=\tau_0}, \quad \tau \geq \tau_0. \tag{5.55}$$

Let Ω_{max} be the maximum missile turning rate. Then,

$$\frac{d\theta}{d\tau} = -\Omega_{max}, \quad \tau \geq \tau_0, \tag{5.56}$$

which can be integrated as

$$\theta = \beta_0 - \Omega_{max}(\tau - \tau_0), \tag{5.57}$$

where we have used the boundary condition $\theta(\tau_0) = \beta_0$. The fundamental equations of homing in Cartesian coordinates and dimensionless form (5.52) imply

$$\begin{aligned}\frac{dx}{d\tau} &= \cos\theta_T - \gamma\cos\theta \\ &= 1 - \gamma\cos\left(\beta_0 - \Omega_{max}(\tau - \tau_0)\right) \\ \frac{dy}{d\tau} &= \sin\theta_T - \gamma\sin\theta \\ &= -\gamma\sin\left(\beta_0 - \Omega_{max}(\tau - \tau_0)\right).\end{aligned} \tag{5.58}$$

Integrate this last equation with respect to $(\tau - \tau_0)$ with boundary conditions

$$\begin{aligned}x(\tau_0) &= \cos\beta_0 \\ y(\tau_0) &= \sin\beta_0,\end{aligned} \tag{5.59}$$

which stem from $r_0 = 1$. We obtain

$$\begin{aligned}x &= \cos\beta_0 + (\tau - \tau_0) + \frac{\gamma}{\Omega_{max}}\left(\sin\left(\beta_0 - \Omega_{max}(\tau - \tau_0)\right) - \sin\beta_0\right) \\ y &= \sin\beta_0 - \frac{\gamma}{\Omega_{max}}\left(\cos\left(\beta_0 - \Omega_{max}(\tau - \tau_0)\right) - \cos\beta_0\right).\end{aligned} \tag{5.60}$$

Furthermore, we have

$$r = \sqrt{x^2 + y^2}. \tag{5.61}$$

Hence, we can compute numerically the dimensionless range r as a function of the dimensionless time $(\tau - \tau_0)$ for given β_0, Ω_{max}, and γ. The miss distance is then given by the minimum value of r.

Note that the miss distance can also be obtained analytically by solving the algebraic equation

$$x\frac{dx}{d\tau} + y\frac{dy}{d\tau} = 0 \tag{5.62}$$

for the unknown $(\tau - \tau_0)$, and then substituting in (5.60) and (5.61). In practice, the approximate method gives an estimate of miss that is within 5% of the exact value whenever $|\beta_0| < 30^\circ$. We use the same approximate analysis technique in studying proportional navigation.

5.3 Fixed Lead Guidance

In this guidance procedure, the missile always heads ahead of the target by a fixed lead angle; that is, referring to Figure 5.1, we have

$$\theta = \beta - \theta_0, \tag{5.63}$$

where θ_0 is the constant lead angle. For a given β_0 and a nonmaneuvering target, there exists a specific $\beta - \theta$ that yields a nonmaneuvering collision course. However, if β_0 changes, the missile must maneuver during the mission. As a consequence, fixed-lead guidance is only useful if one knows the target trajectory.

5.4 Constant Bearing Guidance

In this guidance procedure, we require that the line of sight should not turn, that is,

$$\dot{\beta} = 0. \tag{5.64}$$

Referring to Figure 5.1, this is achieved if

$$v_{\alpha M} = v_{\alpha T}, \tag{5.65}$$

which implies that

$$\dot{v}_{\alpha M} = \dot{v}_{\alpha T}. \tag{5.66}$$

In other words, the missile must match the components of target velocity and acceleration orthogonal to the line of sight. The missile acceleration need not exceed that of the target, as long as the components orthogonal to the line of sight are matched. Constant bearing is difficult to implement for a maneuvering target because, as indicated in (5.66), the missile must react instantaneously to accelerations of the target.

Referring to Figure 5.1, we note that, for a nonmaneuvering target, the value of θ_0 that yields a nonmaneuvering interception satisfies

$$\frac{v_T}{\sin(\beta_0 - \theta_0)} = \frac{v_M}{\sin(\beta_0 - \theta_T)}. \tag{5.67}$$

This formula is used for aiming at moving targets (see Problem 5.6).

In the presence of launching error and target maneuvers, one way to attempt to implement constant bearing is to let $\dot{\theta} = \lambda\dot{\beta}$, which leads to the idea of proportional navigation.

5.5 Proportional Navigation

Here we use the guidance law

$$\dot{\theta} = \lambda\dot{\beta}, \tag{5.68}$$

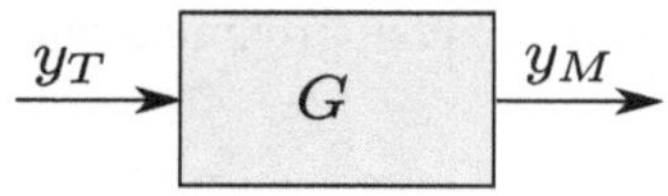

Figure 5.4. Linear time varying dynamic system for linearized proportional navigation.

where λ is the **navigation constant**. Integrating (5.68) yields

$$\theta = \lambda\beta + \theta_0, \tag{5.69}$$

where θ_0 is a constant of integration. This indicates that all the previously studied guidance laws are particular cases of proportional navigation. Indeed, in (5.69),

1. $\lambda = 1$ with $\theta_0 = 0$ constitutes pursuit guidance
2. $\lambda = 1$ with $\theta_0 \neq 0$ constitutes fixed lead guidance
3. $\lambda \to \infty$, which implies $\dot{\beta} = 0$, constitutes constant bearing guidance

For a nonmaneuvering target, the initial heading angle θ_0 yielding a constant bearing interception satisfies

$$\sin(\beta_0 - \theta_0) = \frac{\sin(\beta_0 - \theta_T)}{\gamma}. \tag{5.70}$$

However, for launch errors, the missile trajectory has some curvature. Henceforth, we study the effects of the navigation constant, the missile autopilot dynamics, target maneuvers, and noisy measurements.

5.6 Linearized Proportional Navigation

We consider a constant bearing reference trajectory, and we linearize around it. We focus on the displacements of the target and missile in the direction orthogonal to the nominal line of sight. More specifically, we model how such motions of the target cause motions of the missile. In the resulting linear time varying dynamic system shown in Figure 5.4, the input is the motion of the target in the direction orthogonal to the nominal line of sight, and the output is the displacement of the missile. This system can then be used to analyze the performance of the guidance law.

REMARK 5.4 *In the analysis of linearized proportional navigation, we assume conveniently that the speed of the missile is constant. In practice, this is far from being the case – the speed of the missile may change substantially between the initial boost and the terminal coasting phase. However, under the assumption that the speed of the missile is a known function of time, the analysis method is still applicable, yielding a linear time varying system such as in Figure 5.4.*

Referring to Figure 5.5, the reference trajectory starts at launching time $t = 0$, has a missile velocity v_M^* with heading angle θ^*, and has a target velocity v_T^*. On this reference trajectory, the range satisfies

$$\dot{R}^* = v_{\beta T}^* - v_{\beta M}^* = \text{constant}, \tag{5.71}$$

which can be integrated as

$$R = R_0 - |v_{\beta T}^* - v_{\beta M}^*| t, \tag{5.72}$$

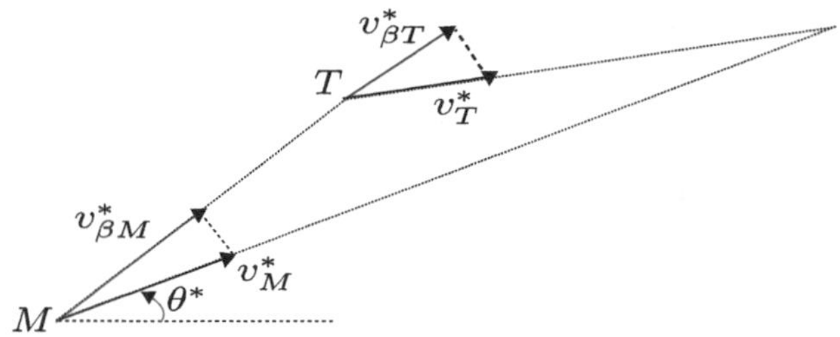

Figure 5.5. Reference trajectory for linearized proportional navigation.

where R_0 is the initial range. Therefore, the reference time to impact is

$$t_f^* = \frac{R_0}{|v^*_{\beta T} - v^*_{\beta M}|}. \tag{5.73}$$

Referring to Figure 5.6, for the linearized equations, β is the angle between the reference line of sight and the actual line of sight, where we assume that $|\beta| \ll 1$; θ is the angle between the reference missile velocity and the actual missile velocity, where we also assume that $|\theta| \ll 1$; y_M is the lateral displacement of the missile relative to the reference line of sight; and y_T is the lateral displacement of the target relative to the reference line of sight. From Figure 5.6, we have

$$\dot{y}_M = v^*_M \theta \cos\alpha. \tag{5.74}$$

Now, the factor:

$$v^*_M \cos\alpha = v^*_{\beta M} \tag{5.75}$$

represents the component of missile velocity that is parallel to the line of sight in the reference trajectory. Hence,

$$\dot{y}_M = v^*_{\beta M}\theta, \tag{5.76}$$

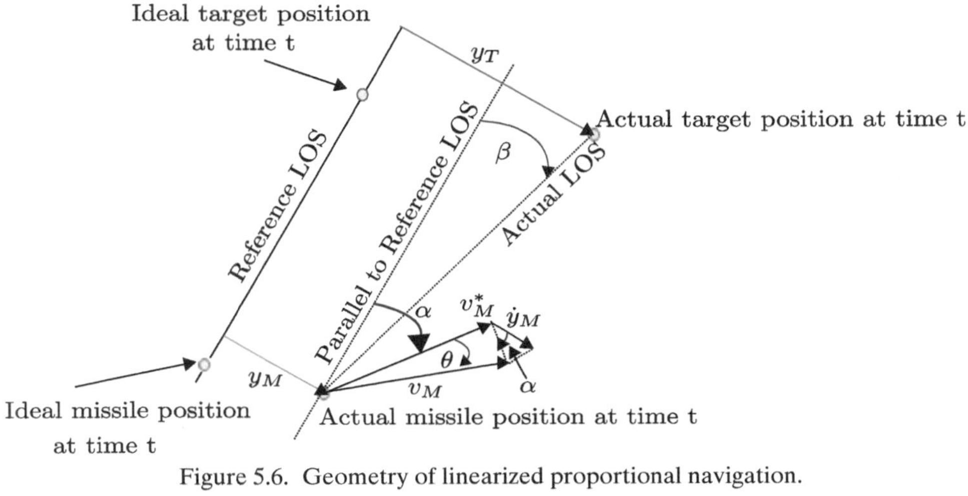

Figure 5.6. Geometry of linearized proportional navigation.

which we can solve for θ as

$$\theta = \frac{\dot{y}_M}{v^*_{\beta M}}. \tag{5.77}$$

From here on, we take into account autopilot dynamics as follows. Whereas ideally, proportional navigation is represented by $\dot{\theta} = \lambda\dot{\beta}$, actually the law that is implemented is

$$\dot{\theta} = \lambda Y(s)\dot{\beta}, \tag{5.78}$$

where

$$Y(s) = \frac{N(s)}{D(s)} \tag{5.79}$$

is the transfer function of the autopilot, s represents differentiation with respect to time, and $N(s)$ and $D(s)$ are polynomials satisfying $\deg(D(s)) > \deg(N(s))$. Integrate (5.79) to obtain

$$\theta = \lambda Y(s)\beta, \tag{5.80}$$

where we have assumed that $\theta(0) = \beta(0) = 0$, that is, a perfect collision-course launch. (For an imperfect launch, we let the target perform a fictitious maneuver in the form of a step velocity.) From Figure 5.6, we have

$$\beta = \frac{y_T - y_M}{R}. \tag{5.81}$$

Also, proximity between the reference and actual trajectories implies

$$R \approx R^*. \tag{5.82}$$

Hence,

$$\beta = \frac{y_T - y_M}{R^*}. \tag{5.83}$$

Introduce (5.77) and (5.83) into (5.80) to obtain

$$\frac{\dot{y}_M}{v^*_{\beta m}} = \lambda Y(s)\left(\frac{y_T - y_M}{R^*(t)}\right), \tag{5.84}$$

or equivalently,

$$D(s)\dot{y}_M = v^*_{\beta M}\lambda N(s)\left(\frac{y_T - y_M}{R^*(t)}\right). \tag{5.85}$$

Now, (5.71) together with the boundary condition $R^*(t^*_f) = 0$ imply

$$R^*(t) = (t^*_f - t)|v^*_{\beta T} - v^*_{\beta M}|. \tag{5.86}$$

Hence, (5.85) becomes

$$D(s)\dot{y}_M = \Lambda N(s)\left(\frac{y_T - y_M}{t^*_f - t}\right), \tag{5.87}$$

where

$$\Lambda = \frac{v^*_{\beta M}\lambda}{|v^*_{\beta T} - v^*_{\beta M}|} \tag{5.88}$$

is called the **effective navigation constant**. Equation (5.87) is the mathematical representation of the linear time varying dynamic system of Figure 5.4, where the input is the target lateral displacement y_T and the output is the missile lateral displacement y_M. For a given time history of target maneuvers, the miss distance is then approximated as

$$M = y_T(t_f^*) - y_M(t_f^*). \tag{5.89}$$

We typically use the method of adjoints of Section 2.4 to evaluate $y_M(t_f^*)$.

REMARK 5.5 *Linearized proportional navigation can be viewed as follows. Assume that a missile, initially located on the x-axis of a plane with Cartesian coordinates, engages a target initially located at the origin, so that the nominal trajectory is along the x-axis. Let the speed of the missile be given by* (5.71). *Assume that the target is allowed to move only along the y-axis by an amount $y_T(t)$, and that the missile is allowed to deviate from the x-axis, in response to target maneuvers, by an amount $y_M(t)$. Then, the line of sight angle is given by* (5.83), (5.82) *and* (5.72)*; proportional navigation implies* (5.87) *and* (5.88)*; and the miss distance is the difference of ordinates at the final time, as given by* (5.89).

EXAMPLE 5.1 *Here, we model the lateral displacements of a target and proportionally navigated missile, assuming that the autopilot has first-order dynamics. In this case, the transfer function is*

$$\begin{aligned} Y(s) &= \frac{1}{sT+1} = \frac{N(s)}{D(s)}, \\ N(s) &= 1, \\ D(s) &= sT+1, \end{aligned} \tag{5.90}$$

where T is the time constant. In practice, this means that the step response of the autopilot, that is, its response to a unit step input, is as depicted in Figure 5.7. Hence, the time constant, T, quantifies how "sluggish" the autopilot is.

Equation (5.87) *becomes*

$$(sT+1)\ddot{y}_M = \frac{\Lambda}{t_f^* - t}(y_T - y_M) \tag{5.91}$$

yielding:

$$\ddot{y}_M = -\frac{\Lambda}{(t_f^* - t)T}y_M - \frac{1}{T}\dot{y}_M + \frac{\Lambda}{(t_f^* - t)T}y_T. \tag{5.92}$$

Note that this equation can be solved in closed form in terms of Kummer functions [73]. We can write it in standard state-space form as follows: let

$$\begin{aligned} x_1 &= y_M \\ x_2 &= \dot{y}_M. \end{aligned} \tag{5.93}$$

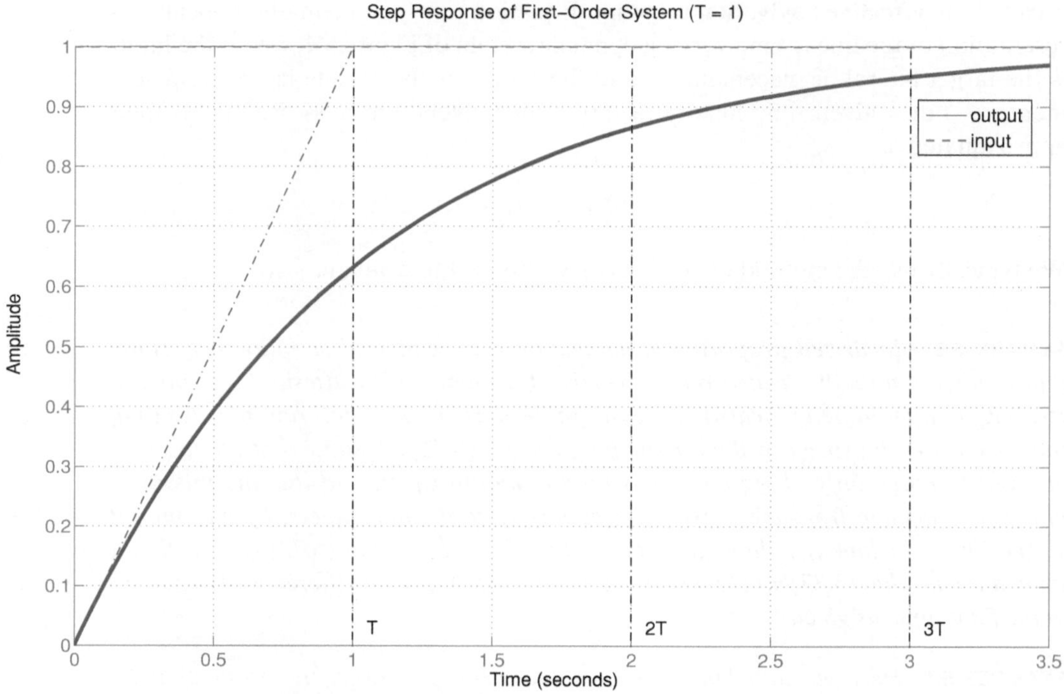

Figure 5.7. Step response of first-order autopilot.

Then, we have

$$\begin{bmatrix} \dot{x}_1 \\ \dot{x}_2 \end{bmatrix} = \begin{bmatrix} 0 & 1 \\ -\frac{\Lambda}{(t_f^*-t)T} & -\frac{1}{T} \end{bmatrix} \begin{bmatrix} x_1 \\ x_2 \end{bmatrix} + \begin{bmatrix} 0 \\ \frac{\Lambda}{(t_f^*-t)T} \end{bmatrix} y_T$$

$$y_M = \begin{bmatrix} 1 & 0 \end{bmatrix} \begin{bmatrix} x_1 \\ x_2 \end{bmatrix}, \tag{5.94}$$

which is a state-space model for the system in Figure 5.4.

In Section 2.4, we showed that the output at time t_f^* of the linear dynamic system in (5.94) could be evaluated using the method of adjoints as

$$y_M(t_f^*) = \int_0^{t_f^*} p^T(\tau)B(\tau)y_T(\tau)\,\mathrm{d}\tau, \tag{5.95}$$

where the adjoint vector p satisfies

$$\begin{bmatrix} \dot{p}_1 \\ \dot{p}_2 \end{bmatrix} = \begin{bmatrix} 0 & \frac{\Lambda}{(t_f^*-t)T} \\ -1 & \frac{1}{T} \end{bmatrix} \begin{bmatrix} p_1 \\ p_2 \end{bmatrix}$$

$$\begin{bmatrix} p_1 \\ p_2 \end{bmatrix}(t_f^*) = \begin{bmatrix} 1 \\ 0 \end{bmatrix}. \tag{5.96}$$

Equations (5.94) and (5.95) imply that

$$y_M(t_f^*) = \frac{\Lambda}{T} \int_0^{t_f^*} \frac{p_2(\tau)y_T(\tau)}{t_f^* - \tau}\,\mathrm{d}\tau. \tag{5.97}$$

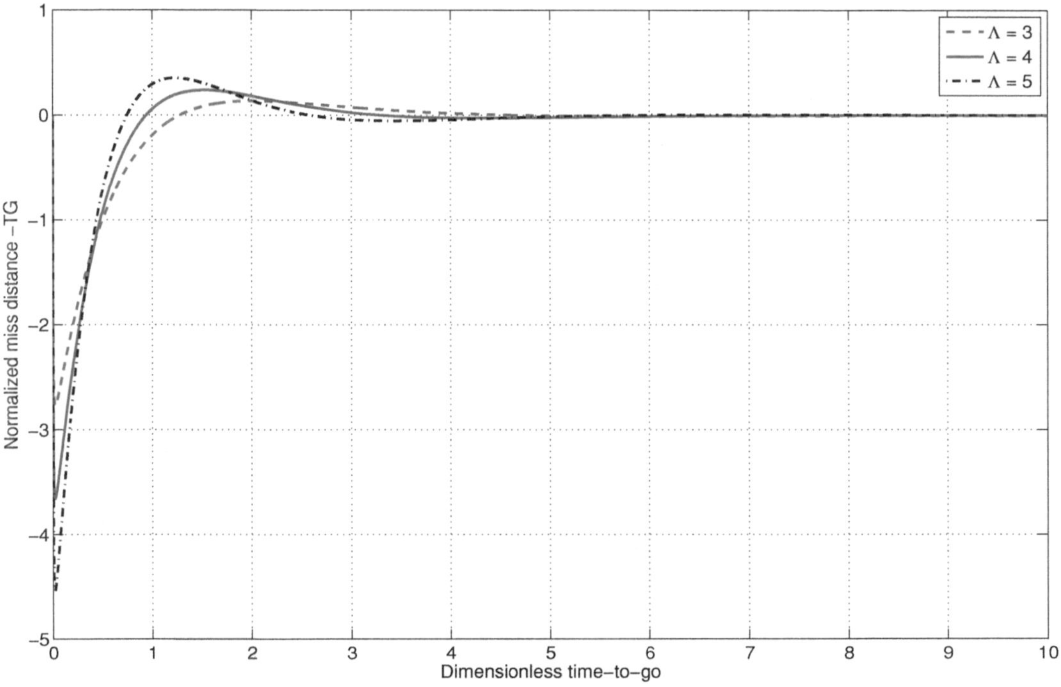

Figure 5.8. Normalized miss distance as a function of dimensionless time-to-go.

Now, recall that the input and output of the linear dynamic system in (5.94) *are related by*

$$y_M(t_f^*) = \int_0^{t_f^*} G(t_f^*, \tau) y_T(\tau)\, d\tau, \tag{5.98}$$

where $G(t, \tau)$ *is the impulse response. Identifying* (5.97) *with* (5.98) *yields*

$$G(t_f^*, \tau) = \frac{\Lambda p_2(\tau)}{T(t_f^* - \tau)}. \tag{5.99}$$

This impulse response is plotted in Figure 5.8 as a function of the dimensionless time-to-go. It represents the normalized miss distance of an engagement where the target performs an impulsive maneuver, as a function of the time at which the target performs the maneuver. The underlying assumptions are that the missile is terminally guided using proportional navigation and that the missile's autopilot has first-order dynamics. From Figure 5.8, the largest miss distance is obtained when the maneuver is performed at time $(t_f^*)^-$*, that is, just before impact. Earlier impulsive maneuvers lead to a smaller miss. When* $|\tau - t_f^*| \gg T$*, the miss distance tends to zero, indicating that the autopilot has enough time to react to the maneuver.*

5.6.1 Miss due to Launch Error

We account for launch errors using two methods. As shown in Figure 5.9, the first method is based on the fact that launching the missile on an erroneous path

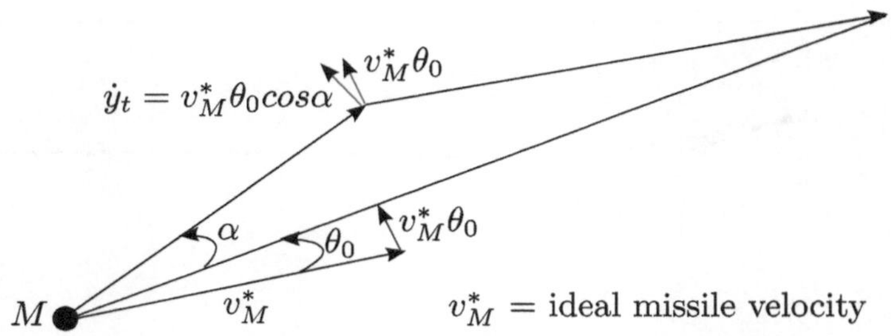

Figure 5.9. Geometry for miss due to launch error analysis.

($\theta_0 \neq 0$) is equivalent to launching it on the correct path, but letting the target have a lateral velocity. From the figure, this yields a lateral target displacement satisfying

$$\dot{y}_T = v_M^* \theta_0 \cos\alpha. \tag{5.100}$$

Now, we have

$$v_M^* \cos\alpha = v_{\beta M}^*. \tag{5.101}$$

Hence,

$$\dot{y}_T = v_{\beta M}^* \theta_0, \tag{5.102}$$

which can be integrated as

$$y_T = v_{\beta M}^* \theta_0 t, \tag{5.103}$$

where we have used the initial condition $y_T(0) = 0$ at launch. We still have that the miss distance is obtained as

$$M = y_T(t_f^*) - \int_0^{t_f^*} G(t_f^*, \tau) y_T(\tau)\, d\tau, \tag{5.104}$$

where $G(t, \tau)$ is the impulse response of the linear time varying system relating the target lateral maneuvers y_T to the missile lateral maneuvers y_M.

The idea for the second method is to perform the modeling by accounting correctly for nonzero initial conditions. In this case, (5.87) becomes

$$D(s)\dot{y}_M = \Lambda N(s) \left(\frac{y_T - y_M}{t_f^* - t} \right) + constant, \tag{5.105}$$

which, by appropriate choice of state variables, can be transformed into

$$\begin{aligned} \dot{x}(t) &= A(t)x(t) + B(t)y_T(t) + d \\ y_M(t) &= C(t)x(t), \end{aligned} \tag{5.106}$$

where d is a constant vector. Let

$$y_T \equiv 0, \tag{5.107}$$

which accounts for the fact that the target does not maneuver, and

$$\begin{aligned} y_M(0) &= 0 \\ \dot{y}_M(0) &= v_{\beta M}^* \theta \end{aligned} \tag{5.108}$$

for initial conditions. Using the adjoint equation

$$\dot{p}(t) = -A^T(t)p(t), \tag{5.109}$$

we have

$$x^T(t_f)p(t_f) = x^T(0)p(0) + \int_0^{t_f} p^T(\tau)d\,\mathrm{d}\tau. \tag{5.110}$$

Then, letting

$$p(t_f) = C^T(t_f), \tag{5.111}$$

we can compute the miss distance as

$$y_M(t_f) = x^T(0)p(0) + \int_0^{t_f} p^T(\tau)d\,\mathrm{d}\tau = -M. \tag{5.112}$$

EXAMPLE 5.2 *Let us use the second method discussed earlier to compute the miss distance due to a launch error for the same first-order autopilot as in Example 5.1; that is,*

$$\frac{N(s)}{D(s)} = \frac{1}{sT+1}. \tag{5.113}$$

Equation (5.105) *yields*

$$(sT+1)\dot{y}_M = \frac{\Lambda}{t_f^* - t}(y_T - y_M) + constant. \tag{5.114}$$

With the same choice of state variables as in (5.93), (5.114) *becomes*

$$\begin{aligned}
\begin{bmatrix}\dot{x}_1\\ \dot{x}_2\end{bmatrix} &= \begin{bmatrix}0 & 1\\ -\frac{\Lambda}{(t_f^*-t)T} & -\frac{1}{T}\end{bmatrix}\begin{bmatrix}x_1\\ x_2\end{bmatrix} + \begin{bmatrix}0\\ \frac{\Lambda}{(t_f^*-t)T}\end{bmatrix} y_T + \begin{bmatrix}0\\ \frac{v_{\beta M}^*\theta_0}{T}\end{bmatrix}\\
y_M &= \begin{bmatrix}1 & 0\end{bmatrix}\begin{bmatrix}x_1\\ x_2\end{bmatrix}.
\end{aligned} \tag{5.115}$$

Let $y_T \equiv 0$. *With the boundary conditions*

$$\begin{aligned}
y_M(0) &= x_1(0) = 0,\\
\dot{y}_M(0) &= x_2(0) = v_{\beta M}^*\theta_0,\\
p(t_f) &= \begin{bmatrix}p_1(t_f)\\ p_2(t_f)\end{bmatrix} = \begin{bmatrix}1\\ 0\end{bmatrix},
\end{aligned} \tag{5.116}$$

the method of adjoints yields

$$\begin{aligned}
y_M(t_f) &= \begin{bmatrix}0 & v_{\beta M}^*\theta_0\end{bmatrix}\begin{bmatrix}p_1(0)\\ p_2(0)\end{bmatrix} + \int_0^{t_f} \begin{bmatrix}p_1(\tau) & p_2(\tau)\end{bmatrix}\begin{bmatrix}0\\ \frac{v_{\beta M}^*\theta_0}{T}\end{bmatrix}\mathrm{d}\tau\\
&= v_{\beta M}^*\theta_0 p_2(0) + \int_0^{t_f} \frac{v_{\beta M}^*\theta_0 p_2(\tau)}{T}\,\mathrm{d}\tau = -M,
\end{aligned} \tag{5.117}$$

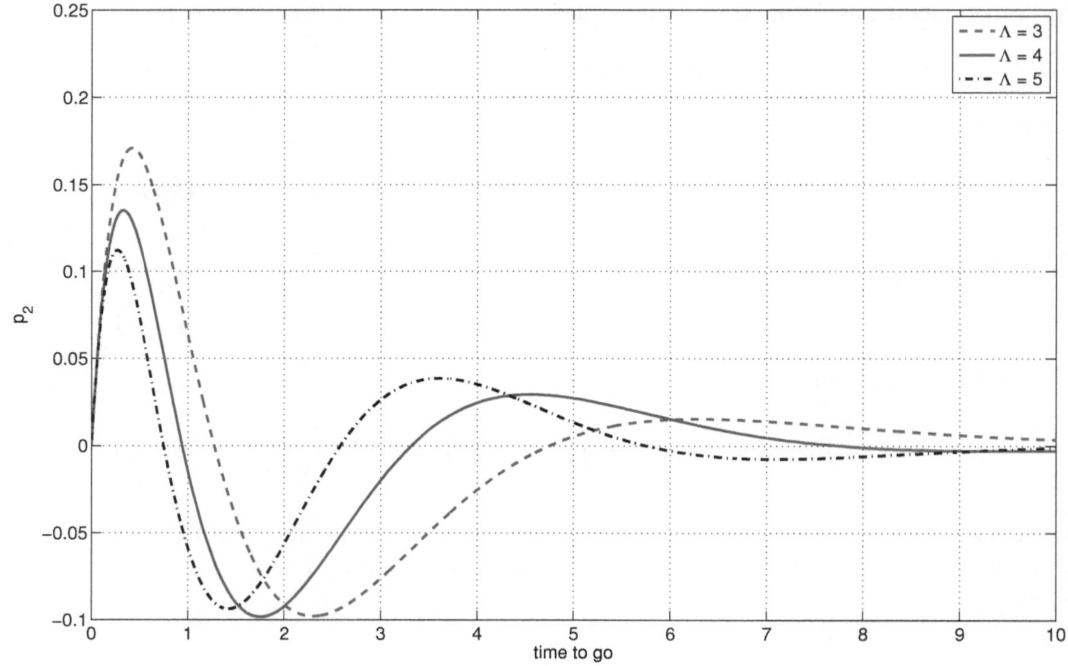

Figure 5.10. Normalized miss distance $\frac{M}{v^*_{\beta M} T\theta_0}$ due to launch error θ_0 as a function of dimensionless time-to-go $\frac{t^*_f}{T}$.

where p satisfies

$$\begin{bmatrix} \dot{p}_1 \\ \dot{p}_2 \end{bmatrix} = \begin{bmatrix} 0 & \frac{\Lambda}{(t^*_f - t)T} \\ -1 & \frac{1}{T} \end{bmatrix} \begin{bmatrix} p_1 \\ p_2 \end{bmatrix}$$

$$\begin{bmatrix} p_1 \\ p_2 \end{bmatrix}(t_f) = \begin{bmatrix} 1 \\ 0 \end{bmatrix}. \tag{5.118}$$

Figure 5.10 shows a plot of p_2 for this example.

5.6.2 Miss due to Step Target Acceleration

Here we assume that the target is subjected to a step lateral acceleration applied at time t_0. Therefore, the system with input the target acceleration and with output the target displacement is a double integrator. In this particular case, we have

$$y_T(t) = \begin{cases} 0 & \text{if } 0 \leq t < t_0 \\ \frac{a_T(t-t_0)^2}{2} & \text{if } t \geq t_0. \end{cases} \tag{5.119}$$

The miss distance can still be computed using (5.104). Alternatively, we can simply integrate the function of Figure 5.10.

Figure 5.11 suggests an evasive maneuver for the target as follows: first, at time t^*_1, apply the maximum positive acceleration. This maneuver in itself yields

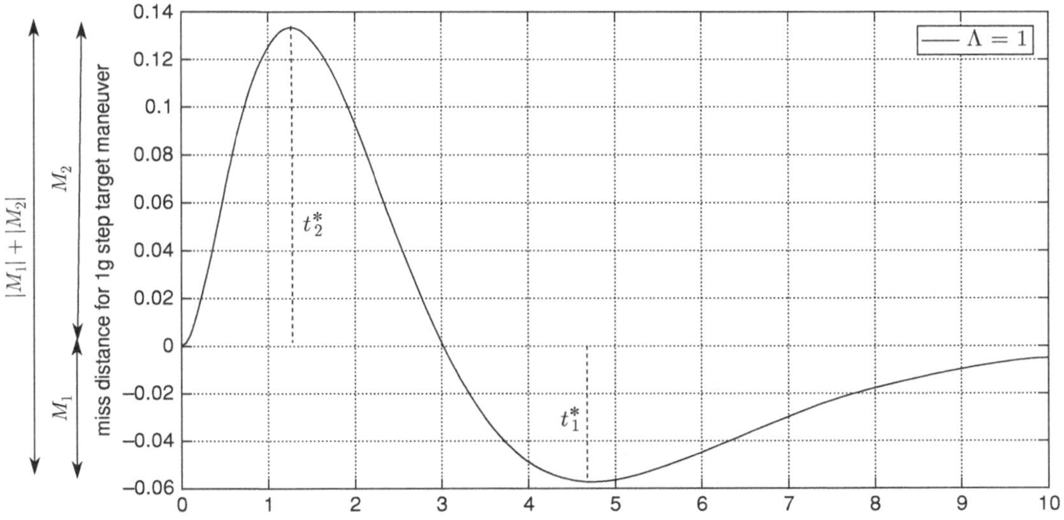

Figure 5.11. Normalized miss distance $\frac{M}{T^2 a_T}$ due to step target acceleration a_T as a function of dimensionless time-to-go $\frac{t_f^* - t_0}{T}$.

a miss $|M_1|a_T$. Then, at time t_2^*, apply the maximum negative acceleration. This second maneuver yields a miss $|M_2|a_T$. By superposition, the total miss is then $a_T(|M_1| + |M_2|)$. This maneuver is called **jinking** in air combat. It is very similar to **juking**, which, in interception sports, consists of faking a move to one side, then turning to the other side. Note, from Figure 5.11, how jinking or juking must be perfectly timed to be effective. These maneuvers are revisited in Chapter 9.

5.6.3 Miss due to Target Sinusoidal Motion

Here we assume that the target undergoes a sinusoidal lateral motion of the form

$$y_T(t) = A \sin(\omega t + \phi), \tag{5.120}$$

where A, ω, and ϕ are the amplitude, frequency, and phase, respectively. The miss distance is still evaluated by (5.104). In terms of ϕ, the miss distance is

$$M(\phi) = A \sin(\omega t_f^* + \phi) - A \int_0^{t_f^*} G(t_f^*, \tau) \sin(\omega \tau + \phi) \, \mathrm{d}\tau. \tag{5.121}$$

Assume that ϕ is a random variable that is uniformly distributed on $[0, 2\pi]$, that is, it has probability density function

$$f(\phi) = \begin{cases} \frac{1}{2\pi} & \text{if } \phi \in [0, 2\pi] \\ 0 & \text{if } \phi \notin [0, 2\pi]. \end{cases} \tag{5.122}$$

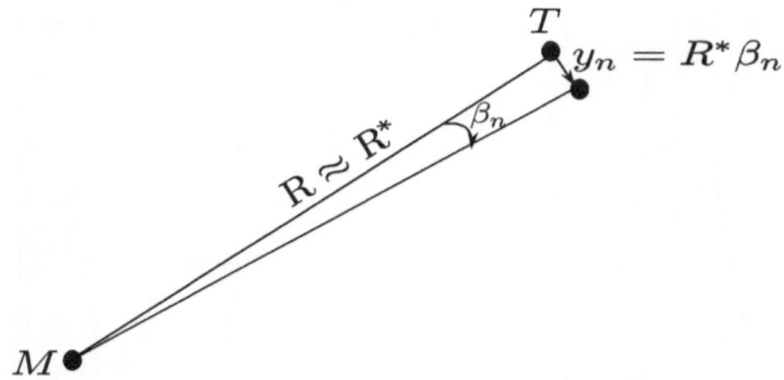

Figure 5.12. Geometry for miss due to noise.

Then, the root-mean-square (RMS) of the miss distance is

$$\sqrt{E[M^2(\phi)]} = \sqrt{\int_{-\infty}^{+\infty} M^2(\phi) f(\phi)\,\mathrm{d}\phi}$$
$$= \sqrt{\frac{1}{2\pi}\int_0^{2\pi} M^2(\phi)\,\mathrm{d}\phi}. \tag{5.123}$$

5.6.4 Miss due to Noise

In this section, we take into account uncertainties in the reading of the sensors. More specifically, we assume that the measurement of β is corrupted by a measurement noise $\beta_n(t)$. Referring to Figure 5.12, the target appears to have a lateral displacement

$$y_T(t) = y_n(t) = R^*(t)\beta_n(t), \tag{5.124}$$

and the miss distance can be computed as if it were the result of this fictitious target maneuver, using (5.104). Because $R^*(t_f^*) = 0$ implies that $y_T(t_f^*) = 0$, we obtain

$$M_n = -\int_0^{t_f^*} G(t_f^*, \tau) y_n(\tau)\,\mathrm{d}\tau. \tag{5.125}$$

We now use the theory presented in Chapter 3 to quantify the miss distance. In the simplest case, assume that the apparent target displacement $y_n(t)$ is a zero-mean Gaussian white process, that is,

$$E[y_n(t)] = 0$$
$$E[y_n(t)y_n(\tau)] = \phi^2(t)\delta(t-\tau) \tag{5.126}$$

for some known function $\phi(t)$. Then, from (5.125), the square of the miss distance is

$$M_n^2 = \int_0^{t_f^*}\int_0^{t_f^*} G(t_f^*, \tau)G(t_f^*, \sigma)y_n(\tau)y_n(\sigma)\,\mathrm{d}\tau\,\mathrm{d}\sigma. \tag{5.127}$$

Hence, its expected value is

$$E[M_n^2] = \int_0^{t_f^*}\int_0^{t_f^*} G(t_f^*, \tau)G(t_f^*, \sigma)\phi^2(\tau)\delta(\tau-\sigma)\,\mathrm{d}\tau\,\mathrm{d}\sigma$$
$$= \int_0^{t_f^*} G^2(t_f^*, \tau)\phi^2(\tau)\,\mathrm{d}\tau. \tag{5.128}$$

The root-mean-square of the miss distance is then

$$\sqrt{E[M_n^2]} = \sqrt{\int_0^{t_f^*} G^2(t_f^*, \tau)\phi^2(\tau)\,\mathrm{d}\tau}. \tag{5.129}$$

In a more complicated case, assume that the apparent target motion $y_n(t)$ is a zero-mean Gaussian colored process with a known covariance kernel

$$E[y_n(t)y_n(\tau)] = \phi(t, \tau). \tag{5.130}$$

Then, using the time varying version of the stochastic realization theory outlined in Remark 3.8, it is possible to find a linear time varying filter that, when driven by a white noise process $n(t)$, realizes $y_n(t)$, that is,

$$\begin{aligned} \dot{x}_f(t) &= A_f(t)x_f(t) + B_f(t)n(t) \\ y_n(t) &= C_f(t)x_f(t). \end{aligned} \tag{5.131}$$

Recall, as demonstrated in Example 5.2, that it is in general possible to model how the missile responds to target maneuvers by a standard linear time varying system, that is,

$$\begin{aligned} \dot{x}(t) &= A(t)x(t) + B(t)y_T(t) \\ y_M(t) &= C(t)x(t). \end{aligned} \tag{5.132}$$

Now, the target maneuver is the apparent motion given by (5.131). Therefore, the missile response satisfies

$$\begin{aligned} \begin{bmatrix} \dot{x}(t) \\ \dot{x}_f(t) \end{bmatrix} &= \begin{bmatrix} A(t) & B(t)C_f(t) \\ 0 & A_f(t) \end{bmatrix} \begin{bmatrix} x(t) \\ x_f(t) \end{bmatrix} + \begin{bmatrix} 0 \\ B_f(t) \end{bmatrix} n(t) \\ y_M(t) &= \begin{bmatrix} C(t) & 0 \end{bmatrix} \begin{bmatrix} x(t) \\ x_f(t) \end{bmatrix}. \end{aligned} \tag{5.133}$$

Because the input $n(t)$ is now white, we can use familiar analysis methods to quantify the miss distance.

REMARK 5.6 *From Figure 5.8, it is clear that increasing the navigation constant λ improves the interception time. It can also be shown that increasing λ worsens the response to measurement noise. In practice, it has been recommended to use $\lambda \in [3, 4]$ for the best trade-off between response to noise and intercept time.*

5.6.5 Use of Power Series Solution

In this section, we consider the singularity that happens in typical homing guidance problems at the time of intercept. This singularity is inherent to homing problems and was anticipated in Remark 5.1. More specifically, we show how to solve the linearized equations of proportional guidance in the presence of this singularity.

Consider the missile of Example 5.1, with a first-order autopilot:

$$Y(s) = \frac{1}{sT + 1}. \tag{5.134}$$

We had obtained the linearized equations

$$\begin{bmatrix}\dot{x}_1\\ \dot{x}_2\end{bmatrix}=\begin{bmatrix}0 & 1\\ -\frac{\Lambda}{(t_f^*-t)T} & -\frac{1}{T}\end{bmatrix}\begin{bmatrix}x_1\\ x_2\end{bmatrix}+\begin{bmatrix}0\\ \frac{\Lambda}{(t_f^*-t)T}\end{bmatrix}y_T+\begin{bmatrix}0\\ \frac{v_{\beta M}^*\theta_0}{T}\end{bmatrix}$$

$$y_M=\begin{bmatrix}1 & 0\end{bmatrix}\begin{bmatrix}x_1\\ x_2\end{bmatrix}, \tag{5.135}$$

relating the lateral target displacement y_T to the lateral missile displacement y_M. The miss distance can be computed using the method of adjoints. However, the corresponding adjoint equations,

$$\begin{bmatrix}\dot{p}_1\\ \dot{p}_2\end{bmatrix}=\begin{bmatrix}0 & \frac{\Lambda}{(t_f^*-t)T}\\ -1 & \frac{1}{T}\end{bmatrix}\begin{bmatrix}p_1\\ p_2\end{bmatrix}$$

$$\begin{bmatrix}p_1\\ p_2\end{bmatrix}(t_f^*)=\begin{bmatrix}1\\ 0\end{bmatrix}, \tag{5.136}$$

have an obvious singularity at t_f^*. We can use a power series expansion to integrate (5.136) beyond the singularity. First, we reformulate the problem in dimensionless form.

Starting from

$$\ddot{y}_M=-\frac{\Lambda}{(t_f^*-t)T}y_M-\frac{1}{T}\dot{y}_M+\frac{\Lambda}{(t_f^*-t)T}y_T, \tag{5.137}$$

let:

$$\hat{t}=\frac{t}{T}$$

$$\hat{t}_f^*=\frac{t_f^*}{T} \tag{5.138}$$

be the dimensionless time and nominal interception time, respectively. Then, the differential operator is transformed as

$$\frac{\mathrm{d}}{\mathrm{d}t}=\frac{\mathrm{d}\hat{t}}{\mathrm{d}t}\frac{\mathrm{d}}{\mathrm{d}\hat{t}}=\frac{1}{T}\frac{\mathrm{d}}{\mathrm{d}\hat{t}}. \tag{5.139}$$

Using (5.139) in (5.137) yields

$$\frac{\mathrm{d}^2y_M}{\mathrm{d}\hat{t}^2}=-\frac{\Lambda}{\hat{t}_f^*-\hat{t}}y_M-\frac{dy_M}{\mathrm{d}\hat{t}}+\frac{\Lambda}{\hat{t}_f^*-\hat{t}}y_T. \tag{5.140}$$

As in Example 5.1, choose the state variables

$$x_1=y_M,\ x_2=\frac{\mathrm{d}}{\mathrm{d}\hat{t}}y_M. \tag{5.141}$$

Then, (5.140) becomes

$$\frac{d}{d\hat{t}}\begin{bmatrix}x_1\\ x_2\end{bmatrix}=\begin{bmatrix}0 & 1\\ -\frac{\Lambda}{(\hat{t}_f^*-\hat{t})} & -1\end{bmatrix}\begin{bmatrix}x_1\\ x_2\end{bmatrix}+\begin{bmatrix}0\\ \frac{\Lambda}{(\hat{t}_f^*-\hat{t})}\end{bmatrix}y_T. \tag{5.142}$$

The adjoint equation and its boundary condition are now

$$\frac{d}{d\hat{t}}\begin{bmatrix}p_1\\p_2\end{bmatrix}=\begin{bmatrix}0 & \frac{\Lambda}{\hat{t}_f^*-\hat{t}}\\-1 & 1\end{bmatrix}\begin{bmatrix}p_1\\p_2\end{bmatrix}$$

$$\begin{bmatrix}p_1\\p_2\end{bmatrix}(\hat{t}_f^*)=\begin{bmatrix}1\\0\end{bmatrix}. \tag{5.143}$$

We still have

$$y_M(\hat{t}_f^*)=\int_0^{\hat{t}_f^*}\frac{\Lambda}{\hat{t}_f^*-\sigma}p_2(\sigma)y_T(\sigma)\,\mathrm{d}\sigma$$

$$=\int_0^{\hat{t}_f^*}G(\hat{t}_f^*,\sigma)y_T(\sigma)\,\mathrm{d}\sigma. \tag{5.144}$$

Hence,

$$G(\hat{t}_f^*,\sigma)=\frac{\Lambda}{\hat{t}_f^*-\sigma}p_2(\sigma). \tag{5.145}$$

Let

$$\tau=\hat{t}_f^*-\hat{t}=\frac{t_f^*-t}{T} \tag{5.146}$$

be the dimensionless time-to-go to intercept. We have

$$\frac{\mathrm{d}}{\mathrm{d}\tau}=\frac{\mathrm{d}\hat{t}}{\mathrm{d}\tau}\frac{\mathrm{d}}{\mathrm{d}\hat{t}}=-\frac{\mathrm{d}}{\mathrm{d}\hat{t}}. \tag{5.147}$$

Hence, the adjoint equation (5.143) becomes

$$\frac{\mathrm{d}}{\mathrm{d}\tau}\begin{bmatrix}p_1\\p_2\end{bmatrix}=\begin{bmatrix}0 & \frac{\Lambda}{\tau}\\1 & -1\end{bmatrix}\begin{bmatrix}p_1\\p_2\end{bmatrix}$$

$$\begin{bmatrix}p_1\\p_2\end{bmatrix}(\hat{t}_f^*)=\begin{bmatrix}1\\0\end{bmatrix}, \tag{5.148}$$

which clearly has a singularity for $\tau=0$, and we use a power series method to integrate (5.148) beyond it. First, rewrite it as

$$\frac{\mathrm{d}^2p_2}{\mathrm{d}\tau^2}+\frac{\mathrm{d}p_2}{\mathrm{d}\tau}+\frac{\Lambda}{\tau}p_2=0. \tag{5.149}$$

We postulate that (5.149) has a solution of the form

$$p_2=a_1\tau+a_2\tau^2+a_3\tau^3+a_4\tau^4+\cdots, \tag{5.150}$$

where we have used our knowledge that $p_2(0)=0$. Differentiating (5.150) twice yields

$$\frac{\mathrm{d}p_2}{\mathrm{d}\tau}=a_1+2a_2\tau+3a_3\tau^2+4a_4\tau^3+\cdots$$

$$\frac{\mathrm{d}^2p_2}{\mathrm{d}\tau^2}=2a_2+6a_3\tau+12a_4\tau^2+\cdots \tag{5.151}$$

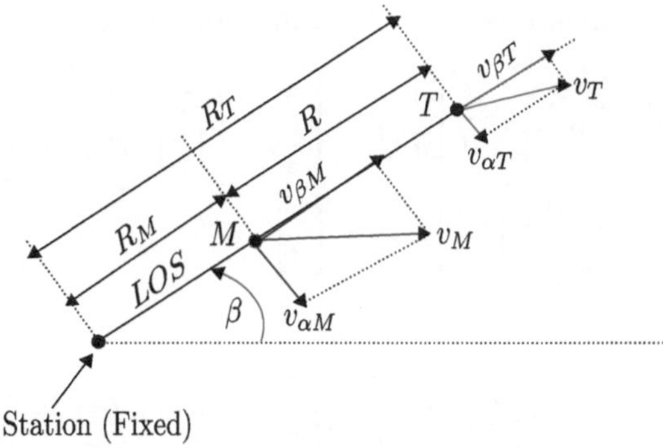

Figure 5.13. Geometry for beam rider guidance.

Substitute (5.150) and (5.151) into (5.149) and identify the coefficients of like powers to obtain

$$a_k = -\frac{a_{k-1}}{k}\left(1 + \frac{\Lambda}{k-1}\right), \quad k \geq 2. \tag{5.152}$$

Furthermore,

$$\left(\frac{\mathrm{d}p_2}{\mathrm{d}\tau}\right)_{\tau=0} = p_1(0) - p_2(0) = 1 = a_1, \tag{5.153}$$

which can be used to initialize the recursion in (5.152). Hence, all the coefficients a_i in (5.150) are known, and p_2 is also known. Also, from (5.148), we have

$$\frac{\mathrm{d}p_2}{\mathrm{d}\tau} = p_1 - p_2. \tag{5.154}$$

Hence,

$$\begin{aligned} p_1 &= p_2 + \frac{\mathrm{d}p_2}{\mathrm{d}\tau} \\ &= 1 + (1 + 2a_2)\tau + (a_2 + 3a_3)\tau^2 + (a_3 + 4a_4)\tau^3 + \cdots, \end{aligned} \tag{5.155}$$

and p_1 is known. We can therefore use this method to compute $p_1(\tau)$ and $p_2(\tau)$ for $\tau \ll 1$, then resume standard numerical integration away from the singularity.

5.7 Beam Rider Guidance

In beam rider guidance, a (typically fixed) tracking station aims an electromagnetic beam at the target. The missile is equipped with a sensor that detects its position with respect to the beam. The missile autopilot is programmed to use feedback to cause the missile to "ride the beam."

As shown in Figure 5.13, the range R is the missile-target distance, R_T is the tracking station-target distance, and R_M is the tracking station-missile distance. For beam riding, the line-of-sight angular rate must satisfy

$$\dot{\beta} = \frac{v_{\alpha M}}{R_M} = \frac{v_{\alpha T}}{R_T}, \tag{5.156}$$

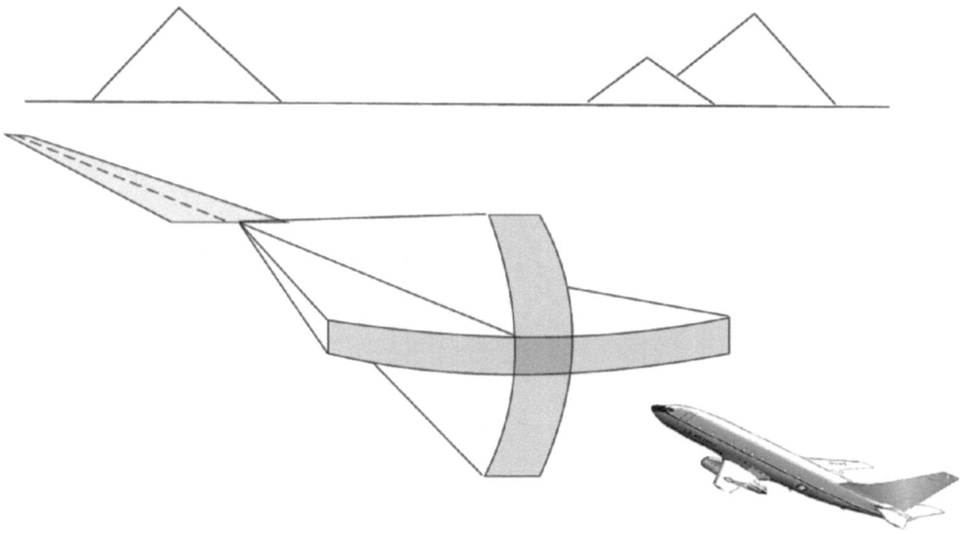

Figure 5.14. Instrument landing system.

implying that the components of velocity orthogonal to the line of sight must satisfy

$$v_{\alpha M} = \frac{R_M}{R_T} v_{\alpha T}. \tag{5.157}$$

This last equation can be used to interpret beam rider guidance in terms of previously studied guidance strategies. Indeed, at launch, we have $R_M \ll 1$, which, from (5.13), implies that $v_{\alpha M} \approx 0$. This turns out to be nearly pursuit guidance, or, equivalently, proportional navigation with $\lambda = 1$. However, near intercept, we have $R_M \approx R_T$, which, from (5.156), implies that $v_{\alpha M} \approx v_{\alpha T}$. This turns out to be nearly constant bearing guidance, or, equivalently, proportional navigation with $\lambda \to \infty$. As a result, we can think of beam rider guidance as a particular case of proportional navigation where the navigation constant is boosted up from 1 to ∞ during the engagement.

The obvious advantage of beam rider guidance compared to other guidance schemes is its simplicity; the missile only needs to sense its relative position with respect to the beam, as opposed to "seeing" the target with a seeker. However, beam rider guidance suffers from several disadvantages. First, it is rather inaccurate at long range. This is due to separate contributions from errors in pointing the beam and riding it. Second, it is inherently inefficient. This is because the missile tends to fly in a pursuit course (which requires turning) in the lower atmosphere, where drag is important.

A form of beam rider guidance is used in the instrument landing system (ILS) to help aircraft land in poor visibility, as depicted in Figure 5.14. A station located at the beginning of the runway broadcasts two beams: one in a vertical plane aligned with the axis of the runway, and one in an inclined plane with elevation equal to the ideal flight path angle for approach. The aircraft instruments can detect whether the aircraft is located to the left of, on, or to the right of the vertical plane. They can also detect whether the aircraft is located above, on, or below the inclined plane.

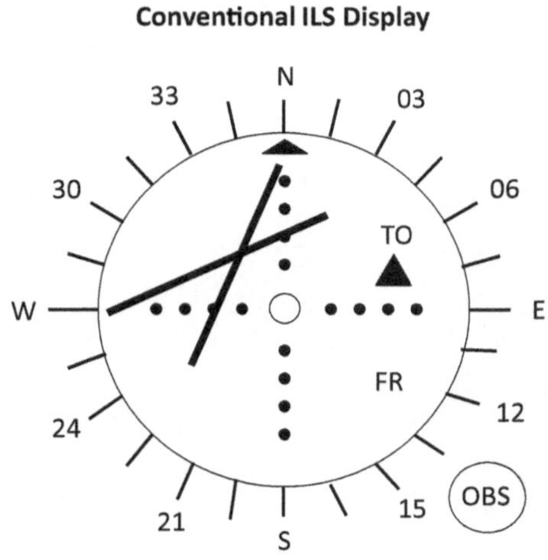

Figure 5.15. ILS display.

Based on these very simple readings, illustrated in Figure 5.15, the pilot can guide the aircraft trajectory to ride the intersection of the planes to touchdown.

5.8 Summary of Key Results

The key results in Chapter 5 are as follows:

1. Equations (5.1), (5.51), and (5.52), which are the fundamental equations of planar homing
2. The constant bearing principle of Proposition 5.1, illustrated in Figure 5.2, which allows a navigator to detect imminent collisions
3. Equation (5.10), which models pursuit guidance against a nonmaneuvering target
4. Equations (5.15), (5.18), and (5.19), which delineate limitations of pursuit guidance
5. Equation (5.44), which quantifies the miss distance in pursuit guidance due to a turn rate limitation
6. Equation (5.67), which provides the lead angle for a constant bearing course against a nonmaneuvering target
7. Equation (5.68), which governs all proportional navigation methods
8. Equations (5.87)–(5.89), which model linearized proportional navigation, accounting for autopilot dynamics
9. Equations (5.102) and (5.105), which model proportional navigation with launch errors
10. Equation (5.157), which allows an interpretation of beam rider guidance as proportional navigation with a time varying gain

5.9 Bibliographic Notes for Further Reading

Our presentation of terminal homing guidance, unified under proportional navigation, closely follows [38], which is out of print. For further reading on homing guidance, the reader may consult [49], [10], [66].

The technology of terminally guided missiles, and in particular the idea of proportional navigation, were under development during World War II [25]. This technology was first used in armed conflict during the Korean War and subsequently during the Vietnam War [25]. Beam rider guidance is used in a variety of short-range missile systems aided by highly portable laser designators [72]. Instrument landing systems are standard equipment in modern airports [1].

5.10 Homework Problems

PROBLEM 5.1 *For pursuit guidance with $\theta_T \equiv 0$ and $\beta(0) \neq \pi$, derive conditions on γ that ensure that*

$$\left(\frac{\mathrm{d}^k\theta}{\mathrm{d}t^k}\right)_{t=t_f}$$

is finite, for $k = 1, 2, 3, \ldots$

PROBLEM 5.2 *A target flies with fixed velocity $v_T = 1{,}000$ ft/s and constant heading $\theta_T = 0$. Assume a pursuit-guided missile, with fixed velocity $v_M = 3{,}000$ ft/s, is launched with initial line-of-sight angle $\beta(0) = 90°$, at an initial range of 20,000 ft. Derive the time histories of the Cartesian coordinates of the missile and target. What is the time of impact? Plot the trajectories.*

PROBLEM 5.3 *Assume that the missile of Problem 5.2 has a maximum lateral acceleration capability of 40 g (where $g = 32.2$ ft/s^2). When does the missile reach its turning rate limitation? What is the radius of its trajectory beyond that point? Give an estimate of the miss distance. Plot the corresponding trajectory on the same diagram as in Problem 5.2.*

PROBLEM 5.4 *A missile using ideal constant bearing guidance is launched on a beam attack ($\beta(0) = 90°$) against a target with constant velocity of 1,000 ft/s. The missile has a constant velocity of 3,000 ft/s. The initial range is 20,000 ft. Determine the initial missile heading angle for a collision course. Determine the time-to-impact. Plot missile and target trajectories.*

PROBLEM 5.5 *In linearized proportional navigation, for a missile autopilot with dynamics*

$$Y(s) = \frac{1}{a_m s^m + a_{m-1} s^{m-1} + \cdots + a_1 s + a_0},$$

use the method of adjoints to obtain the response of the missile to impulsive lateral target maneuvers.

PROBLEM 5.6 *For a nonmaneuvering target with constant velocity, show that the initial lead angle $\beta_0 - \theta_0$ that yields a constant bearing interception is determined by measurements of R, $\dot{\beta}$, and v_M only. Give a formula. Note that this analysis yields the principle of operation of the gyro gunsight, as shown, for example, in Figure 5.16.*

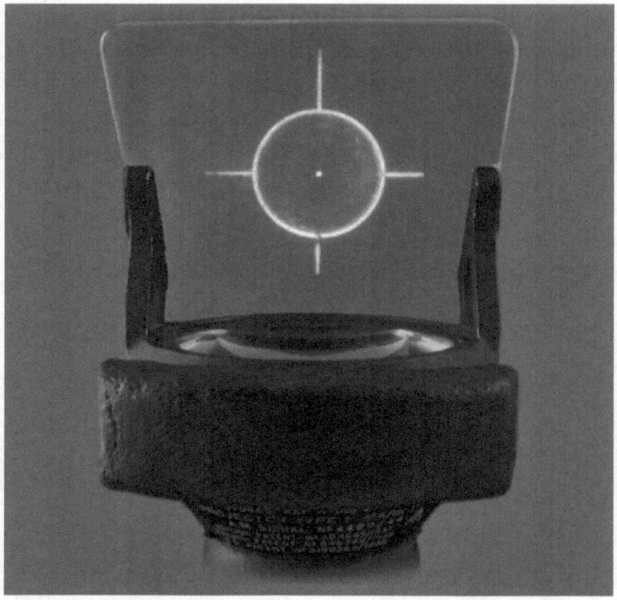

Figure 5.16. Spitfire gunsight. Image courtesy of Philippe Agulhon, Le Comptoir de l'Aviation.

PROBLEM 5.7 *In the Cartesian plane Oxy, let (x_M, y_M) and (x_T, y_T) be the coordinates of a missile and target, respectively, with velocities v_M and v_T and heading angles θ and θ_T, respectively. Define*

$$x = x_T - x_M$$

$$y = y_T - y_M.$$

Derive differential equations for x and y. These could be viewed as the fundamental equations of homing in Cartesian coordinates.

PROBLEM 5.8 *Consider a missile homing toward a fixed target located at the origin of the Cartesian plane. Assume that the missile is subject to currents with velocity components v_x, v_y. Show that the kinematic equations of motion are the same as in Problem 5.7. As a consequence, the effect of currents on the missile trajectory is the same as that of target maneuvers.*

PROBLEM 5.9 *Consider motions of a target and missile controlled using proportional navigation with constant $\lambda = 3$. The target has constant velocity 1,000 ft/s and flies at heading angle $\theta_T(t) = 0.01\pi \sin(2\pi t)$ rad. The missile is launched on a beam attack ($\beta(0) = \pi/2$) with constant velocity 3,000 ft/s and initial range 20,000 ft. Assume that the initial missile heading is $\theta(0) = 0.4\pi$ rad. Assume that the autopilot transfer function is*

$$Y(s) = \frac{1}{1 + 0.5s}.$$

Develop a linearized equation for y_M, the missile displacement orthogonal to the line of sight.

PROBLEM 5.10 *Consider a homing missile launched at $t = 0$ on a collision course against a nonmaneuvering target flying at constant velocity v_T. Assume that the missile velocity is*

$$v_M = v_{M0} - a_D t;$$

that is, due to drag deceleration, the missile velocity decreases with time.

1. *Determine an expression for the required missile heading $\theta^*(t)$ such that the missile remains on a constant bearing collision course with the target.*
2. *Assuming small drag deceleration, determine expressions for $R^*(t)$ and t_f^*, the nominal range and time of intercept.*

PROBLEM 5.11 *For the same missile as in Problem 5.10,*

1. *Develop equations for y_M in terms of y_T assuming an autopilot with transfer function*

$$Y(s) = \frac{N(s)}{D(s)}$$

2. *Assume $Y(s) = (sT + 1)^{-1}$. Using the method of adjoints, describe how to estimate the miss distance due to target maneuvers y_T*

6 Ballistic Guidance

In this chapter, we present the theory that is used in the analysis and design of ballistic guidance systems, with an emphasis on the principles rather than on the implementation. The trajectory of a ballistic missile typically consists of three phases: a powered lift-off, a free flight (this typically lasts about 80% of the engagement), and an aerodynamic reentry (see Figure 6.1). The purpose of ballistic guidance is then to determine the boundary conditions between the first and second phases that lead to a hit, and to analyze the miss due to navigation errors. Note that the optimization of the first phase leads to interesting "rocket staging" problems that can be treated with methods introduced in Chapter 8.

Section 6.1 describes the restricted two-body problem. Section 6.2 deals with the two-dimensional hit equation, whereas Section 6.3 contains the in-plane error analysis. Section 6.4 deals with three-dimensional error analysis, Section 6.5 with accounting for effects of the Earth's rotation. Section 6.6 considers the effect of the Earth's oblateness and geophysical uncertainties. Section 6.7 presents a general framework for numerical solution of general ballistic guidance problems. Sections 6.8, 6.9, and 6.10 present a summary of the key results in the chapter, bibliographic notes for further reading, and homework problems, respectively.

6.1 The Restricted Two-Body Problem

Assume that a ballistic missile, modeled as a particle, moves above the sensible atmosphere (i.e., above an altitude of 80 km). We assume that the Earth is nonrotating and perfectly spherical. We neglect the attraction from other heavenly bodies such as the Moon, the Sun, and the other planets. Finally, we also neglect the aerodynamic forces. (Note that such approximations are not arbitrary but can be justified using the theory of perturbations [19].)

For a conservative mechanical system with generalized coordinates $q \in \mathbb{R}^n$, kinetic energy $T(q, \dot{q})$, and potential energy $V(q)$, the equations of motion can be written in terms of the Lagrange function

$$L(q, \dot{q}) = T(q, \dot{q}) - V(q) \tag{6.1}$$

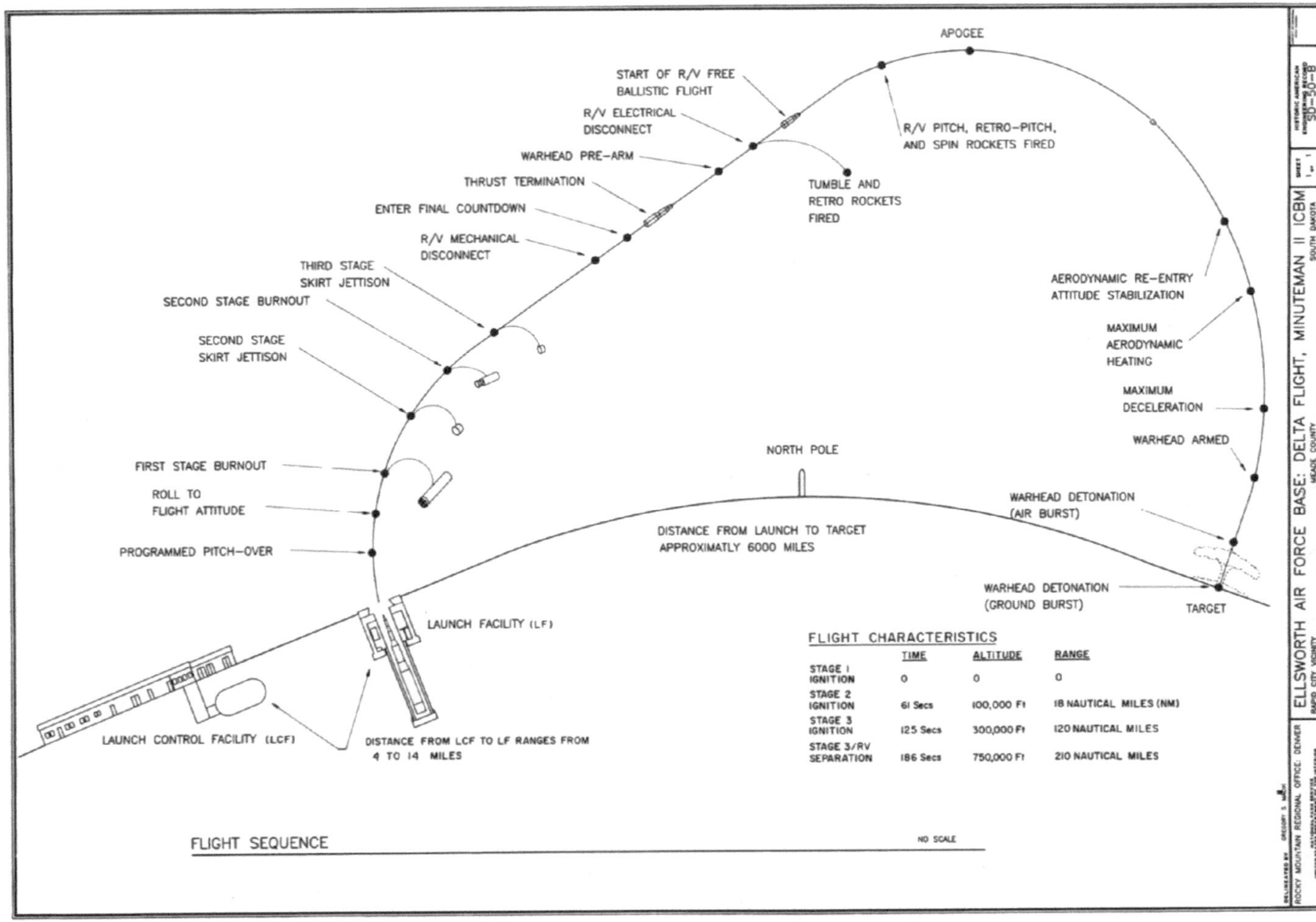

Figure 6.1. Ballistic flight sequence for Minuteman II missile. Image courtesy of National Museum of the U.S. Air Force.

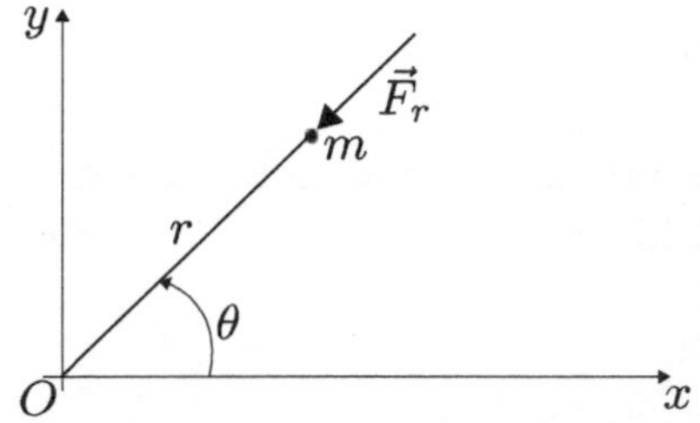

Figure 6.2. Geometry for the restricted two-body problem.

as [30]

$$\frac{\mathrm{d}}{\mathrm{d}t}\left(\frac{\partial L}{\partial \dot{q}}\right) - \frac{\partial L}{\partial q} = 0. \tag{6.2}$$

Let $\{O, x, y\}$ be an inertial frame centered at the center of the Earth, and let (r, θ) be the polar coordinates of the missile. Referring to Figure 6.2, the Lagrange function (6.1) is

$$L(r, \theta, \dot{r}, \dot{\theta}) = \frac{m}{2}\left(\dot{r}^2 + r^2\dot{\theta}^2\right) + \frac{mk}{r}, \tag{6.3}$$

where $k = GM = g_0 r_0^2$ is the gravitational constant of the Earth, G is the universal gravitational constant, M is the mass of the Earth, g_0 is the gravity acceleration at the surface of the Earth, r_0 is the radius of the Earth, and m is the mass of the missile. The equations of motion are then

$$\begin{aligned} \frac{\mathrm{d}}{\mathrm{d}t}\left(\frac{\partial L}{\partial \dot{r}}\right) - \frac{\partial L}{\partial r} &= -\frac{\mathrm{d}}{\mathrm{d}t}(m\dot{r}) - mr\dot{\theta}^2 + \frac{mk}{r^2} = 0 \\ \frac{\mathrm{d}}{\mathrm{d}t}\left(\frac{\partial L}{\partial \dot{\theta}}\right) - \frac{\partial L}{\partial \theta} &= \frac{\mathrm{d}}{\mathrm{d}t}(mr^2\dot{\theta}) = 0. \end{aligned} \tag{6.4}$$

These equations yield

$$\begin{aligned} \ddot{r} - r\dot{\theta}^2 &= -\frac{k}{r^2} \\ r^2\dot{\theta} &= h, \end{aligned} \tag{6.5}$$

where h is a constant that represents the angular momentum per unit mass of the missile.

We now integrate (6.5) in closed form to obtain r as a function of θ. First, from (6.5) (line 2), we have

$$\dot{\theta} = \frac{h}{r^2}. \tag{6.6}$$

Substitute this result in (6.5), (line 1) to obtain

$$\ddot{r} - r\left(\frac{h}{r^2}\right)^2 = -\frac{k}{r^2}, \tag{6.7}$$

which can be simplified as

$$\ddot{r} - \frac{h^2}{r^3} = -\frac{k}{r^2}. \tag{6.8}$$

Define the new dependent variable

$$u = \frac{1}{r}. \tag{6.9}$$

Successive differentiations of (6.9) yield

$$\begin{aligned} \dot{u} &= -\frac{\dot{r}}{r^2} \\ \ddot{u} &= -\frac{\ddot{r}}{r^2} + 2\frac{\dot{r}^2}{r^3}. \end{aligned} \tag{6.10}$$

Solving (6.10) (line 2) for $\ddot{r}$ and substituting in (6.8) and (6.9) yields

$$\left(-r^2\ddot{u} + \frac{\dot{r}^2}{r}\right) - h^2u^3 = -ku^2, \tag{6.11}$$

which, using (6.9) again, can be rewritten as

$$-\frac{\ddot{u}}{u^2} + 2\frac{\dot{u}^2}{u^3} - h^2u^3 = -ku^2. \tag{6.12}$$

Now, from (6.6), we have

$$r^2\frac{\mathrm{d}\theta}{\mathrm{d}t} = h. \tag{6.13}$$

Hence, the time derivatives of u can be expressed as

$$\begin{aligned} \dot{u} &= \frac{\mathrm{d}u}{\mathrm{d}\theta}\frac{\mathrm{d}\theta}{\mathrm{d}t} = \frac{\mathrm{d}u}{\mathrm{d}\theta}\frac{h}{r^2} = hu^2\frac{\mathrm{d}u}{\mathrm{d}\theta} \\ \ddot{u} &= \frac{\mathrm{d}\dot{u}}{\mathrm{d}\theta}\frac{\mathrm{d}\theta}{\mathrm{d}t} = \left(\frac{\mathrm{d}^2u}{\mathrm{d}\theta^2}hu^2 + 2hu\left(\frac{\mathrm{d}u}{\mathrm{d}\theta}\right)^2\right)\frac{h}{r^2}, \end{aligned} \tag{6.14}$$

or equivalently

$$\begin{aligned} \dot{u} &= hu^2u' \\ \ddot{u} &= h^2u^4u'' + 2h^2u^3(u')^2, \end{aligned} \tag{6.15}$$

where $(.)'$ denotes differentiation with respect to θ. Therefore, (6.12) yields

$$-h^2u^2u'' - h^2u^3 = -ku^2. \tag{6.16}$$

Assuming that $h^2u^2 \neq 0$ and simplifying by that quantity, we obtain the differential equation

$$u'' + u = \frac{k}{h^2}, \tag{6.17}$$

which is a second-order linear equation with constant coefficients. Combining the general solution of the homogeneous version of (6.17) with a particular solution of the forced equation, we obtain

$$u(\theta) = \frac{1}{r(\theta)} = c_1\cos\theta + c_2\sin\theta + \frac{k}{h^2}, \tag{6.18}$$

where the constants of integration c_1 and c_2 depend on the initial conditions. It is well known, from analytic geometry in polar coordinates, that (6.18) represents an arc of conic section, that is, an ellipse, a parabola, or a hyperbola. For ballistic guidance, the interesting case is that of bounded orbits, that is, ellipses.

6.2 The Two-Dimensional Hit Equation

In this section, we characterize the cutoff conditions (in position and velocity) that yield ballistic trajectories that reach a target located, without loss of generality, at polar coordinates $(r_0, 0)$. Note that this target may be on the surface of the Earth (which is consistent with neglecting aerodynamic forces) or may be at nonzero altitude and denote the beginning of a reentry trajectory. This reentry trajectory may itself be terminally guided.

Let r_1, θ_1, $\dot{r}_1$, and $\dot{\theta}_1$ be the polar coordinates of the missile at cutoff and their time derivatives. Define the radial and tangential components of velocity as

$$\begin{aligned} v_r &= \dot{r} \\ v_\theta &= r\dot{\theta}, \end{aligned} \tag{6.19}$$

respectively, so that the angular momentum of (6.5) has the expression

$$h = r v_\theta. \tag{6.20}$$

Because the orbit equation (6.18) is satisfied at cutoff, at polar coordinates (r_1, θ_1), we have

$$\frac{1}{r_1} = c_1 \cos\theta_1 + c_2 \sin\theta_1 + \frac{k}{h_1^2}. \tag{6.21}$$

Moreover, differentiating (6.18) with respect to time and evaluating at cutoff yields

$$\dot{u}_1 = -\frac{\dot{r}_1}{r_1^2} = -c_1\dot{\theta}_1 \sin\theta_1 + c_2\dot{\theta}_1 \cos\theta_1. \tag{6.22}$$

Because $r^2\dot{\theta} = h$, (6.22) yields

$$-\dot{r}_1 = -c_1 h_1 \sin\theta_1 + c_2 h_1 \cos\theta_1. \tag{6.23}$$

Equations (6.21) and (6.23) constitute a system of two linear algebraic equations for the two unknowns c_1 and c_2. The solution of this system of equations is

$$\begin{aligned} c_1 &= \left(\frac{1}{r_1} - \frac{k}{h_1^2}\right)\cos\theta_1 + \frac{\dot{r}_1}{h_1}\sin\theta_1 \\ c_2 &= \left(\frac{1}{r_1} - \frac{k}{h_1^2}\right)\sin\theta_1 - \frac{\dot{r}_1}{h_1}\cos\theta_1, \end{aligned} \tag{6.24}$$

or, in terms of coordinates and velocity components at cutoff,

$$\begin{aligned} c_1 &= \left(\frac{1}{r_1} - \frac{k}{(r_1 v_{\theta_1})^2}\right)\cos\theta_1 + \frac{v_{r_1}}{r_1 v_{\theta_1}}\sin\theta_1 \\ c_2 &= \left(\frac{1}{r_1} - \frac{k}{(r_1 v_{\theta_1})^2}\right)\sin\theta_1 + \frac{v_{r_1}}{r_1 v_{\theta_1}}\cos\theta_1. \end{aligned} \tag{6.25}$$

Finally, substituting (6.25) into the orbit equation (6.18), we obtain

$$\frac{1}{r} = \left(\left(\frac{1}{r_1} - \frac{k}{(r_1 v_{\theta_1})^2}\right)\cos\theta_1 + \frac{v_{r_1}}{r_1 v_{\theta_1}}\sin\theta_1\right)\cos\theta$$
$$+ \left(\left(\frac{1}{r_1} - \frac{k}{(r_1 v_{\theta_1})^2}\right)\sin\theta_1 - \frac{v_{r_1}}{r_1 v_{\theta_1}}\cos\theta_1\right)\sin\theta + \frac{k}{(r_1 v_{\theta_1})^2}. \tag{6.26}$$

Equation (6.26) represents a two-parameter bundle of orbits, parametrized by (v_{r_1}, v_{θ_1}) and passing through the location where cutoff occurs, that is, the point of polar coordinates (r_1, θ_1).

We now require that the orbit of the missile pass through the location of the target, that is, that $r = r_0$ when $\theta = 0$, yielding

$$\frac{1}{r_0} - \left(\frac{1}{r_1} - \frac{k}{(r_1 v_{\theta_1})^2}\right)\cos\theta_1 - \frac{v_{r_1}}{r_1 v_{\theta_1}}\sin\theta_1 - \frac{k}{(r_1 v_{\theta_1})^2} = 0. \tag{6.27}$$

This is the **hit equation**, a necessary and sufficient condition that the cutoff coordinates, r_1 and θ_1, and velocity components, v_{r_1} and v_{θ_1}, must satisfy for the missile ballistic trajectory to hit a target located at $(r_0, 0)$.

The hit equation is nonlinear, of the form

$$F(r_1, \theta_1, v_{r_1}, v_{\theta_1}) = \frac{1}{r_0} - \left(\frac{1}{r_1} - \frac{k}{(r_1 v_{\theta_1})^2}\right)\cos\theta_1 - \frac{v_{r_1}}{r_1 v_{\theta_1}}\sin\theta_1 - \frac{k}{(r_1 v_{\theta_1})^2} = 0. \tag{6.28}$$

It may be useful to obtain a linearized form of that equation. Assume that the nominal cutoff conditions, $(r_1^0, \theta_1^0, v_{r_1}^0, v_{\theta_1}^0)$, satisfy it. Also assume that the actual cutoff conditions, $(r_1, \theta_1, v_{r_1}, v_{\theta_1})$, are close to their nominal values. Expand (6.28) in a Taylor series around its nominal conditions:

$$F(r_1, \theta_1, v_{r_1}, v_{\theta_1}) = F(r_1^0, \theta_1^0, v_{r_1}^0, v_{\theta_1}^0) + \left(\frac{\partial F}{\partial r_1}\right)_0 (r_1 - r_1^0)$$
$$+ \left(\frac{\partial F}{\partial \theta_1}\right)_0 (\theta_1 - \theta_1^0) + \left(\frac{\partial F}{\partial v_{r_1}}\right)_0 (v_{r_1} - v_{r_1}^0)$$
$$+ \left(\frac{\partial F}{\partial v_{\theta_1}}\right)_0 (v_{\theta_1} - v_{\theta_1}^0) + \text{H.O.T.} \tag{6.29}$$

If the actual cutoff conditions also satisfy the hit equation, then, to first order, we have

$$\left(\frac{\partial F}{\partial r_1}\right)_0 (r_1 - r_1^0) + \left(\frac{\partial F}{\partial \theta_1}\right)_0 (\theta_1 - \theta_1^0)$$
$$+ \left(\frac{\partial F}{\partial v_{r_1}}\right)_0 (v_{r_1} - v_{r_1}^0) + \left(\frac{\partial F}{\partial v_{\theta_1}}\right)_0 (v_{\theta_1} - v_{\theta_1}^0) = 0, \tag{6.30}$$

which is a linear equation of the form

$$A_0 + A_1 r_1 + A_2 \theta_1 + A_3 v_{r_1} + A_4 v_{\theta_1} = 0, \tag{6.31}$$

where

$$A_0 = -\left(\frac{\partial F}{\partial r_1}\right)_0 r_1^0 - \left(\frac{\partial F}{\partial \theta_1}\right)_0 \theta_1^0 - \left(\frac{\partial F}{\partial v_{r_1}}\right)_0 v_{r_1}^0 - \left(\frac{\partial F}{\partial v_{\theta_1}}\right)_0 v_{\theta_1}^0,$$

$$A_1 = \left(\frac{\partial F}{\partial r_1}\right)_0, \quad A_2 = \left(\frac{\partial F}{\partial \theta_1}\right)_0,$$

$$A_3 = \left(\frac{\partial F}{\partial v_{r_1}}\right)_0, \quad A_4 = \left(\frac{\partial F}{\partial v_{\theta_1}}\right)_0. \tag{6.32}$$

Condition (6.30) is the **linearized hit equation** and can be checked by the guidance system. When it is satisfied, the engines are turned off, guaranteeing a hit of the target to within first order.

EXAMPLE 6.1 (Least Energetic Ballistic Shot) *To illustrate the use of the hit equation, consider a case where cutoff is known to occur at polar coordinates* $r_1 = r_0$, $\theta_1 = \pi/2$. *For these values, the hit equation* (6.27) *becomes*

$$\frac{1}{r_0} - \frac{v_{r_1}}{r_0 v_{\theta_1}} - \frac{k}{r_0^2 v_{\theta_1}^2} = 0, \tag{6.33}$$

which can be solved for v_{r_1} *as*

$$v_{r_1} = v_{\theta_1} - \frac{k}{r_0 v_{\theta_1}}. \tag{6.34}$$

Figure 6.3 shows the relation between v_{r_1} *and* v_{θ_1} *specified by* (6.34). *Each point of this locus represents a combination of components of cutoff velocity that yields a ballistic trajectory that hits the target. The resulting bundle of trajectories is illustrated in Figure 6.4.*

Because many trajectories accomplish the mission, we need a subsidiary criterion to choose a particular trajectory out of the bundle. Here we minimize the kinetic energy per unit mass at cutoff, leading to the ***least energetic ballistic shot****. The constrained optimization problem is*

$$\min_{v_{r_1}, v_{\theta_1}} T(v_{r_1}, v_{\theta_1}) = \frac{1}{2}(v_{r_1}^2 + v_{\theta_1}^2)$$

$$\textit{subject to } (6.34). \tag{6.35}$$

To solve this constrained optimization problem, we eliminate v_{r_1} *in favor of* v_{θ_1} *using* (6.34) *and obtain the specific kinetic energy (i.e., the kinetic energy per unit mass) as a function of* v_{θ_1} *alone:*

$$T(v_{\theta_1}) = \frac{1}{2}\left(v_{\theta_1} - \frac{k}{r_0 v_{\theta_1}}\right)^2 + \frac{1}{2}v_{\theta_1}^2 = v_{\theta_1}^2 + \frac{1}{2}\frac{k^2}{r_0^2 v_{\theta_1}^2} - \frac{k}{r_0}. \tag{6.36}$$

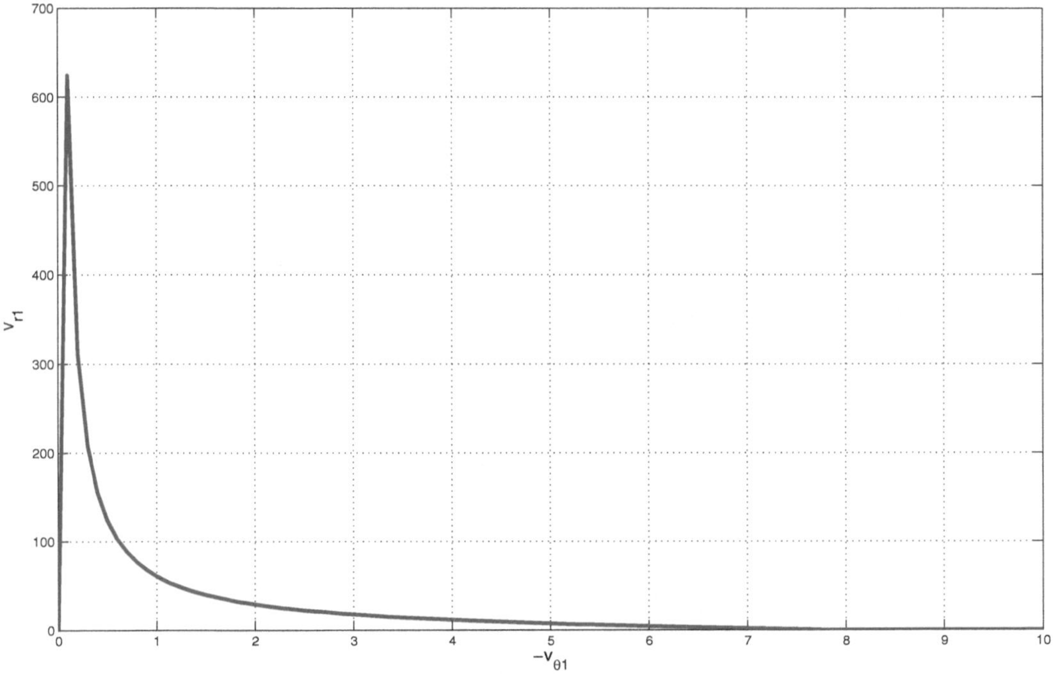

Figure 6.3. Relation between v_{r_1} and v_{θ_1} for a ballistic shot.

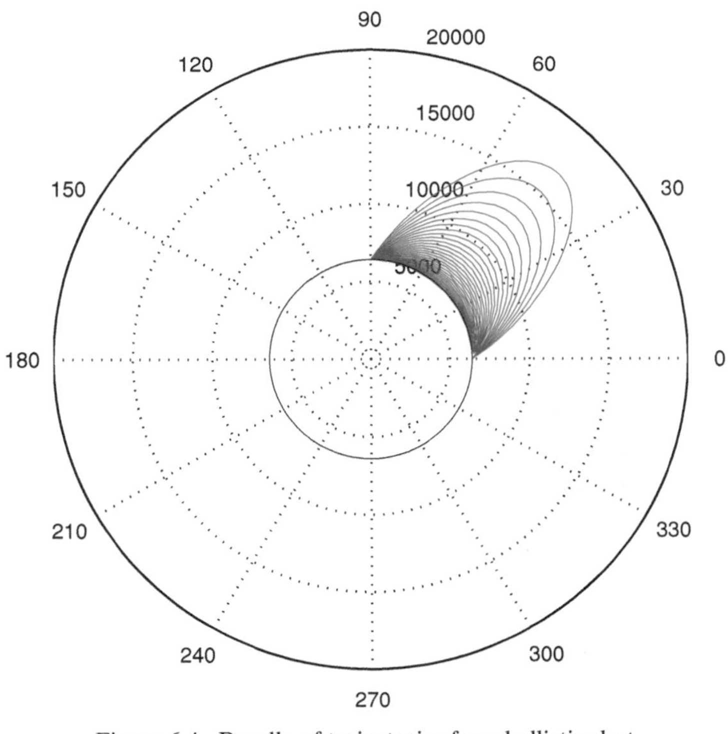

Figure 6.4. Bundle of trajectories for a ballistic shot.

Then, we minimize $T(v_{\theta_1})$ with respect to its argument. The first derivative is

$$\frac{\mathrm{d}}{\mathrm{d}v_{\theta_1}} T(v_{\theta_1}) = 2v_{\theta_1} - \frac{k^2}{r_0^2 v_{\theta_1}^3}. \tag{6.37}$$

Setting the first derivative to zero yields

$$v_{\theta_1} = \pm\sqrt{\frac{k}{r_0\sqrt{2}}}. \tag{6.38}$$

Recalling that v_{r_1} is obtained through (6.34)*, it is easily seen that we must take the negative alternative in* (6.38) *to have $v_{r_1} \geq 0$, that is, a launch that clears the surface of the planet. Moreover, the second derivative of $T(v_{\theta_1})$,*

$$\frac{\mathrm{d}^2}{\mathrm{d}v_{\theta_1}^2} T(v_{\theta_1}) = 2 + \frac{3k^2}{r_0^2 v_{\theta_1}^4}, \tag{6.39}$$

is always strictly positive, which guarantees that the values of v_{θ_1} in (6.38) *are global minimizers. Hence, for the given cutoff and target locations, the combination of components of cutoff velocity that achieves a hit while minimizing energy is*

$$\begin{aligned} v_{r_1} &= (\sqrt{2}-1)\sqrt{\frac{k}{r_0\sqrt{2}}} \\ v_{\theta_1} &= -\sqrt{\frac{k}{r_0\sqrt{2}}}. \end{aligned} \tag{6.40}$$

Note that the resulting ballistic trajectory has a flight path angle at launch:

$$\begin{aligned} \gamma_{\text{launch}} &= \arctan\left(\frac{v_{r_1}}{|v_{\theta_1}|}\right) \\ &= \arctan(\sqrt{2}-1) \\ &= 22.5 \text{ deg}. \end{aligned} \tag{6.41}$$

6.3 In-Plane Error Analysis

In this section, we quantify the miss that occurs when cutoff conditions are a small perturbation away from nominal conditions that satisfy the hit equation. For that purpose, write the general relation of (6.26) between the six quantities $r, \theta, r_1, \theta_1, v_{r_1}$, and v_{θ_1} as

$$H(r, \theta, r_1, \theta_1, v_{r_1}, v_{\theta_1}) = 0, \tag{6.42}$$

where

$$\begin{aligned} H(r, \theta, r_1, \theta_1, v_{r_1}, v_{\theta_1}) = &-\frac{1}{r} + B(r_1, \theta_1, v_{r_1}, v_{\theta_1})\cos\theta \\ &+ C(r_1, \theta_1, v_{r_1}, v_{\theta_1})\sin\theta + \frac{k}{(r_1 v_{\theta_1})^2}, \end{aligned} \tag{6.43}$$

and

$$B(r_1, \theta_1, v_{r_1}, v_{\theta_1}) = \left(\frac{1}{r_1} - \frac{k}{(r_1 v_{\theta_1})^2}\right)\cos\theta_1 - \frac{v_{r_1}}{r_1 v_{\theta_1}}\sin\theta_1$$

$$C(r_1, \theta_1, v_{r_1}, v_{\theta_1}) = \left(\frac{1}{r_1} - \frac{k}{(r_1 v_{\theta_1})^2}\right)\sin\theta_1 - \frac{v_{r_1}}{r_1 v_{\theta_1}}\cos\theta_1. \quad (6.44)$$

Note that, in the formalism of (6.42) through (6.44), the hit equation for a target located at polar coordinates $(r, \theta) = (r_0, 0)$ can be rewritten as

$$H(r_0, 0, r_1^0, \theta_1^0, v_{r_1}^0, v_{\theta_1}^0) = 0, \quad (6.45)$$

where the superscript "0" pertains to nominal cutoff conditions that satisfy the hit equation. Under the assumption of a perfectly spherical Earth, if the cutoff conditions do not satisfy the hit equation, the missile impacts the planet at polar coordinates (r_0, θ), where the angular coordinate θ satisfies

$$H(r_0, \theta, r_1^0, \theta_1^0, v_{r_1}^0, v_{\theta_1}^0) = 0, \quad (6.46)$$

or, more explicitly,

$$-\frac{1}{r_0} + B(r_1, \theta_1, v_{r_1}, v_{\theta_1})\cos\theta + C(r_1, \theta_1, v_{r_1}, v_{\theta_1})\sin\theta + \frac{k}{(r_1 v_{\theta_1})^2} = 0. \quad (6.47)$$

For given cutoff conditions $(r_1, \theta_1, v_{r_1}, v_{\theta_1})$, (6.46) and (6.47) define the angular coordinate of the location of impact. For a target located at angular coordinate $\theta = 0$, the **in-plane**, or **down-range**, **miss distance** is then quantified as

$$M_{DR} = -r_0\theta. \quad (6.48)$$

To obtain a first-order approximation of this miss distance, we evaluate the partial derivatives of the implicit function defined in (6.46) around the nominal condition as defined in (6.45), that is,

$$\left(\frac{\partial\theta}{\partial r_1}\right)_0, \left(\frac{\partial\theta}{\partial \theta_1}\right)_0, \left(\frac{\partial\theta}{\partial v_{r_1}}\right)_0, \left(\frac{\partial\theta}{\partial v_{\theta_1}}\right)_0. \quad (6.49)$$

To do so, we use the **implicit function theorem**, described in Appendix A.5, to evaluate the partial derivatives of (6.49), where θ is implicitly defined in (6.46). Here, $n = 1, m = 4, x = \theta, y = (r_1, \theta_1, v_{r_1}, v_{\theta_1})^T$, and $F(x, y) = H(r_0, \theta, r_1, \theta_1, v_{r_1}, v_{\theta_1})$. According to the theorem, if $(\frac{\partial H}{\partial\theta})_0 \neq 0$, then, in the vicinity of 0, θ can be expressed explicitly as

$$\theta = G(r_1, \theta_1, v_{r_1}, v_{\theta_1}), \quad (6.50)$$

where the partial derivatives of G, evaluated at $(r_1^0, \theta_1^0, v_{r_1}^0, v_{\theta_1}^0)$, are

$$\left(\frac{\partial G}{\partial r_1}\right)_0 = \frac{-\left(\frac{\partial H}{\partial r_1}\right)_0}{\left(\frac{\partial H}{\partial\theta}\right)_0}, \quad \left(\frac{\partial G}{\partial \theta_1}\right)_0 = \frac{-\left(\frac{\partial H}{\partial \theta_1}\right)_0}{\left(\frac{\partial H}{\partial\theta}\right)_0}$$

$$\left(\frac{\partial G}{\partial v_{r_1}}\right)_0 = \frac{-\left(\frac{\partial H}{\partial v_{r_1}}\right)_0}{\left(\frac{\partial H}{\partial\theta}\right)_0}, \quad \left(\frac{\partial G}{\partial v_{\theta_1}}\right)_0 = \frac{-\left(\frac{\partial H}{\partial v_{\theta_1}}\right)_0}{\left(\frac{\partial H}{\partial\theta}\right)_0}. \quad (6.51)$$

Define the perturbations in cutoff conditions as

$$\begin{aligned}
\delta r_1 &= r_1 - r_1^0, \\
\delta\theta_1 &= \theta_1 - \theta_1^0, \\
\delta v_{r_1} &= v_{r_1} - v_{r_1}^0, \\
\delta v_{\theta_1} &= v_{\theta_1} - v_{\theta_1}^0, \\
\delta y &= (\delta r_1, \delta\theta_1, \delta v_{r_1}, \delta v_{\theta_1})^T.
\end{aligned} \tag{6.52}$$

Also define the miss coefficients and the miss coefficient vector as

$$\begin{aligned}
M_{r_1} &= -r_0 \frac{\left(\frac{\partial H}{\partial r_1}\right)_0}{\left(\frac{\partial H}{\partial \theta}\right)_0}, \quad M_{\theta_1} = -r_0 \frac{\left(\frac{\partial H}{\partial \theta_1}\right)_0}{\left(\frac{\partial H}{\partial \theta}\right)_0}, \\
M_{v_{r_1}} &= -r_0 \frac{\left(\frac{\partial H}{\partial v_{r_1}}\right)_0}{\left(\frac{\partial H}{\partial \theta}\right)_0}, \quad M_{v_{\theta_1}} = -r_0 \frac{\left(\frac{\partial H}{\partial v_{\theta_1}}\right)_0}{\left(\frac{\partial H}{\partial \theta}\right)_0}, \\
M &= (M_{r_1}, M_{\theta_1}, M_{v_{r_1}}, M_{v_{\theta_1}})^T.
\end{aligned} \tag{6.53}$$

Then, in terms of cutoff perturbations and miss coefficients, the down-range miss of (6.48) has the form

$$\begin{aligned}
M_{DR} &= -r_0\theta = M_{r_1}\delta r_1 + M_{\theta_1}\delta\theta_1 + M_{v_{r_1}}\delta v_{r_1} + M_{v_{\theta_1}}\delta v_{\theta_1} \\
&= M^T \delta y.
\end{aligned} \tag{6.54}$$

If the cutoff perturbations are Gaussian with zero mean and known covariance, that is, $\delta y = \mathcal{N}(0, P)$, then the expected square miss can be evaluated as

$$E(M_{DR}^2) = M^T P M, \tag{6.55}$$

which can be used as a figure of merit for the accuracy of the ballistic shot.

EXAMPLE 6.2 *Here we illustrate the computation of the miss coefficients for the least energetic ballistic shot of Example 6.1. With the components of velocity in* (6.40) *and* H *defined in* (6.43) *and* (6.44), (6.53) *yields*

$$\begin{aligned}
|M_{r_1}| &= 5,850 \text{ ft/ft}, \\
|M_{\theta_1}| &= r_0 \text{ ft/rad}, \\
|M_{v_{r_1}}| &= 2,310 \text{ ft/\{ft/s\}}, \\
|M_{v_{\theta_1}}| &= 5,590 \text{ ft/\{ft/s\}}.
\end{aligned} \tag{6.56}$$

Note, for instance, that a velocity error in v_{θ_1} *of 1 ft/s at cutoff implies a down-range miss of more than a mile. (There are 5,280 feet in a mile.)*

EXAMPLE 6.3 (Most Accurate Ballistic Shot) *Here we illustrate the use of the expected square miss given in* (6.55) *as a figure of merit. For the sake of simplicity, we assume flat Earth with constant gravity, in addition to the absence of aerodynamic forces.*

Consider a vertical plane with Cartesian coordinates (x_1, x_2), where the x_1 axis is horizontal and the x_2 axis is vertical ascending. Assume that a canon, located at the origin, shoots a cannonball against a target located on the x_1 axis at a known range R.

Also assume that the components of initial velocity are subject to small uncertainties that are jointly Gaussian, with zero mean and known covariance:

$$\text{cov}\left(\begin{bmatrix}\delta V_1\\ \delta V_2\end{bmatrix}\right) = \begin{bmatrix}\sigma_{11} & \sigma_{12}\\ \sigma_{12} & \sigma_{22}\end{bmatrix}, \tag{6.57}$$

where V_1 and V_2 are the horizontal and vertical components of initial velocity, respectively. We determine, among all the ballistic trajectories that nominally hit the target, which one is the most accurate in that it minimizes the expected square miss.

The equations of motion with initial conditions are

$$\begin{aligned}\ddot{x}_1 &= 0,\\ \ddot{x}_2 &= -g,\\ (x_1(0), x_2(0)) &= (0, 0),\\ (\dot{x}_1(0), \dot{x}_2(0)) &= (V_1, V_2),\end{aligned} \tag{6.58}$$

where g is the acceleration of gravity. These equations are easily integrated as

$$\begin{aligned}x_1(t) &= V_1 t\\ x_2(t) &= -\frac{gt^2}{2} + V_2 t.\end{aligned} \tag{6.59}$$

The ballistic flight terminates at a time $t_f > 0$ satisfying $x_2(t_f) = 0$. Hence,

$$t_f = \frac{2V_2}{g}. \tag{6.60}$$

The range achieved by the shot is then $r = x_1(t_f)$, that is,

$$r = \frac{2V_1V_2}{g}. \tag{6.61}$$

Therefore, the condition for hitting the target, that is, the hit equation, is

$$V_1V_2 = \frac{Rg}{2}. \tag{6.62}$$

This relationship between V_1 and V_2 is illustrated in Figure 6.5, which should be compared with Figure 6.3. Here also, each point on the locus represents a combination of components of launch velocity that yields a ballistic trajectory that hits the target. The resulting bundle of trajectories is illustrated in Figure 6.6, which should be compared with Figure 6.4.

Because many trajectories accomplish the mission, we need a subsidiary criterion to select one from the bundle in Figure 6.6. Although we could use energy as in Example 6.1, here we use accuracy. From (6.61), the miss coefficients and miss

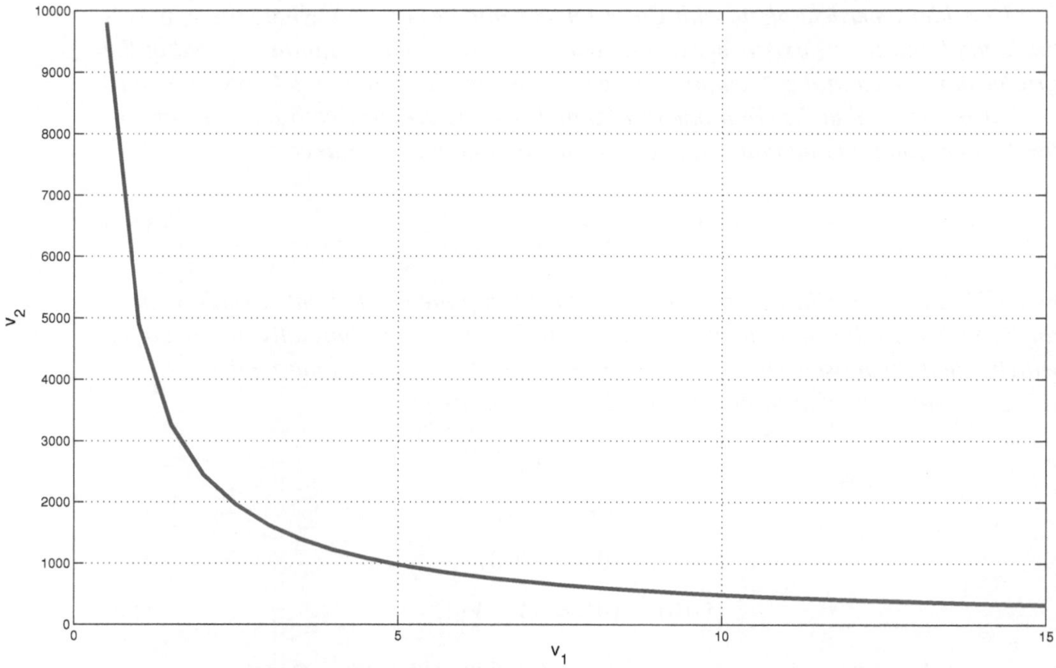

Figure 6.5. Relation between V_1 and V_2 for a ballistic shot.

coefficient vector are

$$M_1 = \frac{\partial r}{\partial V_1} = \frac{2V_2}{g},$$
$$M_2 = \frac{\partial r}{\partial V_2} = \frac{2V_1}{g},$$
$$M = (M_1, M_2)^T. \tag{6.63}$$

In terms of the miss coefficients, the down-range miss is

$$M_{DR} = M_1 \delta V_1 + M_2 \delta V_2. \tag{6.64}$$

From (6.55), (6.57), *and* (6.63), *the expected square miss is*

$$E(M_{DR}^2) = \begin{bmatrix} \frac{2V_2}{g} & \frac{2V_1}{g} \end{bmatrix} \begin{bmatrix} \sigma_{11} & \sigma_{12} \\ \sigma_{12} & \sigma_{22} \end{bmatrix} \begin{bmatrix} \frac{2V_2}{g} \\ \frac{2V_1}{g} \end{bmatrix}$$
$$= \frac{4}{g^2} \left(\sigma_{11} V_2^2 + 2\sigma_{12} V_1 V_2 + \sigma_{22} V_1^2 \right). \tag{6.65}$$

The most accurate hit is obtained by minimizing expression (6.65) *with respect to* V_1 *and* V_2, *subject to the constraint* (6.62). *Eliminating* V_2 *in favor of* V_1, *we obtain the expected square miss in terms of* V_1 *alone:*

$$E(M_{DR}^2) = \frac{4}{g^2} \left(\sigma_{11} \frac{R^2 g^2}{4V_1^2} + \sigma_{12} R g + \sigma_{22} V_1^2 \right). \tag{6.66}$$

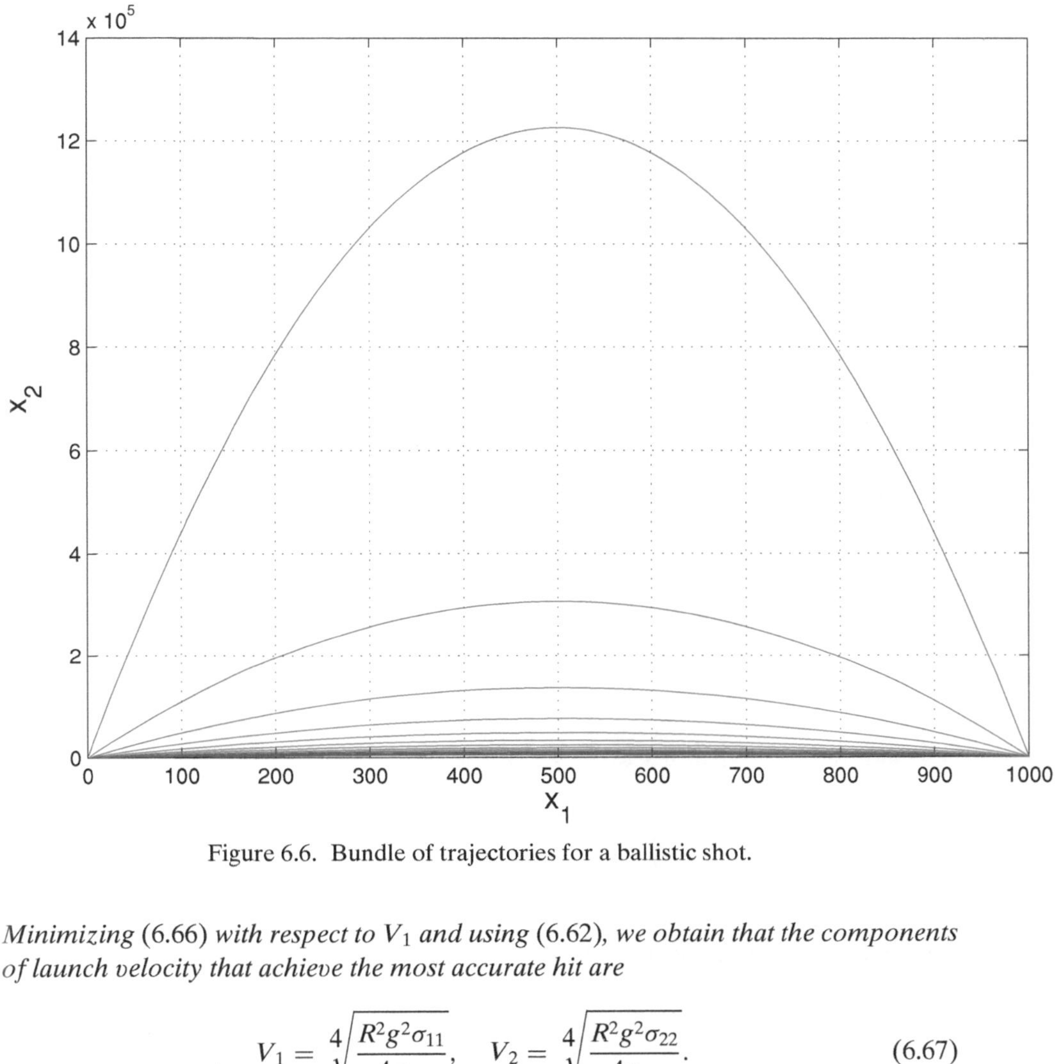

Figure 6.6. Bundle of trajectories for a ballistic shot.

Minimizing (6.66) *with respect to* V_1 *and using* (6.62), *we obtain that the components of launch velocity that achieve the most accurate hit are*

$$V_1 = \sqrt[4]{\frac{R^2 g^2 \sigma_{11}}{4\sigma_{22}}}, \quad V_2 = \sqrt[4]{\frac{R^2 g^2 \sigma_{22}}{4\sigma_{11}}}. \tag{6.67}$$

Note that this ballistic trajectory has a flight path angle at launch:

$$\begin{aligned} \gamma_{\text{launch}} &= \arctan\left(\frac{V_2}{V_1}\right) \\ &= \arctan\left(\sqrt{\frac{\sigma_{22}}{\sigma_{11}}}\right), \end{aligned} \tag{6.68}$$

which depends only on the ratio of the variances of V_2 *and* V_1.

6.4 Three-Dimensional Error Analysis

Three-dimensional error analysis is more typical of real life than two-dimensional analysis yet does not require a closed-form solution. We consider motion of a missile in a typical cylindrical coordinate system, as shown in Figure 6.7. Let $\{O, x, y\}$ be the

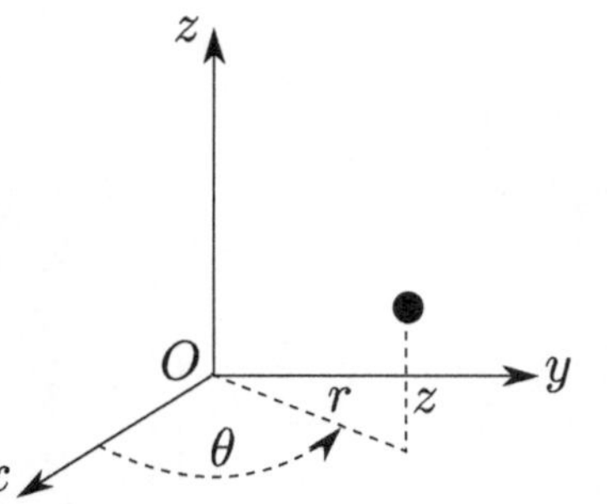

Figure 6.7. Setup for three-dimensional error analysis.

plane containing the center of Earth, the powered flight cutoff point, and the desired impact point. Then, the trajectory is always near the $\{O, x, y\}$ plane.

Consider ballistic flight using cylindrical (r, θ, z) coordinates to allow for out-of-orbital-plane motions. Assume that the nominal orbit is in the plane $z = 0$. Here the Lagrange function is

$$L(r, \theta, z, \dot{r}, \dot{\theta}, \dot{z}) = \frac{m}{2}(\dot{r}^2 + \dot{\theta}^2 + \dot{z}^2) + \frac{mk}{\sqrt{r^2 + z^2}}. \tag{6.69}$$

The equations of motion are given by

$$\frac{\mathrm{d}}{\mathrm{d}t}\left(\frac{\partial L}{\partial(\dot{r}, \dot{\theta}, \dot{z})^T}\right) - \frac{\partial L}{\partial(r, \theta, z)^T} = 0, \tag{6.70}$$

yielding dynamics of order $n = 6$:

$$\begin{aligned}
\dot{r} &= v_r \\
\dot{\theta} &= \frac{v_\theta}{r} \\
\dot{z} &= v_z \\
\dot{v}_r &= \frac{v_\theta^2}{r} - \frac{kr}{(r^2 + z^2)^{3/2}} \\
\dot{v}_\theta &= -\frac{v_r v_\theta}{r} \\
\dot{v}_z &= \frac{-kz}{(r^2 + z^2)^{3/2}}.
\end{aligned} \tag{6.71}$$

Let $(r_1, \theta_1, 0, v_{r1}, v_{\theta_1}, 0)$ be the cutoff conditions of powered flight. Let $(r_0, \theta_0, 0)$ be the desired impact point.

Assume that $x^0 = (r^0(t), \theta^0(t), 0, v_r^0(t), v_\theta^0(t), 0)$ is a nominal trajectory, that is, a trajectory that satisfies the equations of motion and the boundary conditions.

Assume further that the true trajectory is close to nominal, as shown in Figure 6.8. Let

$$
\begin{aligned}
\delta r(t) &= r(t) - r^0(t),\\
\delta\theta(t) &= \theta(t) - \theta^0(t),\\
\delta z(t) &= z(t) - z^0(t) = z(t),\\
\delta v_r(t) &= v_r(t) - c_r^0(t),\\
\delta v_\theta(t) &= v_\theta(t) - v_\theta^0(t),\\
\delta v_z(t) &= v_z(t) - v_z^0(t) = v_z(t),
\end{aligned}
\tag{6.72}
$$

where $x = (r, \theta, z, v_r, v_\theta, v_z)$ are for the true trajectory, and $z^0(t) = v_z^0(t) = 0$ by choice.

The equations of motion have the form

$$\dot{x} = f(x), \tag{6.73}$$

where $x^0(t)$ is a solution. We linearize these equations about $x^0(t)$ to obtain

$$\frac{\mathrm{d}}{\mathrm{d}t}(x^0 + \delta x) = f(x^0 + \delta x), \tag{6.74}$$

or

$$\dot{x}^0 + \delta\dot{x} = f(x^0) + \left(\frac{\partial f}{\partial x}\right)_0^T \delta x + \text{H.O.T.} \tag{6.75}$$

Now, $\dot{x}^0 = f(x^0)$, hence, to first order,

$$\delta\dot{x} = \left(\frac{\partial f}{\partial x}\right)_0^T \delta x = A(t)\delta x. \tag{6.76}$$

This is a standard linear, time varying perturbation equation. For this example,

$$
\begin{aligned}
\delta\dot{r} &= \delta v_r\\
\delta\dot{\theta} &= -\left(\frac{v_\theta}{r^2}\right)_0 \delta r + \left(\frac{1}{r}\right)_0 \delta\dot{\theta}\\
\delta\dot{z} &= \delta v_z\\
\delta\dot{v}_r &= \left(-\frac{v_\theta^2}{r^2} - \frac{k}{(r^2+z^2)^{3/2}} + \frac{3kr^2}{(r^2+z^2)^{5/2}}\right)_0 \delta r + \left(\frac{3krz}{(r^2+z^2)^{5/2}}\right)_0 \delta z + \left(\frac{2v_\theta}{r}\right)_0 \delta v_\theta\\
\delta\dot{v}_\theta &= \left(\frac{v_r v_\theta}{r^2}\right)_0 \delta r + \left(-\frac{v_\theta}{r}\right)_0 \delta v_r + \left(-\frac{v_r}{r}\right)_0 \delta v_\theta\\
\delta\dot{v}_z &= \left(\frac{3krz}{(r^2+z^2)^{5/2}}\right)_0 \delta r + \left(\frac{-k}{(r^2+z^2)^{3/2}} + \frac{3kz^2}{(r^2+z^2)^{5/2}}\right)_0 \delta z,
\end{aligned}
\tag{6.77}
$$

which indeed has the form $\delta\dot{x}(t) = A(t)\delta x(t)$.

We are concerned with the miss due to errors at the powered flight cutoff point. Let t_f^0 be the final time of the reference trajectory that satisfies impact conditions and lies in the $\{O, x, y\}$ plane. We proceed in two steps, first computing an approximation of the actual flight time, t_f, and then computing down-range and cross-range miss distances.

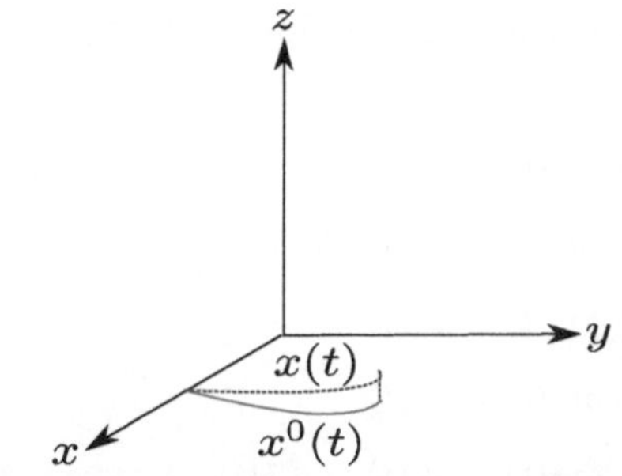

Figure 6.8. Nominal and actual trajectories for three-dimensional error analysis.

6.4.1 Actual Flight Time Approximation

The effect on miss of using nominal termination time instead of actual in three-dimensional error analysis for ballistic guidance is shown in Figure 6.9. To compute an approximation of the actual flight time, assume that z is small, and consider the terminal radius

$$r_E = \sqrt{r^2(t_f) + z^2(t_f)} \approx \sqrt{r^2(t_f)} = r(t_f) = r_0. \tag{6.78}$$

Then, expansion of $r(t)$ yields

$$r(t) = r(t_f^0) + \dot{r}\Big|_{t_f^0} (t - t_f^0) + \text{H.O.T.} \tag{6.79}$$

Now, at the actual impact time t_f, from (6.78), $r(t_f) = r_0$, implying

$$r_0 = r(t_f^0) + v_r(t_f^0)(t_f - t_f^0), \tag{6.80}$$

yielding

$$t_f \approx t_f^0 - \frac{r(t_f^0) - r_0}{v_r(t_f^0)}. \tag{6.81}$$

Using the fact that $r_0 = r^0(t_f^0)$ and the notation $\delta r(t_f^0) = r(t_f^0) - r^0(t_f^0)$, we obtain

$$t_f = t_f^0 - \frac{\delta r(t_f^0)}{v_r(t_f^0)}. \tag{6.82}$$

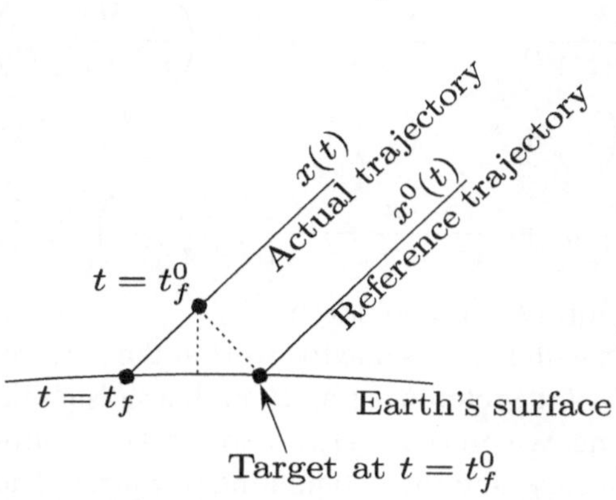

Figure 6.9. Effect of using nominal termination time instead of actual in three-dimensional error analysis for ballistic guidance.

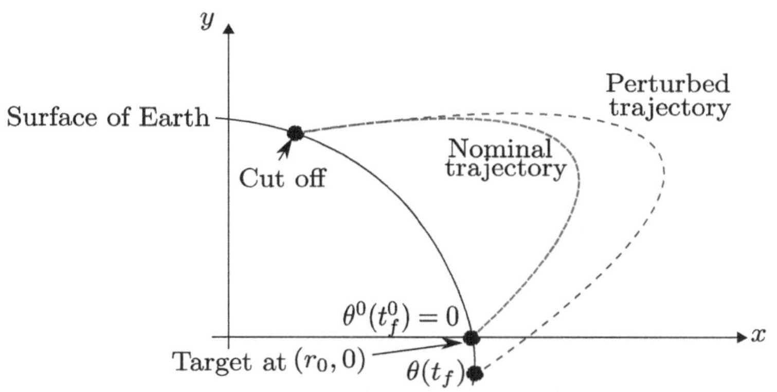

Figure 6.10. Setup for down-range miss distance analysis.

Note that (6.82) requires knowledge of the perturbed trajectory at t_f^0 because $r(t_f^0)$ and $v_r(t_f^0)$ are required. These can be determined using the method of adjoints of Section 2.4.

For the rest of the analysis, we assume that t_f is known, and we determine the down-range and cross-range miss distances.

6.4.2 Down-Range Miss Distance, M_{DR}

The down-range miss is the miss in the $\{O, x, y\}$ plane. The setup and geometry are shown in Figure 6.10.

We have

$$\begin{aligned} M_{DR} &= r_0(\theta(t_f) - \theta^0(t_f^0)) \\ &= r_0(\theta(t_f) - \theta_0). \end{aligned} \tag{6.83}$$

Let us first obtain an expression for $\theta(t_f)$:

$$\begin{aligned} \theta(t) &= \theta(t_f^0) + \dot{\theta}(t_f^0)(t - t_f^0) + \text{H.O.T} \\ \Rightarrow \theta(t_f) &\approx \theta(t_f^0) + \dot{\theta}(t_f^0)(t_f - t_f^0). \end{aligned} \tag{6.84}$$

Substituting in (6.83),

$$\begin{aligned} M_{DR} &= r_0\left(\theta(t_f^0) + \dot{\theta}(t_f^0)\left(\frac{-\delta r(t_f^0)}{v_r(t_f^0)}\right)\right) \\ &= r_0\delta\theta(t_f^0) - \frac{r_0\dot{\theta}(t_f^0)}{v_r(t_f^0)}\delta r(t_f^0). \end{aligned} \tag{6.85}$$

We use the method of adjoints to compute M_{DR}. We have

$$\begin{aligned} \delta\dot{x} &= A(t)\delta x \\ M_{DR} &= C\delta x(t_f^0), \end{aligned} \tag{6.86}$$

so

$$M_{DR} = C\Phi(t_f, t_1)\delta x(t_1), \tag{6.87}$$

where $C\Phi(t_f, t_1)$ is a vector of miss coefficients. We want an expression for the miss coefficients as a function of t_1. Let us define an adjoint vector λ satisfying

$$\dot{\lambda} = -A^T(t)\lambda, \tag{6.88}$$

so that

$$\lambda^T \delta x = \text{ constant } = \lambda^T(t_1)\delta x(t_1) = \lambda^T(t_f)\delta x(t_f). \tag{6.89}$$

Then, choosing $\lambda(t_f^0) = C^T$,

$$M_{DR} = C\delta x(t_f^0) = \lambda^T(t_f^0)\delta x(t_f^0) = \lambda^T(t_1)\delta x(t_1). \tag{6.90}$$

Here $\lambda^T(t_1)$ contains the time history of miss coefficients. A difficulty, however, is that the matrix C depends on the perturbed trajectory.

There are two ways to deal with this difficulty. The first is a computationally inexpensive but inaccurate computation of M_{DR}. Assume that

$$\begin{aligned} \dot{\theta}(t_f^0) &\approx \dot{\theta}^0(t_f^0) \\ v_r(t_f^0) &\approx v_r^0(t_f^0). \end{aligned} \tag{6.91}$$

Then,

$$M_{DR} = r_0\delta\theta(t_f^0) - \frac{v_\theta^0(t_f^0)}{v_r^0(t_f^0)}\delta r(t_f^0), \tag{6.92}$$

where the quantities r_0 and $v_r^0(t_f^0)$ are known from the nominal trajectory. Then, let

$$\lambda_1(t_f^0) = \frac{-v_\theta^0(t_f^0)}{v_r^0(t_f^0)}, \quad \lambda_2(t_f^0) = r_0, \quad \lambda_i(t_f^0) = 0, i \in [3, 6], \tag{6.93}$$

and integrate the vector equation $\dot{\lambda} = -A^T(t)\lambda$ backward in time until t_1, to obtain

$$\sum_{i=1}^{6} \lambda_i(t_1)\delta x_i(t_1) = -\frac{v_\theta^0(t_f^0)}{v_r^0(t_f^0)}\delta r(t_f^0) + r_0\delta\theta(t_f^0) + 0 + \cdots + 0. \tag{6.94}$$

In (6.94), the left-hand side gives the down-range miss distance due to perturbations:

$$(\delta r(t_1), \delta\theta(t_1), \delta z(t_1), \delta v_r(t_1), \delta v_\theta(t_1), \delta v_z(t_1)) = (\delta x_1(t_1), \ldots, \delta x_6(t_1)). \tag{6.95}$$

The second method to deal with the fact that C depends on the perturbed trajectory is more computationally expensive than the method we just presented but yields a more accurate computation of M_{DR}. As before,

$$M_{DR} = r_0\delta\theta(t_f^0) - \frac{r_0\dot{\theta}(t_f^0)}{v_r(t_f^0)}\delta r(t_f^0). \tag{6.96}$$

The coefficients of $\delta\theta(t_f^0)$ and $\delta r(t_f^0)$ depend on the perturbed final state. Four integrations of the adjoint equation can be performed:

$$\begin{aligned} \dot{\lambda}^{(i)} &= -A^T\lambda^{(i)}, 1 \le i \le 4 \\ \lambda^{(i)}(t_f^0) &= (1, 0, 0, 0, 0, 0, 0); (0, r_0, 0, 0, 0, 0); (0, 0, 0, 1, 0, 0); (0, 0, 0, 0, 1, 0) \\ &= \lambda^{(1)}(t_f^0); \lambda^{(2)}(t_f^0); \lambda^{(3)}(t_f^0); \lambda^{(4)}(t_f^0). \end{aligned} \tag{6.97}$$

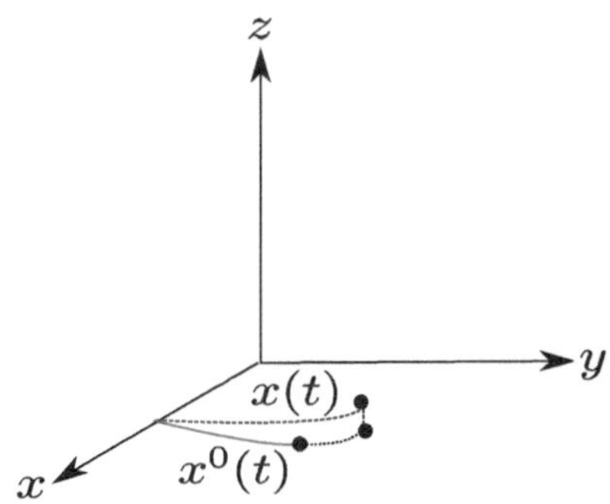

Figure 6.11. Setup for cross-range miss distance analysis.

Recall that $\lambda^T \delta x$ is a constant. The respective results of the four integrations yield

$$\begin{aligned} \delta r(t_f^0) &= r(t_f^0) - r^0(t_f^0 *) = \lambda^{(1)}(t_1)^T \delta x(t_1), \\ r_0 \delta\theta(t_f^0) &= \lambda^{(2)}(t_1)^T \delta x(t_1), \\ \delta v_r(t_f^0) &= v_r(t_f^0) - v_r^0(t_f^0) = \lambda^{(3)}(t_1)^T \delta x(t_1), \\ \delta v_\theta(t_f^0) &= v_\theta(t_f^0) - v_\theta^0(t_f^0) = \lambda^{(4)}(t_1)^T \delta x(t_1). \end{aligned} \tag{6.98}$$

These equations give the quantities necessary to compute M_{DR}.

6.4.3 Cross-Range Miss Distance, M_{CR}

Assuming that z and $\dot{z}$ are small, the out-of-plane miss distance, $z(t_f)$, is given by

$$\begin{aligned} z(t_f) &= z(t_f^0) + \dot{z}(t_f^0)(t_f - t_f^0) + \text{H.O.T.} \\ &= z(t_f^0) + \dot{z}^0(t_f^0)(t_f - t_f^0), \end{aligned} \tag{6.99}$$

as illustrated in Figure 6.11.

In addition, assume that $v_r(t_f^0) \approx v_r^0(t_f^0)$. The cross-range miss distance is then given by the expression

$$M_{CR} = z(t_f) - z^0(t_f^0) = \delta z(t_f^0) + v_z^0(t_f^0) \left(\frac{-\delta r(t_f^0)}{v_r^0(t_f^0)} \right), \tag{6.100}$$

or equivalently,

$$M_{CR} = \delta z(t_f^0) - \frac{v_z^0(t_f^0)}{v_r^0(t_f^0)} \delta r(t_f^0). \tag{6.101}$$

Let us introduce a new adjoint system:

$$\dot{\mu} = -A^T \mu \tag{6.102}$$

with

$$\mu(t_f^0) = \left(-\frac{v_z^0(t_f^0)}{v_r^0(t_f^0)}, 0, 1, 0, 0, 0 \right)^T. \tag{6.103}$$

Integrate (6.102) with (6.103) backward until t_1, to yield

$$M_{CR} = \mu^T(t_1) \delta x(t_1). \tag{6.104}$$

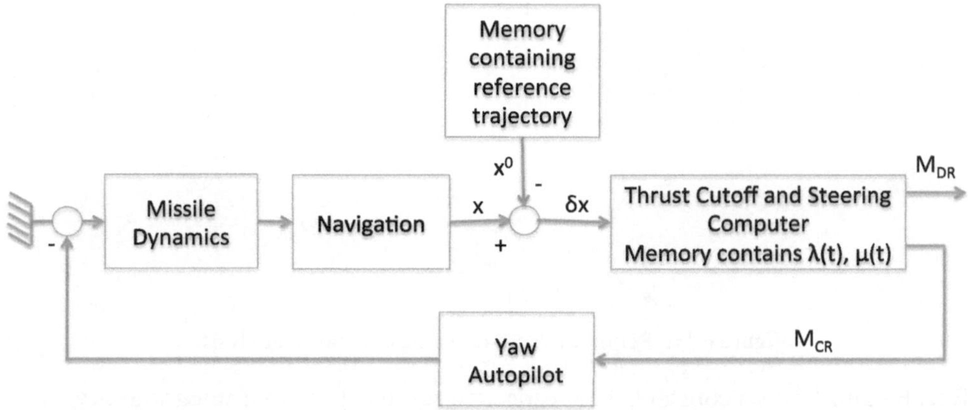

Figure 6.12. Mechanization of guidance scheme.

REMARK 6.1 *Note that here we have* $z^0(t_f^0) = v_z(t_f^0) = 0$. *We have included them in anticipation of the rotating Earth case.*

In conclusion, using the first (computationally inexpensive but less accurate) method to obtain M_{DR}, we have

$$\begin{aligned} M_{DR} &= \lambda^T(t)\delta x(t) \\ M_{CR} &= \mu^T(t)\delta x(t), \end{aligned} \tag{6.105}$$

and both expressions hold for $t \leq t_f^0$. This leads to a possible **mechanization of the guidance scheme**, as shown in block diagram form in Figure 6.12.

In Figure 6.12, the quantities x, x^0, and δx are as follows:

$$\begin{aligned} x &= (r, \theta, z, v_r, v_\theta, v_z)^T, \\ x^0 &= \left(r^0, \theta^0, z^0, v_r^0, v_\theta^0, v_z^0\right)^T, \\ \delta x &= \left(\delta r^0, \delta\theta^0, \delta z^0, \delta v_r^0, \delta v_\theta^0, \delta v_z^0\right)^T. \end{aligned} \tag{6.106}$$

The guidance computer's memory contains the time histories of x^0, λ, and μ. These time histories can be stored as polynomials (using curve fitting), or digitally (using sampling).

The basic idea of the mechanization of the guidance scheme is to use feedback control to drive M_{CR} to zero. Then, cut thrust off when M_{DR} is zero, ensuring a hit.

6.5 Effects of the Earth's Rotation

The effects of the Earth's rotation on ballistic trajectories can cause miss distances of up to 600 kilometers. The Earth's rotation has two basic effects, due to the circumferential motion of the target during the flight, and to the initial tangential velocity of the missile.

In Figure 6.13, $\{O, \hat{x}, \hat{y}, \hat{z}\}$ is a frame fixed at the center of the Earth, the $\hat{z}$-axis is the polar axis of Earth, $\vec{\omega} = \omega\hat{z}$ is the angular velocity of Earth, $\hat{x}$ is determined by the longitude of the powered flight cutoff point, and α is the latitude of the powered

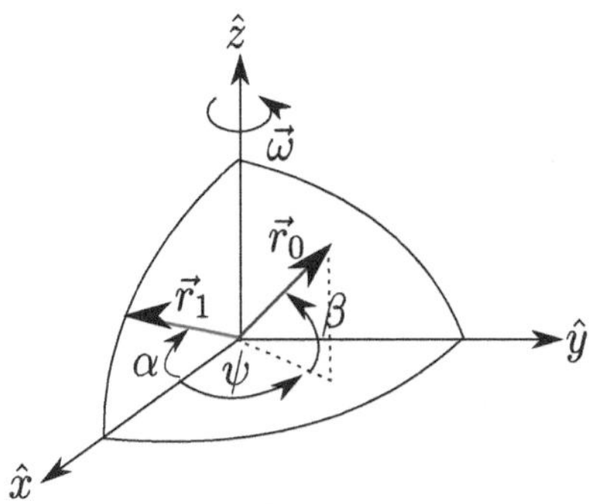

Figure 6.13. Geometry to study the effect of the Earth's rotation on ballistic flight.

flight cutoff point. In addition, $r_1 = r_0 + h$, where r_0 is the radius of the Earth and h is the altitude of the missile above Earth.

Let $t = 0$ at the powered flight cutoff. At $t = 0$, the target is at latitude $\beta = \beta_0$ and longitude $\psi = \psi_1$. During free flight, the target's longitude changes, but its latitude remains the same. Let T be the free flight time. At impact,

$$\psi_0 = \psi_1 + \omega T$$
$$\beta = \beta_0. \tag{6.107}$$

The motion of free flight is still planar with respect to the inertial system but does not appear planar to an observer on Earth. As a consequence, our previous analyses are still valid.

Let $\vec{r}_1$ be the powered flight cutoff position vector and $\vec{r}_2$ be the impact position vector. Then, $|\vec{r}_1| = r_0 + h$ and $|\vec{r}_2| = r_0$. The motion is in the $(\vec{r}_1, \vec{r}_2)$ plane.

Assume that the latitude of the launch is close to the latitude of the powered flight cutoff. The true powered flight cutoff velocity is given by

$$\vec{u} = \vec{v} + \omega r_0 \cos\alpha \hat{y}, \tag{6.108}$$

where the first term on the right-hand side is the relative velocity, and the second is the entrainment velocity, that is, the velocity of the launch site. Let

$$\vec{v} = lv\hat{x} + mv\hat{y} + nv\hat{z} = v\left(l\hat{x} + m\hat{y} + n\hat{z}\right), \tag{6.109}$$

where l, m, and n are direction cosines. Then,

$$\vec{u} = lv\hat{x} + (mv + \omega r_0 \cos\alpha)\hat{y} + nv\hat{z}. \tag{6.110}$$

Also,

$$\vec{r}_1 = r_1 \cos\alpha \hat{x} + r_1 \sin\alpha \hat{z}$$
$$\vec{r}_2 = r_0 \cos\beta \cos\psi \hat{x} + r_0 \cos\beta \sin\psi \hat{y} + r_0 \sin\beta \hat{z}. \tag{6.111}$$

We need to determine equations between l, m, n, and T to ensure a hit, taking the Earth's rotation into account. Evidently,

$$l^2 + m^2 + n^2 = 1. \tag{6.112}$$

Also, the motion must take place in the $(\vec{r}_1, \vec{r}_2)$ plane. This implies that $\vec{u}$ must also lie in this plane, that is,

$$\vec{u} \cdot (\vec{r}_1 \times \vec{r}_2) = 0. \tag{6.113}$$

REMINDER 6.1 *Recall that, given three vectors*

$$\begin{aligned} \vec{a} &= a_x\hat{x} + a_y\hat{y} + a_z\hat{z}, \\ \vec{b} &= b_x\hat{x} + b_y\hat{y} + b_z\hat{z}, \\ \vec{c} &= c_x\hat{x} + c_y\hat{y} + c_z\hat{z}, \end{aligned} \tag{6.114}$$

the cross-product of $\vec{b}$ and $\vec{c}$ is given by

$$\vec{b} \times \vec{c} = \hat{x}(b_yc_z - c_yb_z) + \hat{y}(c_xb_z - b_xc_z) + \hat{z}(b_xc_y - c_xb_y) = X\hat{x} + Y\hat{y} + Z\hat{z}, \tag{6.115}$$

and the box product of the three vectors is given by

$$\vec{a} \cdot (\vec{b} \times \vec{c}) = a_xX + a_yY + a_zZ = \det \begin{bmatrix} a_x & a_y & a_z \\ b_x & b_y & b_z \\ c_x & c_y & c_z \end{bmatrix}. \tag{6.116}$$

*This is called a **box product** because it represents the (signed) volume of the parallelepiped constructed on the three vectors. Three vectors are coplanar if and only if their box product is zero.*

According to Reminder 6.1, (6.110), (6.111), and (6.113) yield

$$\det \begin{bmatrix} lv & mv + \omega r_0 \cos\alpha & nv \\ r_1 \cos\alpha & 0 & r_1 \sin\alpha \\ r_0 \cos\beta \cos\psi & r_0 \cos\beta \sin\psi & r_0 \sin\beta \end{bmatrix} = 0. \tag{6.117}$$

Simplifying by vr_1r_0, we obtain

$$\begin{aligned} &\left(m + \frac{\omega r_0 \cos\alpha}{v}\right)(\cos\beta\cos\psi\sin\alpha - \cos\alpha\sin\beta) \\ &\quad -l\sin\alpha\cos\beta\sin\psi + n\cos\alpha\cos\beta\sin\psi = 0. \end{aligned} \tag{6.118}$$

Now recall the planar (2D) hit equation, derived from $H(r_2, \theta_2, r_1, \theta_1, v_{r1}, v_{\theta 1}) = 0$:

$$\begin{aligned} &\frac{1}{r_2} + \left(\left(\frac{1}{r_1} - \frac{k}{(r_1v_{\theta1})^2}\right)\cos\theta_1 + \frac{v_{r1}}{r_1v_{\theta1}}\sin\theta_1\right)\cos\theta_2 \\ &\quad + \left(\left(\frac{1}{r_1} - \frac{k}{(r_1v_{\theta1})^2}\right)\sin\theta_1 - \frac{v_{r1}}{r_1v_{\theta1}}\cos\theta_1\right)\sin\theta_2 + \frac{k}{(r_1v_{\theta1})^2} = 0. \end{aligned} \tag{6.119}$$

We need to interpret (6.119) in terms of l, m, n, α, and ψ.

Consider the case of polar coordinates, as shown in Figure 6.14. In this case, $\vec{r}_1$ is along $\hat{x}'$, and $\vec{r}_2$ is in the $(\hat{x}', \hat{y}')$ plane, a positive angle ϕ away from $\vec{r}_1$. Therefore, $\vec{u}$ lies in the $(\hat{x}', \hat{y}')$ plane, and

$$\begin{aligned} v_{r1} &= |\vec{u}|\cos\gamma \\ v_{\theta1} &= |\vec{u}|\sin\gamma, \end{aligned} \tag{6.120}$$

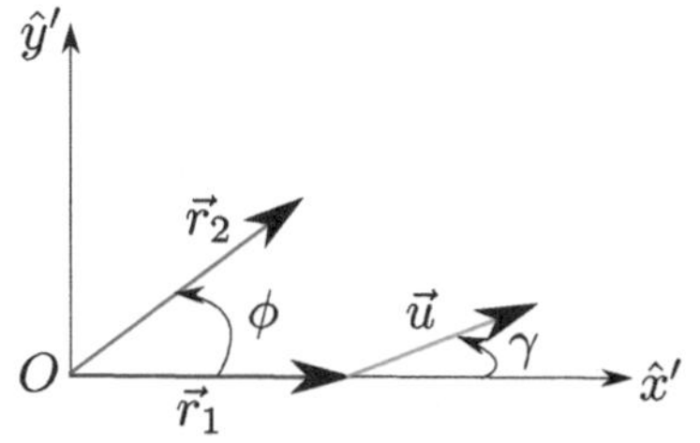

Figure 6.14. Polar coordinates to study the hit equation for ballistic flight considering the Earth's rotation.

where

$$\cos\gamma = \frac{\vec{u}\cdot\vec{r}_1}{|\vec{u}||\vec{r}_1|}. \tag{6.121}$$

For the hit equation, we have

$$\begin{aligned} r_1 &= r_0 + h, \quad \theta_1 = 0, \\ v_{r1} &= |\vec{u}|\cos\gamma, \\ v_{\theta 1} &= |\vec{u}|\sin\gamma, \\ r_2 &= r_0, \quad \theta_2 = \phi, \end{aligned} \tag{6.122}$$

and

$$\frac{1}{r_0} - \left(\frac{1}{r_1}\frac{k}{(r_1 v_{\theta 1})^2}\right)\cos\phi + \frac{v_{r1}}{r_1 v_{\theta 1}}\sin\phi - \frac{k}{(r_1 v_{\theta 1})^2} = 0. \tag{6.123}$$

If α and $r_1 = r_0 + h$ are known, then (6.112), (6.118), and (6.123) are three equations with unknowns l, m, n, and T. T is determined by the fact that, for a Keplerian orbit,

$$r v_\theta = \text{ constant}, \tag{6.124}$$

implying that

$$r^2\dot{\phi} = r_1 v_{\theta 1}. \tag{6.125}$$

Therefore,

$$\mathrm{d}t = \frac{r^2\,\mathrm{d}\phi}{r_1 v_{\theta 1}}. \tag{6.126}$$

Recall, in addition, that $r = r(\phi)$, hence

$$T = \int_0^{\phi} \frac{r^2\,\mathrm{d}\phi}{r_1 v_{\theta 1}}. \tag{6.127}$$

Now, (6.112), (6.118), (6.123), and (6.127) are four algebraic-integral equations for l, m, n, and T, to be solved iteratively using, for example, Newton's method as discussed in Appendix A.4. As an initial condition, the solution for nonrotating Earth can be used.

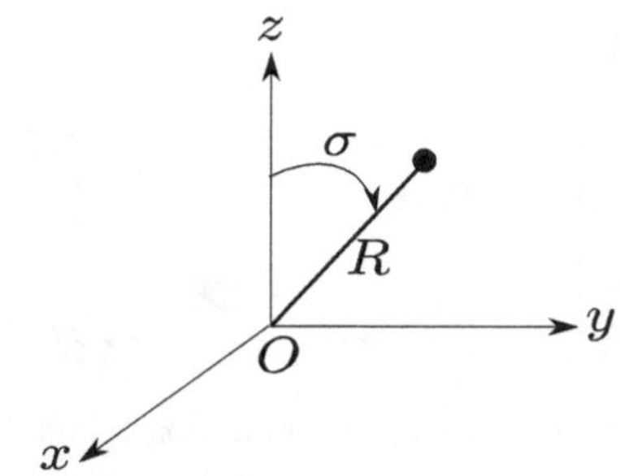

Figure 6.15. Colatitude σ.

REMARK 6.2 *Note that the miss coefficients are similar for the rotating and nonrotating Earth cases.*

REMARK 6.3 *The previous three-dimensional error analysis can be modified for the rotating Earth case by adjoining the in-plane and out-of-plane components of the target velocity, and multiplying by* $(t_f - t_f^0)$ *to obtain* M_{DR} *and* M_{CR}:

$$\begin{aligned} M_{DR} &= r_0 \Delta\theta + v_{DR}(t_f - t_f^0) \\ M_{CR} &= \Delta z + v_{CR}(t_f - t_f^0). \end{aligned} \tag{6.128}$$

6.6 Effects of Earth's Oblateness and Geophysical Uncertainties

Exact computation of effects due to the Earth's oblateness and geophysical uncertainties requires spherical trigonometry and is not attempted here. Instead, we use an approximation that yields an order of magnitude analysis.

Assume that the shape of the Earth is that of an oblate spheroid. The potential of an oblate spheroid, given by potential theory, is

$$V(r, \sigma) = -\frac{mk}{r_e}\left(\frac{r_e}{R} + J\frac{r_e^3}{R^3}\left(\frac{1}{3} - \cos^2\sigma\right) + \frac{8}{35}D\frac{r_e^5}{R^7}P_4(\sigma) + \cdots\right), \tag{6.129}$$

where

$$P_4(\sigma) = \frac{1}{8}(35\cos^4\sigma - 30\cos^2\sigma + 3) \tag{6.130}$$

is the Legendre polynomial of the first kind, $R = \sqrt{r^2 + z^2}$, and J and D are spherical harmonics.

As shown in Figure 6.15, σ is the colatitude. For Earth,

$$\begin{aligned} J &= 1.637 \times 10^{-3} \pm 0.25\% \\ D &= 1.07 \times 10^{-5}. \end{aligned} \tag{6.131}$$

A crude approximation of the effect of J at intercontinental ballistic missile distances is

$$\Delta \approx J r_e \phi, \tag{6.132}$$

where $r_e\phi$ is the arc length along Earth. For our purposes, $\phi \approx 90°$, so $\Delta \approx 10$ miles. As such, the effect of Δ is approximately one-tenth of a mile.

6.6.1 Effects of Other Perturbations

The following table gives the maximum miss distance due to uncertainty in the knowledge of a number of parameters.

Parameter	Max. Miss
k	0.2 miles
r_e	0.06 miles
ω	inches
J	0.02 miles

Additional effects are due to inhomogeneity of the Earth's surface (gravitational anomaly), which yield maximum miss distances on the order of 0.1 mile; the attraction of the Sun and the Moon, which yield maximum miss distances on the order of feet; and relativistic effects, which yield maximum miss distances on the order of less than one inch. All these effects combined yield maximum miss distances of less than half a mile.

6.7 General Solution of Ballistic Guidance Problems

In this section, we remove the restrictive assumptions that have allowed closed form treatment of ballistic guidance problems in Sections 6.1 through 6.6: restricted two-body problem, absence of aerodynamic forces, nonrotating Earth, and point-mass gravity field. The treatment here is a better reflection of practice and is mostly numerical, often resorting to iterative computations that can be initialized using the results of Sections 6.1 through 6.6.

6.7.1 General Framework

Assume that the *dynamics* of the ballistic missile in free flight are described by the general equations

$$\dot{x}(t) = f(x(t), t)$$
$$x(t_0) = x_0, \tag{6.133}$$

where $x(t) \in \mathbb{R}^n$ is the state vector, containing the components of position and velocity, t_0 is the cutoff time, and x_0 contains the cutoff conditions.

Assume that we have a *flight termination condition* of the form

$$g(x(t), t) = 0, \tag{6.134}$$

where $g(.,.)$ is a scalar function of its arguments. Let t_f be the termination time, that is, the first time $t > t_0$ at which condition (6.134) is satisfied, and let $x(t_f)$ be the state vector at that time.

Finally, assume that we are given an m-dimensional *miss vector* of the form

$$h(x(t_f), t_f), \tag{6.135}$$

where $m \leq n$ and the components of (6.135) quantify the miss when flight termination occurs.

6.7.2 Problem Formulation

Within the preceding general framework, we consider the following two problems:

1. **Targeting Problem:** Given t_0, find x_0 such that if $t_f > t_0$ is defined by $g(x(t_f), t_f) = 0$, we have $h(x(t_f), t_f) = 0$.
2. **Miss Analysis Problem:** Given a nominal trajectory that solves the targeting problem, evaluate the miss due to slightly off-nominal cutoff conditions.

Note that in the Targeting Problem, x_0 may be restricted, for example, if some components of position are specified but some components of velocity are unknown. Also, note that the solution of the targeting problem may not be unique, which allows the optimization of a subsidiary criterion such as energy or accuracy.

6.7.3 Examples

In this subsection, we illustrate how several ballistic guidance problems fit within the general framework, and, in each instance, what the dynamics, flight termination condition, and miss vector are.

EXAMPLE 6.4 (Cannonball on Flat Earth) *Consider again the cannonball shot of Example 6.3. In this case, the dynamic order is $n = 4$, and the dynamics are*

$$\begin{aligned}
\dot{x}_1 &= x_3, \\
\dot{x}_2 &= x_4, \\
\dot{x}_3 &= 0, \\
\dot{x}_4 &= -g, \\
(x_1(0), x_2(0)) &= (0, 0), \text{ given}, \\
(x_3(0), x_4(0)) &= (V_1, V_2), \text{ unknown}.
\end{aligned} \tag{6.136}$$

The flight termination condition is

$$g(x(t), t) = x_2(t) = 0. \tag{6.137}$$

Moreover, the miss is a scalar quantity, that is, $m = 1$. Finally, when the flight terminates, the miss is quantified as

$$h(x(t_f), t_f) = x_1(t_f) - R. \tag{6.138}$$

EXAMPLE 6.5 (Ballistic Flight around a Nonrotating Planet) *Consider the ballistic flight treated in Sections 6.1 through 6.6. The dynamic order is $n = 4$. From (6.4), (6.5),* *and* (6.19), *the dynamics are*

$$\begin{aligned}
\dot{r} &= v_r, \\
\dot{\theta} &= \frac{v_\theta}{r}, \\
\dot{v}_r &= \frac{v_\theta^2}{r} - \frac{k}{r^2}, \\
\dot{v}_\theta &= -\frac{v_r v_\theta}{r},
\end{aligned}$$

$$(r(t_0), \theta(t_0), v_r(t_0), v_\theta(t_0)) = (r_1, \theta_1, v_{r_1}, v_{\theta_1}). \tag{6.139}$$

For a target located at polar coordinates $(r_0, 0)$, *the flight termination condition is*

$$g(r, \theta, v_r, v_\theta, t) = r - r_0 = 0. \tag{6.140}$$

Here, also, the miss is scalar, that is, $m = 1$. *From* (6.48), *when the flight terminates, the down-range miss is*

$$h(r(t_f), \theta(t_f), v_r(t_f), v_\theta(t_f), t_f) = -r_0\theta(t_f). \tag{6.141}$$

EXAMPLE 6.6 (Ballistic Flight in Cylindrical Coordinates) *Consider again the ballistic flight treated in Section 6.4, using cylindrical* (r, θ, z) *coordinates to allow for out-of-orbital-plane motions. Assume that the nominal orbit is in the plane* $z = 0$. *The equations of motion are derived in Section 6.4, and the dynamics are of order* $n = 6$:

$$\begin{aligned}
\dot{r} &= v_r, \\
\dot{\theta} &= \frac{v_\theta}{r}, \\
\dot{z} &= v_z, \\
\dot{v}_r &= \frac{v_\theta^2}{r} - \frac{kr}{(r^2 + z^2)^{3/2}}, \\
\dot{v}_\theta &= -\frac{v_r v_\theta}{r}, \\
\dot{v}_z &= \frac{-kz}{(r^2 + z^2)^{3/2}}.
\end{aligned} \tag{6.142}$$

For a target located at cylindrical coordinates $(r, \theta, z) = (r_0, 0, 0)$, *the flight termination condition is*

$$g(r, \theta, z, v_r, v_\theta, v_z, t) = r - r_0 = 0. \tag{6.143}$$

Here the miss vector has two components: a down-range component in the nominal orbital plane and a cross-range component perpendicular to that plane. Hence, $m = 2$ *and*

$$h(r(t_f), \theta(t_f), z(t_f), v_r(t_f), v_\theta(t_f), v_z(t_f), t_f) = \begin{bmatrix} -r_0\theta(t_f) \\ z(t_f) \end{bmatrix}. \tag{6.144}$$

EXAMPLE 6.7 (Ballistic Flight around a Rotating Planet) *Here we write Newton's equations of ballistic motion in a rotating reference frame attached to a rotating planet (for the sake of our example, Earth). We use standard spherical coordinates* (r, θ, ϕ) *where* r *is the distance from O, the center of Earth, to the orbiter,* θ *is the longitude, and* ϕ *is the latitude. Let* ω_E *be the constant spin rate of Earth.*

We use the following reference frames:

1. $\{O, \hat{I}, \hat{J}, \hat{K}\}$*, inertially fixed, with the* $\hat{K}$*-axis along the spin axis of Earth*
2. $\{O, \hat{x}, \hat{y}, \hat{z}\}$*, obtained from* $\{O, \hat{I}, \hat{J}, \hat{K}\}$ *by a rotation about* $\hat{K}$ *of magnitude* $\omega_E t + \theta$
3. $\{O, \hat{\imath}, \hat{\jmath}, \hat{k}\}$*, obtained from* $\{O, \hat{x}, \hat{y}, \hat{z}\}$ *by a rotation about* $\hat{y}$ *of magnitude* $-\phi$

With these reference frames, the position vector of the orbiter takes the simple form

$$\vec{r} = r\hat{\imath}. \tag{6.145}$$

Using the formalism of elementary rotation matrices to transform vectrices, we have

$$\begin{bmatrix} \hat{x} \\ \hat{y} \\ \hat{z} \end{bmatrix} = R_3(\omega_E t + \theta) \begin{bmatrix} \hat{I} \\ \hat{J} \\ \hat{K} \end{bmatrix} = \begin{bmatrix} \cos(\omega_E t + \theta) & \sin(\omega_E t + \theta) & 0 \\ -\sin(\omega_E t + \theta) & \cos(\omega_E t + \theta) & 0 \\ 0 & 0 & 1 \end{bmatrix} \begin{bmatrix} \hat{I} \\ \hat{J} \\ \hat{K} \end{bmatrix} \tag{6.146}$$

$$\begin{bmatrix} \hat{\imath} \\ \hat{\jmath} \\ \hat{k} \end{bmatrix} = R_2(-\phi) \begin{bmatrix} \hat{x} \\ \hat{y} \\ \hat{z} \end{bmatrix} = \begin{bmatrix} \cos\phi & 0 & \sin\phi \\ 0 & 1 & 0 \\ -\sin\phi & 0 & \cos\phi \end{bmatrix} \begin{bmatrix} \hat{x} \\ \hat{y} \\ \hat{z} \end{bmatrix}. \tag{6.147}$$

Because the angular velocity vector for a sequence of rotations is the sum of the angular velocity vectors of the individual rotations, the angular velocity vector is

$$\vec{\omega} = (\omega_E + \dot{\theta})\hat{z} - \dot{\phi}\hat{y}. \tag{6.148}$$

Inverting (6.147)*, we obtain*

$$\hat{z} = \sin\phi\hat{\imath} + \cos\phi\hat{k}. \tag{6.149}$$

Hence, the angular velocity vector, expressed in the $\{\hat{\imath}, \hat{\jmath}, \hat{k}\}$ *basis, is*

$$\vec{\omega} = (\omega_E + \dot{\theta})\sin\phi\hat{\imath} - \dot{\phi}\hat{\jmath} + (\omega_E + \dot{\theta})\cos\phi\hat{k}. \tag{6.150}$$

Using (6.145) *and* (6.150)*, we evaluate the velocity of the orbiter as*

$$\begin{aligned} \vec{v} &= \overset{\circ}{\vec{r}} + \vec{\omega} \times \vec{r} \\ &= \dot{r}\hat{\imath} + r(\omega_E + \dot{\theta})\cos\phi\hat{\jmath} + r\dot{\phi}\hat{k}, \end{aligned} \tag{6.151}$$

where the $\overset{\circ}{\vec{r}}$ *overset indicates the relative derivative of* $\vec{r}$*, and the term* $\vec{\omega} \times \vec{r}$ *is called the entrainment derivative. Both terms are needed to compute the derivative of a vector in a noninertial frame.*

Using (6.150) *and* (6.151)*, we evaluate the acceleration of the orbiter as*

$$\vec{a} = \overset{\circ}{\vec{v}} + \vec{\omega} \times \vec{v}, \tag{6.152}$$

which, after collecting terms, yields

$$\begin{aligned}\vec{a} &= \left(\ddot{r} - r\dot{\phi}^2 - r(\omega_E + \dot{\theta})^2\cos^2\phi\right)\hat{\imath} \\ &\quad + \left(r\ddot{\theta}\cos\phi + 2\dot{r}(\omega_E + \dot{\theta})\cos\phi - 2r\dot{\phi}(\omega_E + \dot{\theta})\sin\phi\right)\hat{\jmath} \\ &\quad + \left(r\ddot{\phi} + 2\dot{r}\dot{\phi} + r(\omega_E + \dot{\theta})^2\sin\phi\cos\phi\right)\hat{k}. \end{aligned} \tag{6.153}$$

For orbital motion, the force acting on the orbiter is purely radial, of the form

$$\vec{F} = -\frac{km}{r^2}\hat{\imath}, \tag{6.154}$$

where k is the gravitational constant of Earth.

Equations (6.153) *and* (6.154) *yield Newton's equations of motion:*

$$\begin{aligned}\ddot{r} &= r\dot{\phi}^2 + r(\omega_E + \dot{\theta})^2\cos^2\phi - \frac{k}{r^2} \\ \ddot{\theta} &= -\frac{2\dot{r}(\omega_E + \dot{\theta})}{r} + 2\dot{\phi}(\omega_E + \dot{\theta})\tan\phi \\ \ddot{\phi} &= -\frac{2\dot{r}\dot{\phi}}{r} - (\omega_E + \dot{\theta})^2\sin\phi\cos\phi. \end{aligned} \tag{6.155}$$

These equations can be written in the standard form (6.133)*, as follows. The dynamic order is* $n = 6$*. Define the state vector to be* $x = (r, \dot{r}, \theta, \dot{\theta}, \phi, \dot{\phi})^T \in \mathbb{R}^6$*. Then, we have*

$$\begin{bmatrix}\dot{r} \\ \ddot{r} \\ \dot{\theta} \\ \ddot{\theta} \\ \dot{\phi} \\ \ddot{\phi}\end{bmatrix} = \begin{bmatrix}\dot{r} \\ r\dot{\phi}^2 + r(\omega_E + \dot{\theta})^2\cos^2\phi - \frac{k}{r^2} \\ \dot{\theta} \\ -\frac{2\dot{r}(\omega_E+\dot{\theta})}{r} + 2\dot{\phi}(\omega_E + \dot{\theta})\tan\phi \\ \dot{\phi} \\ -\frac{2\dot{r}\dot{\phi}}{r} - (\omega_E + \dot{\theta})^2\sin\phi\cos\phi\end{bmatrix}. \tag{6.156}$$

For a target located at spherical coordinates (r_0, θ_0, ϕ_0)*, the flight termination condition is*

$$g(r, \dot{r}, \theta, \dot{\theta}, \phi, \dot{\phi}, t) = r - r_0 = 0. \tag{6.157}$$

The miss vector has two components: a meridial miss along the meridian of the target, and a parallel miss along the parallel of the target:

$$h(r(t_f), \dot{r}(t_f), \theta(t_f), \dot{\theta}(t_f), \phi(t_f), \dot{\phi}(t_f), t_f) = \begin{bmatrix} r_0(\theta(t_f) - \theta_0) \\ r_0\cos\theta_0(\phi(t_f) - \phi_0)\end{bmatrix}. \tag{6.158}$$

6.7.4 Targeting

Consider (6.133) and let $x(t; x_0, t_0)$ denote the solution originating from x_0 at time t_0. The requirement to hit the target can be formulated as simultaneously satisfying

$$\begin{aligned} g(x(t_f; x_0, t_0), t_f) &= 0 \\ h(x(t_f; x_0, t_0), t_f) &= 0, \end{aligned} \tag{6.159}$$

which can be viewed as a system of $m + 1$ equations for the $n + 1$ unknowns (x_0, t_f). Note that, for given values of (x_0, t_f), evaluating the left-hand side of (6.159) requires integrating the differential equation (6.133). Despite this complexity, however, (6.159) can be solved using Newton's iteration as presented in Appendix A.4 for solving nonlinear algebraic equations. Conceptually, the development is as follows.

Assume that at iteration k, the candidates (x_0^k, t_f^k) do not satisfy (6.159), and denote $(g(k), h(k))$ the left-hand sides of (6.159) evaluated at the candidates. Update the candidates by (yet to be determined) increments $(\delta x_0^k, \delta t_f^k)$ and assume that, to first order, the updated candidates satisfy (6.159). Therefore,

$$\begin{bmatrix} g(k) \\ h(k) \end{bmatrix} + \begin{bmatrix} J_{xx}(k) & J_{xt}(k) \\ J_{tx}(k) & J_{tt}(k) \end{bmatrix} \begin{bmatrix} \delta x_0^k \\ \delta t_f^k \end{bmatrix} = \begin{bmatrix} 0 \\ 0 \end{bmatrix}, \tag{6.160}$$

where the matrices $J_{\alpha\beta}(k)$ are matrices of partial derivatives of the left-hand sides of (6.159) with respect to (x_0, t_f) evaluated at the candidates. Equation (6.160) is a system of $m+1$ linear algebraic equations in the $n+1$ unknowns $(\delta x_0^k, \delta t_f^k)$. If this system can be solved, the candidates are incremented and the iteration proceeds until convergence.

The matrices of partial derivatives in (6.160) can be evaluated as follows. First, linearize (6.133) around $x(t; x_0^k, t_0)$ to obtain

$$\begin{aligned} \delta \dot{x}^k(t) &= \left(\frac{\partial f}{\partial x}\right)^T_{x^k(t)} \delta x^k(t) = A^k(t)\delta x^k(t) \\ \delta x^k(t_0) &= \delta x_0^k. \end{aligned} \tag{6.161}$$

Write the solution of the homogeneous linear differential equation (6.161) as

$$\delta x^k(t) = \Phi^k(t, t_0)\delta x_0^k, \tag{6.162}$$

where $\Phi^k(t, t_0)$ is the state transition matrix associated with the state matrix $A^k(t)$. Owing to increments $(\delta x_0^k, \delta t_f^k)$, the left-hand side of (6.159) (line 1) has a first-order variation

$$\delta g\left(x(t_f^k; x_0^k, t_0)\right) = \left(\frac{\partial g}{\partial x}\right)^T_k \delta x^k + \left(\frac{\partial g}{\partial t}\right)_k \delta t_f^k. \tag{6.163}$$

Now,

$$\delta x^k = \left(\frac{\partial x}{\partial t}\right)_{t_f^k} \delta t_f^k + \left(\frac{\partial x}{\partial x_0}\right)_{x_0^k} \delta x_0^k. \tag{6.164}$$

Hence,

$$\delta g\left(x(t_f^k; x_0^k, t_0)\right) = \left(\frac{\partial g}{\partial x}\right)^T_k \Phi^k(t_f^k, t_0)\delta x_0^k + \left[\left(\frac{\partial g}{\partial x}\right)^T_k f(x^k(t_f^k), t_f^k) + \left(\frac{\partial g}{\partial t}\right)_k\right] \delta t_f^k. \tag{6.165}$$

Using a similar argument, we have

$$\delta h\left(x(t_f^k; x_0^k, t_0)\right) = \left(\frac{\partial h}{\partial x}\right)^T_k \Phi^k(t_f^k, t_0)\delta x_0^k + \left[\left(\frac{\partial h}{\partial x}\right)^T_k f(x^k(t_f^k), t_f^k) + \left(\frac{\partial h}{\partial t}\right)_k\right] \delta t_f^k. \tag{6.166}$$

Therefore, the matrices of partial derivatives in (6.160) are

$$
\begin{aligned}
J_{xx}(k) &= \left(\frac{\partial g}{\partial x}\right)_k^T \Phi^k(t_f^k, t_0),\\
J_{xt}(k) &= \left(\frac{\partial g}{\partial x}\right)_k^T f(x^k(t_f^k), t_f^k) + \left(\frac{\partial g}{\partial t}\right)_k,\\
J_{tx}(k) &= \left(\frac{\partial h}{\partial x}\right)_k^T \Phi^k(t_f^k, t_0),\\
J_{tt}(k) &= \left(\frac{\partial h}{\partial x}\right)_k^T f(x^k(t_f^k), t_f^k) + \left(\frac{\partial h}{\partial t}\right)_k.
\end{aligned}
\tag{6.167}
$$

Equations (6.160) should then be solved for $(\delta x_0^k, \delta t_f^k)$ to yield the next iterate according to

$$
\begin{bmatrix} x_0^{k+1} \\ t_f^{k+1} \end{bmatrix} = \begin{bmatrix} x_0^k \\ t_f^k \end{bmatrix} + \begin{bmatrix} \delta x_0^k \\ \delta t_f^k \end{bmatrix}. \tag{6.168}
$$

6.7.5 Miss Analysis

In this subsection, we quantify the miss due to off-nominal cutoff conditions. Assume that we have a nominal solution of (6.133), $x^0(t)$, that hits the target at the nominal termination time t_f^0, that is,

$$
\begin{aligned}
\dot{x}^0(t) &= f(x^0(t), t),\\
x^0(t_0) &= x_0^0,\\
g\left(x^0(t_f^0), t_f^0\right) &= 0,\\
h\left(x^0(t_f^0), t_f^0\right) &= 0.
\end{aligned}
\tag{6.169}
$$

Assume that the cutoff conditions are perturbed away from x_0^0, which generally implies perturbations of both the solution and the termination time. The question is then to quantify the miss distance, that is, quantify $h(x(t_f), t_f)$, where $x(t)$ is the perturbed solution and t_f is the perturbed termination time.

We first evaluate the perturbation in termination time. To do so, we linearize (6.169) around the nominal solution. Letting $\delta x(t) = x(t) - x^0(t)$, we obtain to first order

$$
\begin{aligned}
\delta\dot{x}(t) &= \left(\frac{\partial f}{\partial x}\right)_0^T \delta x(t) = A(t)\delta x(t)\\
\delta x(t_0) &= \delta x_0.
\end{aligned}
\tag{6.170}
$$

The perturbed termination time satisfies

$$
g\left(x(t_f), t_f\right) = 0. \tag{6.171}
$$

A first-order expansion of the left-hand side of (6.171) around nominal conditions yields

$$g\left(x(t_f), t_f\right) = g\left(x^0(t_f^0), t_f^0\right) + \left(\frac{\partial g}{\partial x}\right)_0^T \left(x(t_f) - x^0(t_f^0)\right) + \left(\frac{\partial g}{\partial t}\right)_0 (t_f - t_f^0). \tag{6.172}$$

Note that the left-hand side of (6.172) is zero because of (6.171) and that the first term in the right-hand side is also zero because of (6.169) (line 3). Hence,

$$\left(\frac{\partial g}{\partial x}\right)_0^T \left(x(t_f) - x^0(t_f^0)\right) + \left(\frac{\partial g}{\partial t}\right)_0 (t_f - t_f^0) = 0. \tag{6.173}$$

Also, a first-order expansion of $x(t_f)$ around t_f^0 yields

$$x(t_f) = x(t_f^0) + \dot{x}(t_f^0)(t_f - t_f^0). \tag{6.174}$$

Substituting in (6.173), we obtain

$$\left(\frac{\partial g}{\partial x}\right)_0^T \left(x(t_f^0) + \dot{x}(t_f^0)(t_f - t_f^0) - x^0(t_f^0)\right) + \left(\frac{\partial g}{\partial t}\right)_0 (t_f - t_f^0) = 0. \tag{6.175}$$

Using the familiar notation

$$\delta x(t) = x(t) - x^0(t), \tag{6.176}$$

we write (6.175) as

$$\left(\frac{\partial g}{\partial x}\right)_0^T \delta x(t_f^0) + \left[\left(\frac{\partial g}{\partial x}\right)_0^T \dot{x}(t_f^0) + \left(\frac{\partial g}{\partial t}\right)_0\right](t_f - t_f^0) = 0, \tag{6.177}$$

which can be solved for the perturbation of termination time as

$$t_f - t_f^0 = -\frac{\left(\frac{\partial g}{\partial x}\right)_0^T \delta x(t_f^0)}{\left(\frac{\partial g}{\partial x}\right)_0^T \dot{x}(t_f^0) + \left(\frac{\partial g}{\partial t}\right)_0}. \tag{6.178}$$

We can now evaluate the miss:

$$M = h\left(x(t_f), t_f\right). \tag{6.179}$$

A first-order expansion of the right-hand side of (6.179) around nominal conditions yields

$$M = h\left(x^0(t_f^0), t_f^0\right) + \left(\frac{\partial h}{\partial x}\right)_0^T \left(x(t_f) - x^0(t_f^0)\right) + \left(\frac{\partial h}{\partial t}\right)_0 (t_f - t_f^0). \tag{6.180}$$

Now, the first term of the right-hand side of (6.180) is zero because of (6.169) (line 4). Moreover, using (6.174), we obtain

$$M = \left(\frac{\partial h}{\partial x}\right)_0^T \left(x(t_f^0) + \dot{x}(t_f^0)(t_f - t_f^0) - x^0(t_f^0)\right) + \left(\frac{\partial h}{\partial t}\right)_0 (t_f - t_f^0), \tag{6.181}$$

which, using (6.176), can be rewritten as

$$M = \left(\frac{\partial h}{\partial x}\right)_0^T \delta x(t_f^0) + \left[\left(\frac{\partial h}{\partial x}\right)_0^T \dot{x}(t_f^0) + \left(\frac{\partial h}{\partial t}\right)_0\right](t_f - t_f^0). \tag{6.182}$$

Finally, using (6.178), we obtain

$$M = \frac{\left[\left(\frac{\partial g}{\partial x}\right)_0^T \dot{x}(t_f^0) + \left(\frac{\partial g}{\partial t}\right)_0\right]\left(\frac{\partial h}{\partial x}\right)_0^T - \left[\left(\frac{\partial h}{\partial x}\right)_0^T \dot{x}(t_f^0) + \left(\frac{\partial h}{\partial t}\right)_0\right]\left(\frac{\partial g}{\partial x}\right)_0^T}{\left[\left(\frac{\partial g}{\partial x}\right)_0^T \dot{x}(t_f^0) + \left(\frac{\partial g}{\partial t}\right)_0\right]} \delta x(t_f^0). \tag{6.183}$$

This expression has the form $M = C\delta x(t_f^0)$ (a perturbation of the initial conditions). The difficulty here is that C depends on $\dot{x}(t_f^0)$.

An inexpensive way of overcoming this difficulty consists of approximating x by x^0. We then obtain

$$C \approx \frac{\left[\left(\frac{\partial g}{\partial x}\right)_0^T \dot{x}^0(t_f^0) + \left(\frac{\partial g}{\partial t}\right)_0\right]\left(\frac{\partial h}{\partial x}\right)_0^T - \left[\left(\frac{\partial h}{\partial x}\right)_0^T \dot{x}^0(t_f^0) + \left(\frac{\partial h}{\partial t}\right)_0\right]\left(\frac{\partial g}{\partial x}\right)_0^T}{\left[\left(\frac{\partial g}{\partial x}\right)_0^T \dot{x}^0(t_f^0) + \left(\frac{\partial g}{\partial t}\right)_0\right]}, \tag{6.184}$$

where everything is known from the nominal trajectory.

Consider the adjoint system

$$\begin{aligned} \dot{p} &= -A^T p \\ p(t_f^0) &= C^T. \end{aligned} \tag{6.185}$$

Then,

$$M = p^T(t)\delta x(t), \tag{6.186}$$

where $p(t)$ contains the time history of miss coefficients. Differentiating (6.186) yields

$$\begin{aligned} \dot{M} &= \dot{p}^T \delta x + p^T \delta \dot{x} \\ &= p^T B \delta u. \end{aligned} \tag{6.187}$$

If we enforce exponential decay of the miss distance through midcourse corrections,

$$\dot{M} = -\frac{1}{T} M, \tag{6.188}$$

where $T > 0$ is a chosen time constant, then we must have

$$p^T B \delta u = \frac{p^T \delta x}{T}, \tag{6.189}$$

so that the midcourse corrections, δu, are given by

$$\delta u = -\frac{(p^T B)^{-1} p^T \delta x}{T}. \tag{6.190}$$

A more expensive way of overcoming the difficulty that (6.183) depends on $\dot{x}(t_f^0)$ is as follows. Consider the adjoint system:

$$\begin{aligned} \dot{\psi}(t) &= -A^T(t)\psi(t) \\ \psi(t_f^0) &= I. \end{aligned} \tag{6.191}$$

Then,

$$\psi^T(t)\delta x(t) = \text{constant}. \tag{6.192}$$

Indeed,

$$\begin{aligned}\frac{\mathrm{d}}{\mathrm{d}t}(\psi^T(t)\delta x(t)) &= \dot{\psi}^T(t)\delta x(t) + \psi^T(t)\delta\dot{x}(t) \\ &= -\psi^T(t)A(t)\delta x(t) + \psi^T(t)A(t)\delta x(t) \\ &= 0. \end{aligned} \tag{6.193}$$

Hence,

$$\psi^T(t)\delta x(t) = \delta x(t_f^0), \tag{6.194}$$

which gives the last factor of M in (6.183). Also,

$$\begin{aligned} x(t_f^0) &= x^0(t_f^0) + \delta x(t_f^0) \\ &= x^0(t_f^0) + \psi^T(t)\delta x(t), \end{aligned} \tag{6.195}$$

and

$$\begin{aligned} \dot{x}(t_f^0) &= f(x(t_f^0), t_f^0) \\ &= f(x^0(t_f^0) + \psi^T(t)\delta x(t), t_f^0). \end{aligned} \tag{6.196}$$

Therefore, given a nominal trajectory $x^0(t)$, t_f^0, $A(t)$, (6.194) and (6.196) allow us to compute M in terms of $\delta x(t)$, the current excursion with respect to nominal.

6.8 Summary of Key Results

The key results in Chapter 6 are as follows:

1. Equation (6.18), which gives the orbit in the restricted two-body problem
2. Equation (6.30), which is the hit equation that cutoff conditions must satisfy to hit a target located at polar coordinates $(r, \theta) = (r_0, 0)$
3. Equations (6.53), (6.54), and (6.55), which provide the miss coefficients and expected square miss in planar ballistic guidance
4. Equations (6.71), which provide the orbit, allowing for out-of-plane motion
5. Equation (6.82), which provides the duration of a perturbed ballistic orbit
6. Equation (6.85), which provides the down-range miss
7. Equation (6.101), which provides the cross-range miss
8. Equations (6.112), (6.118), (6.123), and (6.127), which account for Earth's rotation in ballistic guidance
9. Equations (6.160) and (6.167), which allow targeting in general ballistic guidance problems
10. Equation (6.183), which allows miss analysis in general ballistic guidance problems

6.9 Bibliographic Notes for Further Reading

Our presentation of ballistic guidance closely follows [38], which is out of print. For further reading on the restricted two-body problem, the reader may consult [18] and [19]. Lagrange's equations of motion are covered in [30]. The representation of a sequence of rotations using a product of orthogonal matrices (such as in Example 6.7) is covered in [18].

Ballistic missile technology was developed and first used during World War II [33]. Subsequently, it was used to establish space programs in the United States, the USSR, the European Union, China, and India ([33], [76], [47]). The sobering prospect of intercontinental ballistic missiles (ICBMs) carrying nuclear warheads has been a major driver of international diplomacy [56].

6.10 Homework Problems

For this chapter and all following, the problems may require numerical solutions and computer programming. All problems for Chapter 6 relate to the following scenario, **Accurate Ballistic Targeting**.

The purpose of this homework set is to compute the cutoff velocity that achieves the most accurate hit for a missile. The assumptions are as follows:

1. Perfectly spherical, nonrotating planet
2. Point mass missile, subject to no lift, but subject to atmospheric drag
3. Exponential atmosphere

Under these assumptions, the motion is planar, and the equations of motion are, in vector form,

$$\begin{aligned}
\dot{\vec{r}} &= \vec{v} \\
\dot{\vec{v}} &= \vec{g}(\vec{r}) + \vec{d}(\vec{v}) \\
\vec{g}(\vec{r}) &= -k\frac{\vec{r}}{r^3} \\
\vec{d}(\vec{v}) &= -\eta(r)v\vec{v} \\
\eta(r) &= \eta_0 e^{-\alpha(r-R)} \\
r &\geq R, \qquad (6.197)
\end{aligned}$$

where

1. $\vec{r}$ is the position vector
2. $\vec{v}$ is the velocity vector
3. $\vec{g}(\vec{r})$ is the gravity vector
4. $\vec{d}(\vec{v})$ is the drag vector
5. $r = \sqrt{\vec{r} \cdot \vec{r}}$ is the magnitude of $\vec{r}$
6. $v = \sqrt{\vec{v} \cdot \vec{v}}$ is the magnitude of $\vec{v}$
7. k is the gravitational constant of the planet under consideration
8. $\eta(r)$ is the drag factor
9. η_0 is the drag factor on the planet's surface
10. α is the drag decay factor
11. R is the radius of the planet

In your computations, use the following numerical values:

$$k = 4.305 \times 10^{13} \text{ m}^3/\text{s}^2,$$
$$R = 3.38 \times 10^6 \text{ m},$$
$$\eta_0 = 4.69 \times 10^{-7} \text{ 1/m},$$
$$\alpha = 3.609 \times 10^{-5} \text{ 1/m}. \tag{6.198}$$

PROBLEM 6.1 *Obtain the equations of motion in terms of the polar coordinates, (r, θ), and the polar components of velocity, (v_r, v_θ), in nondimensional form.*

PROBLEM 6.2 *Assume that cutoff occurs exactly at the known polar coordinates $(r_0, \theta_0) = (R, 135^\circ)$. Solve the targeting problem for a target located at $(r, \theta) = (R, 0)$. In other words, in the plane (v_r, v_θ), determine the locus of cutoff velocities that achieve a hit.*

PROBLEM 6.3 *Assume that, owing to navigation errors, the previous cutoff velocities are affected by uncertainties that can be modeled as zero-mean Gaussian with known covariance:*

$$P_{v_r v_\theta} = \begin{bmatrix} 10^3 & 0 \\ 0 & 10^3 \end{bmatrix} \text{ m}^2/\text{s}^2. \tag{6.199}$$

For each combination of cutoff velocities that achieve a hit, compute the expected value of the square of the down-range miss.

PROBLEM 6.4 *Using the result of Problem 6.3, obtain the combination of cutoff velocities that yields the most accurate hit. Plot the associated trajectory in the Cartesian plane. Would you recommend using this trajectory in practice?*

7 Midcourse Guidance

In this chapter, we present the fundamentals used in the analysis and design of midcourse guidance systems. These systems are predicated on the following assumptions:

1. In mission planning, a set of nominal trajectories that meet mission specifications is determined.
2. During flight, corrections are continually applied to the trajectory to return it to nominal. This sustained application of control authority is the distinguishing feature of **midcourse guidance**.

We consider three methods for midcourse guidance. In the first method, the set of nominal trajectories is parametrized by initial position and initial time, which specify the required velocity. This leads to the formalism of **velocity-to-be-gained guidance**, also known as **Q-guidance**. In the second method, a single nominal trajectory is determined and control is applied to return the trajectory to it. This leads to the formalism of **state feedback guidance**, also known as **Delta-guidance**. The third method combines state feedback guidance with navigation in that it uses Delta-guidance based on an estimate of the state vector rather than the true state vector.

Midcourse guidance is related to work presented in the previous two chapters. Indeed, in Chapter 5, the primary purpose of homing is to come close to a target – in other words, homing is a "final-value" problem. However, constant bearing guidance can be viewed as a form of midcourse guidance where the nominal trajectory is defined by $\dot{\beta} = 0$. Also, in Chapter 6, in the thrusting phase of two-dimensional ballistic guidance, we do not correct the trajectory to return to some nominal path, but wait until the hit equation is satisfied, then turn the thrusters off. However, in the three-dimensional scenario, we correct M_{CR}, the cross-range miss.

Section 7.1 describes velocity-to-be-gained guidance, Section 7.2 describes guidance by state feedback, and Section 7.3 considers combined navigation and guidance. Sections 7.4, 7.5, and 7.6 present a summary of the key results in the chapter, bibliographic notes for further reading, and homework problems, respectively.

7.1 Velocity-to-Be-Gained Guidance

Consider the equations of motion for a spacecraft:

$$\begin{aligned} \dot{r} &= v \\ \dot{v} &= g(r, t) + a_T, \end{aligned} \tag{7.1}$$

where $r \in \mathbb{R}^3$ contains the Cartesian coordinates of the spacecraft, $v \in \mathbb{R}^3$ contains the components of the spacecraft's velocity vector, $g \in \mathbb{R}^3$ is the local gravity vector, and $a_T \in \mathbb{R}^3$, the thrust acceleration, is the control input.

Assume that for every initial condition r_0 and t_0, there exists a unique $v_r(r_0, t_0)$ such that, starting from initial position r_0, $v_r(r_0, t_0)$ at initial time t_0 and falling freely (i.e., $a_T = 0$), the mission specifications are met. An example of mission specification could be to pass through position r_f at a given time t_f.

REMARK 7.1 *The "required velocity" field $v_r(r, t)$ must be determined before flight; this is a targeting problem. This is similar to the derivation of the hit equation in Chapter 6, which was also a targeting problem.*

In this section, let us assume that the targeting problem has been solved. The main idea of this section is to choose the thrust acceleration, a_T, to drive $v_r - v$ to zero. For this purpose, let us define the **velocity-to-be-gained**, v_d, as

$$v_d = v_r - v. \tag{7.2}$$

Let us set the desired dynamics for v_d as

$$\dot{v}_d(t) + \frac{1}{T(t)} v_d(t) = 0, \ \ T(t) > 0. \tag{7.3}$$

This implies that

$$v_d(t) = \exp\left(-\int_0^t \frac{\mathrm{d}\tau}{T(\tau)}\right) v_d(0). \tag{7.4}$$

Also notice that

$$\frac{\mathrm{d}}{\mathrm{d}t}||v_d||^2 = \frac{\mathrm{d}}{\mathrm{d}t}(v_d^T v_d) = 2v_d^T \dot{v}_d = -\frac{1}{T(t)} v_d^T v_d < 0. \tag{7.5}$$

Equations (7.3) and (7.2) imply that

$$(\dot{v}_r - \dot{v}) + \frac{1}{T}(v_r - v) = 0. \tag{7.6}$$

Also note that, because $v_r = v_r(r, t)$, we have

$$\dot{v}_r = \left(\frac{\partial v_r}{\partial r}\right)^T v + \frac{\partial v_r}{\partial t}. \tag{7.7}$$

Equations (7.6) and (7.1) (line 2) imply that

$$\dot{v}_r - g - a_T + \frac{1}{T}(v_r - v) = 0. \tag{7.8}$$

Let us define

$$b = \dot{v}_r - g. \tag{7.9}$$

Then,

$$a_T = b + \frac{1}{T} v_d. \tag{7.10}$$

REMARK 7.2 *Note that* (7.3) *means that* v_d *and* $\dot{v}_d$ *are collinear, which implies that* $\dot{v}_d \times v_d = 0$. *For this reason, velocity-to-be-gained guidance is sometimes called* ***cross-product guidance****. Also, in* (7.7), *the* 3×3 *matrix* $\left(\frac{\partial v_r}{\partial r}\right)^T$ *is often noted* Q, *hence the name* ***Q-guidance*** *for the guidance law* (7.10).

We consider two versions of the velocity-to-be-gained guidance law as follows.

7.1.1 Velocity-to-Be-Gained Guidance with Unlimited Thrust

In this case, we assume that $T(t)$ is a constant. Then, (7.4) implies that

$$v_d(t) = e^{-\frac{t}{T}} v_d(0), \tag{7.11}$$

which decays to 0 as $t \to \infty$. The guidance law is, then, given r, v, v_r, and T,

$$\begin{aligned} v_d &= v_r - v, \\ b &= \dot{v}_r - g = \left(\frac{\partial v_r}{\partial r}\right)^T v + \frac{\partial v_r}{\partial t} - g(r,t), \\ a_T &= b + \frac{1}{T} v_d. \end{aligned} \tag{7.12}$$

REMARK 7.3 *This guidance law may require a large* a_T, *if, for instance,* v_d *is large, or if* v_d *is aligned with* b.

7.1.2 Velocity-to-Be-Gained Guidance with Limited Thrust

Here we assume that

$$a_T^T a_T = a^2(t), \tag{7.13}$$

where $a^2(t)$ is a known function of time (typically constant) representing the thrusting capability of the engines. The strategy is then to choose T such that a_T in (7.10) satisfies (7.13), that is, use the maximum thrust available to implement (7.3).

Equations (7.13) and (7.10) imply that

$$b^T b + \frac{1}{T^2} v_d^T v_d + \frac{2}{T} b^T v_d = a^2. \tag{7.14}$$

Define

$$b^2 = b^T b \tag{7.15}$$

$$v_d^2 = v_d^T v_d. \tag{7.16}$$

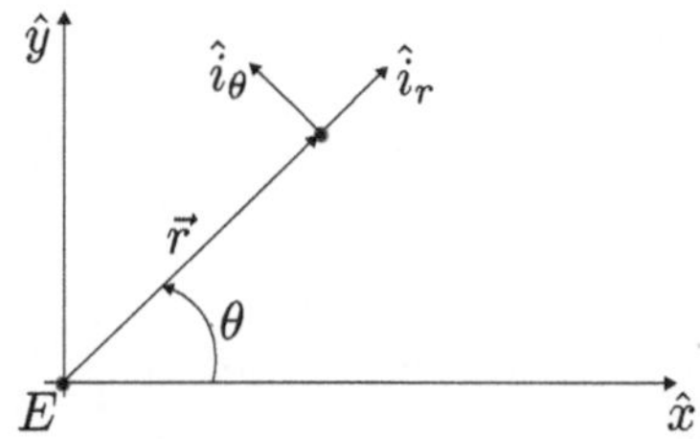

Figure 7.1. Polar coordinates representation for ascent/descent into a circular orbit.

Because (7.14) is a quadratic equation for $1/T$, its only nonnegative solution is

$$\frac{1}{T} = \frac{-b^T v_d + \sqrt{(b^T v_d)^2 - v_d^2(b^2 - a^2)}}{v_d^2}. \tag{7.17}$$

The guidance law is, then, given r, v, v_r, and a^2,

$$\begin{aligned} v_d &= v_r - v, \\ b &= \dot{v}_r - g, \\ v_d^2 &= v_d^T v_d, \\ b^2 &= b^T b, \\ \frac{1}{T} &= \frac{-b^T v_d + \sqrt{(b^T v_d)^2 - v_d^2(b^2 - a^2)}}{v_d^2}, \\ a_T &= b + \frac{1}{T} v_d. \end{aligned} \tag{7.18}$$

REMARK 7.4 *For* (7.17) *to yield a real time constant T, we must have*

$$a^2 \geq b^2 - (b^T v_d)^2 / v_d^2, \tag{7.19}$$

that is, we must have sufficient thrust capability. If we do not have enough thrust, we must use a guidance law where $\dot{v}_d$ is not collinear with v_d.

REMARK 7.5 *Equation* (7.17) *becomes ill-conditioned for v_d small. In practice, we often stop the guidance when $||v_d|| \leq \epsilon$, where ϵ is a given tolerance.*

EXAMPLE 7.1 *Consider the case of a planar ascent or descent into a circular orbit; that is, the main mission specification is to establish a circular orbit, regardless of terminal radius. We use polar coordinates for the trajectory modeling, as shown in Figure 7.1, but the ideas are as presented previously.*

Let $k = GM$ be the gravitational constant of Earth. For a circular orbit, we want

$$\frac{mv^2}{r} = \frac{mk}{r^2}, \tag{7.20}$$

implying that

$$v = \sqrt{\frac{k}{r}}. \tag{7.21}$$

Hence, the required velocity is

$$\vec{v}_r = \sqrt{\frac{k}{r}}\hat{i}_\theta. \tag{7.22}$$

In addition,

$$\vec{g} = -\frac{k}{r^2}\hat{i}_r, \tag{7.23}$$

which implies that

$$\vec{b} = \dot{\vec{v}}_r - \vec{g} = -\frac{1}{2}\frac{\dot{r}}{r}\sqrt{\frac{k}{r}}\hat{i}_\theta + \sqrt{\frac{k}{r}}\frac{\mathrm{d}}{\mathrm{d}t}(\hat{i}_\theta) - \frac{k}{r^2}\hat{i}_r. \tag{7.24}$$

Now,

$$\frac{\mathrm{d}}{\mathrm{d}t}(\hat{i}_\theta) = -\dot{\theta}\hat{i}_r. \tag{7.25}$$

Hence,

$$\vec{b} = -\left(\frac{k}{r^2} + \dot{\theta}\sqrt{\frac{k}{r}}\right)\hat{i}_r - \frac{\dot{r}}{2r}\sqrt{\frac{k}{r}}\hat{i}_\theta. \tag{7.26}$$

Navigation measures $\vec{v} = \dot{r}\hat{i}_r + r\dot{\theta}\hat{i}_\theta$. With that knowledge, we can compute $\vec{v}_d$ as

$$\vec{v}_d = \vec{v}_r - \vec{v} = -\dot{r}\hat{i}_r + \left(\sqrt{\frac{k}{r}} - r\dot{\theta}\right)\hat{i}_\theta. \tag{7.27}$$

If, for example, we choose to use unlimited thrust velocity-to-be-gained guidance, we obtain

$$\begin{aligned}\vec{a}_T &= \vec{b} + \frac{1}{T}v_d,\\ &= -\left(\frac{k}{r^2} + \dot{\theta}\sqrt{\frac{k}{r}} + \frac{\dot{r}}{T}\right)\hat{i}_r + \left(-\frac{\dot{r}}{2r}\sqrt{\frac{k}{r}} + \frac{1}{T}\left(\sqrt{\frac{k}{r}} - r\dot{\theta}\right)\right)\hat{i}_\theta.\end{aligned} \tag{7.28}$$

7.2 Guidance by State Feedback

Let us consider the general equations of motion for a vehicle:

$$\dot{x} = f(x, u, t), \tag{7.29}$$

where $x \in \mathbb{R}^n$ contains the positions and velocities of the vehicle, and $u \in \mathbb{R}^m$ contains the control inputs (e.g., thrusts and torques).

Assume that in mission planning, we have determined a nominal trajectory $x^0(t), u^0(t)$. As in Chapter 2, we linearize the equations of motion of the vehicle about the nominal trajectory, yielding

$$\begin{aligned}
\delta x &= x - x^0, \\
\delta u &= u - u^0, \\
A(t) &= \left(\frac{\partial f}{\partial x}\right)_0^T, \\
B(t) &= \left(\frac{\partial f}{\partial u}\right)_0^T.
\end{aligned} \tag{7.30}$$

We then obtain a linearized system in the standard linear time varying form

$$\delta \dot{x} = A(t)\delta x + B(t)\delta u. \tag{7.31}$$

Assume for now that δx is perfectly measured in flight. The purpose of midcourse guidance is then to find a δu that drives δx to zero. The idea behind guidance by state feedback is to let

$$\delta u = F(t)\delta x, \tag{7.32}$$

which implies that

$$\delta \dot{x} = (A(t) + B(t)F(t))\,\delta x. \tag{7.33}$$

REMARK 7.6 *Because the guidance law (7.32) relies on excursions from nominal for both the state vector and the input vector, it is often called* ***Delta-guidance****.*

In (7.33), $A(t)$ and $B(t)$ are given, and the unknown is $F(t)$. We then choose $F(t)$ so that (7.33) is asymptotically stable, which in turn implies that

$$\lim_{t\to\infty} \delta x(t) = 0. \tag{7.34}$$

Recall that in Section 4.6, we had obtained the equation for the observation error in an asymptotic observer:

$$\dot{\tilde{x}} = (A(t) + G(t)C(t))\,\tilde{x}, \tag{7.35}$$

in which $A(t)$ and $C(t)$ are given, and $G(t)$ is the unknown. As discussed in Section 2.5 on controllability and duality, (7.35) is related to the dual of (7.33). Hence, the same techniques that were used to select $G(t)$ in Section 4.6 can be used to select $F(t)$ for guidance by state feedback.

EXAMPLE 7.2 *Consider again the double integrator, introduced and motivated in Example 2.3. Here it is used to model one-dimensional guidance along a straight line assuming measurements of both position and velocity. The equations of motion are*

$$\delta \dot{x} = \begin{bmatrix} 0 & 1 \\ 0 & 0 \end{bmatrix} \delta x + \begin{bmatrix} 0 \\ 1 \end{bmatrix} \delta u. \tag{7.36}$$

Let

$$\delta u = \begin{bmatrix} f_1 & f_2 \end{bmatrix} \delta x = f_1 \delta x_1 + f_2 \delta x_2. \tag{7.37}$$

Then the closed-loop equation is

$$\delta\dot{x} = \begin{bmatrix} 0 & 1 \\ f_1 & f_2 \end{bmatrix} \delta x = A_c \delta x. \tag{7.38}$$

The closed-loop characteristic polynomial is

$$\det(sI - A_c) = s^2 - f_2 s - f_1. \tag{7.39}$$

For stability, we select $f_1 < 0$ and $f_2 < 0$.

REMARK 7.7 *In Section 4.7, we introduced the Kalman filter as a way of keeping the observer poles from becoming arbitrarily fast. In Chapter 9, we introduce the linear quadratic regulator (LQR) as a way of keeping the controls from becoming arbitrarily large. The LQR is the dual of the Kalman filter.*

REMARK 7.8 *Equations* (7.33) *and* (7.35) *show that observation and control are dual problems. In particular, in the following section, we show that navigation and guidance are dual problems.*

7.3 Combined Navigation and Guidance

In Section 7.2, we assume that δx is perfectly known and that it can be fed back. But in reality, δx is corrupted by navigation errors. In this section, we analyze the effect of those navigation errors on combined navigation and guidance.

Assume that we have planned a nominal trajectory and linearized the equations of motion about it. In the presence of noise, the system equations are, as presented in Chapter 3:

$$\begin{aligned} \delta\dot{x}(t) &= A(t)\delta x(t) + B(t)\delta u(t) + w(t) \\ \delta y(t) &= C(t)\delta x(t) + v(t), \end{aligned} \tag{7.40}$$

where $\delta x(t) \in \mathbb{R}^n$ contains the excursions, with respect to nominal values, of the positions, velocities, filter states, and so on; $\delta u(t) \in \mathbb{R}^m$ contains the excursions, with respect to nominal values, of accelerometer readings or inputs; $\delta y(t) \in \mathbb{R}^p$ contains the excursions, with respect to nominal values, of the readings of position and velocity sensors; $w(t)$ is the random vector of accelerometer noises and state disturbances; and $v(t)$ is the random vector of sensor noises.

As in Chapter 4, we use a linear navigator

$$\delta\dot{\hat{x}} = A(t)\delta\hat{x}(t) + B(t)\delta u(t) + G(t)\left(C(t)\delta\hat{x}(t) - \delta y(t)\right), \tag{7.41}$$

and, like in Section 7.2, we use linear guidance of the form

$$\delta u = F(t)\delta\hat{x}(t). \tag{7.42}$$

Note that, unlike in Section 7.2, we use $\delta\hat{x}(t)$, which is known, in the right-hand side of (7.42). Let

$$\delta\tilde{x} = \delta x - \delta\hat{x} \tag{7.43}$$

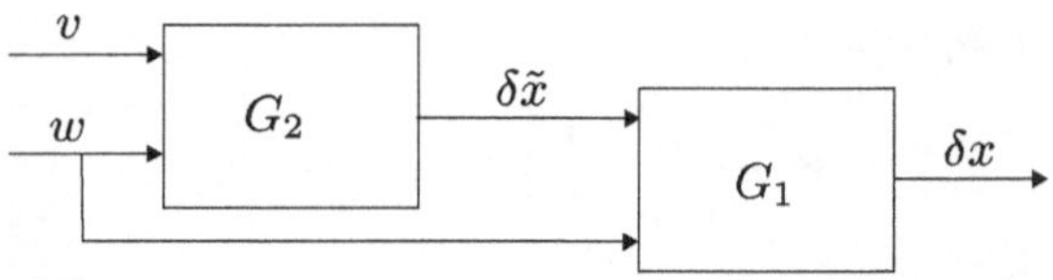

Figure 7.2. Block diagram for analysis of combined navigation and guidance.

be the navigation error. Then, the closed-loop dynamics are given by

$$\delta\dot{x}(t) = A(t)\delta x(t) + B(t)F(t)\delta x - B(t)F(t)\delta\tilde{x} + w(t), \tag{7.44}$$

and the navigation error dynamics are given by

$$\delta\dot{\tilde{x}}(t) = A(t)\delta\tilde{x}(t) + G(t)C(t)\delta\tilde{x} + G(t)v(t) + w(t). \tag{7.45}$$

We can rewrite (7.44) and (7.45) in matrix form:

$$\begin{bmatrix}\delta\dot{x}\\ \delta\dot{\tilde{x}}\end{bmatrix} = \begin{bmatrix}A(t)+B(t)F(t) & -B(t)F(t)\\ 0 & A(t)+G(t)C(t)\end{bmatrix}\begin{bmatrix}\delta x\\ \delta\tilde{x}\end{bmatrix} + \begin{bmatrix}I & 0\\ I & G(t)\end{bmatrix}\begin{bmatrix}w(t)\\ v(t)\end{bmatrix}. \tag{7.46}$$

Equation (7.46) leads to an illuminating extension of inequality (1.2) as follows. First, represent the dynamics of (7.46) as the block diagram in Figure 7.2 and assume that the systems G_1 and G_2 are both BIBO stable. Then, applying inequality (2.39), we obtain

$$||\delta\tilde{x}|| \leq ||G_2|| \left|\left|\begin{bmatrix}v\\ w\end{bmatrix}\right|\right| \tag{7.47}$$

and

$$||\delta x|| \leq ||G_1|| \left|\left|\begin{bmatrix}\delta\tilde{x}\\ w\end{bmatrix}\right|\right|. \tag{7.48}$$

Now,

$$\left|\left|\begin{bmatrix}\delta\tilde{x}\\ w\end{bmatrix}\right|\right| = \left|\left|\begin{bmatrix}\delta\tilde{x}\\ 0\end{bmatrix} + \begin{bmatrix}0\\ w\end{bmatrix}\right|\right|. \tag{7.49}$$

Applying the triangle inequality (2.37) (line 3), we obtain

$$\left|\left|\begin{bmatrix}\delta\tilde{x}\\ w\end{bmatrix}\right|\right| \leq ||\delta\tilde{x}|| + ||w||. \tag{7.50}$$

Therefore, (7.48) and (7.50) yield

$$||\delta x|| \leq ||G_1|| \left(||\delta\tilde{x}|| + ||w||\right). \tag{7.51}$$

Now, in (7.51), the term $||\delta x||$ quantifies the true guidance error, and the term $||\delta\tilde{x}||$ quantifies the navigation error. In the same spirit as (1.2), let us use for an upper bound on the navigation error the notation ϵ_n, where

$$\epsilon_n = ||\delta\tilde{x}||. \tag{7.52}$$

Also, from the block diagram of Figure 7.2, it is clear that the guidance error is due to two causes: the navigation error and w. The superposition principle in linear systems implies that the guidance error is the sum of two contributions: the navigation error assuming that $w = 0$, and w assuming that $\delta\tilde{x} = 0$. The guidance error due to the latter contribution is precisely what we call the estimated guidance error in (1.2).

Its norm is upper-bounded by $||G_1||||w||$, which is a term in (7.51). Hence, using the same notation as in (1.2), ϵ_g, for an upper bound on that error, we have

$$\epsilon_g = ||G_1||||w||. \tag{7.53}$$

Therefore, (7.51)–(7.53) yield the following bound on the true guidance error:

$$||\delta x|| \leq ||G_1||\epsilon_n + \epsilon_g. \tag{7.54}$$

Inequality (7.54) is an extension to dynamic systems of inequality (1.2) and leads to the same conclusion: good navigation combined with good navigation-based guidance guarantees good guidance.

REMARK 7.9 *The attentive reader may ask the question: Why does inequality* (7.54) *contain a factor* $||G_1||$, *whereas inequality* (1.2) *does not? The answer is as follows. On one hand, inequality* (1.2) *pertains to a static situation, and the notions of error are instantaneous. In that case, a navigation error at a particular time causes a guidance error only at that particular time. On the other hand, inequality* (7.54) *pertains to a dynamic situation, and the notions of error pertain to the whole time history of a signal, according to* (2.36). *In that case, a navigation error at a particular time generally causes a guidance error at many other times, with a propagation governed by the differential equation* (7.46). *The factor* $||G_1||$ *in* (7.54) *captures this effect of dynamic propagation of navigation error into guidance error.*

Inequality (7.54) is predicated upon BIBO stability of both G_1 and G_2 in the block diagram of Figure 7.2. However, examining (7.46) reveals that BIBO stability of G_2 depends only on the choice of $G(t)$ and does not depend on the choice of $F(t)$. Moreover, once the boundedness of $\delta\tilde{x}$ is secured by choice of $G(t)$, BIBO stability of G_1 depends only on the choice of $F(t)$ and does not depend on the choice of $G(t)$. Hence, $F(t)$ and $G(t)$ can be chosen separately to ensure that both G_1 and G_2 in Figure 7.2 are BIBO stable. This property, which is due to the upper block-triangular structure of the state matrix in (7.46), is called the **deterministic separation principle**.

REMARK 7.10 *In the time invariant case, the deterministic separation principle takes the following form. Recall that, for time invariant systems, as per Propositions 2.15, 2.16, and 2.17, all notions of stability are determined by the eigenvalues of the state matrix, including the subset of eigenvalues that are poles of the transfer function. When the state matrix in* (7.46) *is constant, its eigenvalues are those of the matrices on its diagonal because of the upper block-triangular structure. Clearly, the eigenvalues of* $A + BF$ *depend on* F *alone and not on* G, *and those of* $A + GC$ *depend on* G *alone and not on* F. *If the pair* (A, B) *is controllable and the pair* (A, C) *is observable, as per Propositions 4.4 and 2.23, the matrices* F *and* G *can be selected separately to cause the state matrix of* (7.46) *to have arbitrary eigenvalues. In particular, the matrices* F *and* G *can then be selected separately so that all the eigenvalues of the state matrix of* (7.46) *are strictly in the left half of the complex plane, which guarantees BIBO stability of* G_1 *and* G_2 *in Figure 7.2.*

The analysis provided by inequality (7.54) can be refined as follows. Let us now assume that, in (7.40), the standard assumptions for linear Gauss–Markov processes are satisfied, as discussed in Section 3.8, namely, that the initial condition $x(t_0)$ is Gaussian, and that $w(t)$ and $v(t)$ are zero-mean, Gaussian, white, and uncorrelated. Then, write (7.46) as a linear system driven by white noise, of the form

$$\dot{\xi} = \mathbb{A}(t)\xi + \Gamma(t)\omega. \tag{7.55}$$

Since (7.55) is of the form (3.85) where assumptions (3.87)–(3.90) are satisfied, Propositions 3.5 and 3.6 apply. In particular, the mean value function of ξ satisfies

$$\dot{\bar{\xi}} = \mathbb{A}(t)\bar{\xi}, \tag{7.56}$$

and, if we denote by $\mathbb{P}(t)$ the covariance of ξ, we have

$$\dot{\mathbb{P}}(t) = \mathbb{A}(t)\mathbb{P}(t) + \mathbb{P}(t)\mathbb{A}^T(t) + \Gamma(t)R_\omega(t)\Gamma^T(t), \tag{7.57}$$

where

$$R_\omega = \text{cov}\begin{bmatrix} w(t) \\ v(t) \end{bmatrix} = \begin{bmatrix} R_w & 0 \\ 0 & R_v \end{bmatrix}. \tag{7.58}$$

The Lyapunov equation (7.57) allows an exact quantification of the navigation and guidance errors in combined navigation and navigation-based guidance under Gauss–Markov assumptions. Specifically, the covariance matrices of the navigation and guidance errors are the two blocks on the diagonal of matrix $\mathbb{P}(t)$.

In stochastic linear optimal control, another separation principle is established, guaranteeing that navigation and guidance can be optimized separately. This result is obtained in the context of the LQG regulator, discussed in Chapter 9. Hence, our treatment of navigation and guidance as separate problems is justified within linear systems theory.

7.4 Summary of Key Results

The key results in Chapter 7 are as follows:

1. Equations (7.9) and (7.10), which provide the thrust acceleration required in space flight guidance, in terms of the velocity-to-be-gained, to achieve its exponential decay with specified time constant
2. Equation (7.33), which provides the guidance error in the case of state feedback guidance
3. Equation (7.46), which provides both the navigation and guidance errors in the case of combined navigation and navigation-based guidance
4. The deterministic separation principle, which results from the upper block triangular structure of the state matrix in (7.46)
5. Inequality (7.54), which extends inequality (1.2) to dynamic situations and justifies our separate treatment of the navigation and guidance problems
6. Equation (7.57), which provides the covariance matrices of the navigation and guidance errors under Gauss–Markov assumptions

7.5 Bibliographic Notes for Further Reading

Midcourse guidance is covered in [8] and [60]. For further reading on guidance by state feedback and combined navigation and guidance, the reader may consult [17], [71], and [46].

Velocity-to-be-gained and state feedback guidance, that is, Q-guidance and Delta-guidance, respectively, have been used on a variety of ballistic systems [50]. The combination of navigation and guidance was extensively used in the Apollo lunar missions [9].

7.6 Homework Problems

For this chapter, as in the last, the problems may require numerical solutions and computer programming.

PROBLEM 7.1 *Show that in velocity-to-be-gained guidance, the* 3×3 *matrix* $\left(\frac{\partial v_r}{\partial r}\right)^T$ *is symmetric.*

Problems 7.2 through 7.4 relate to the scenario **Asteroid Assessment and Targeting.**

The purpose of these problems is threefold. In the first part, you assess the danger posed by an asteroid in terms of probability of impact, accounting for uncertainty in the observation data. In the second part, you assess the accuracy of a shot of a ballistic missile aimed at destroying the asteroid in terms of expected square miss. In the third part, you devise a midcourse guidance law using continuous course corrections to improve the accuracy of the ballistic shot.

The assumptions are as follows:

1. Perfectly spherical, nonrotating planet, with radius R and gravitational constant k
2. Point mass asteroid
3. Point mass missile
4. Asteroid and missile are within the sphere of influence of the planet, and their gravitational interaction is negligible
5. No aerodynamic forces

Under these assumptions, the motions of both asteroid and missile are described by the restricted two-body problem, which has been studied extensively (e.g., [18]). Furthermore, assume that the asteroid and missile have the same orbital plane.

In this homework set, you will choose data to illustrate that your computations work. In your computations, use the following numerical values:

$$k = 4.305 \times 10^{13} \text{ m}^3/\text{s}^2,$$
$$R = 3.38 \times 10^6 \text{ m}. \tag{7.59}$$

PROBLEM 7.2 *Under the standard assumptions of the restricted two-body problem, consider the problem of determining the orbit of the asteroid, based on measurements of its position. The basic equation is*

$$r = \frac{p}{1 + e\cos(\theta - \theta_0)}, \tag{7.60}$$

where (r, θ) are the polar coordinates of the asteroid and θ_0, e, and p are unknown parameters. Assume that we are given a record of measurements of the asteroid position at various times, (r_1, θ_1), (r_2, θ_2), ..., (r_n, θ_n), where $n \geq 3$. As would be typical in this case, we assume that the goniometry data $(\theta_1, \ldots, \theta_n)$ are perfectly reliable but that the telemetry data $(r_1, \ldots, r_n)$ are corrupted by measurement errors.

1. *Suggest a method for computing the unknown parameters θ_0, e, and p based on the available data.*
2. *Assume that we know the statistical properties of the measurement errors that corrupt the telemetry data. Suggest a method for assessing the statistical properties of the estimate of the unknown parameters θ_0, e, and p.*
3. *Use this result to estimate the probability of impact. (Hint: Impact is the event that the radius at periapse is smaller than the sum of the radii of the planet and the asteroid.)*

PROBLEM 7.3 *Assume that the cutoff occurs on the planet surface at exactly known polar coordinates and at a known initial time. Choose a time of intercept.*

1. *Solve the targeting problem to intercept the asteroid at the chosen time of intercept. In other words, determine the locus of cutoff velocities (v_r, v_θ) that would achieve a hit if the asteroid were following its nominal trajectory. (Here you recognize Lambert's problem [18].)*
2. *Quantify the uncertainty in the position of the asteroid due to uncertainty in its orbital elements.*
3. *Quantify the expected square miss of the shot in the targeting problem due to the uncertainty in the position of the asteroid.*

PROBLEM 7.4 *Devise a midcourse guidance law using continuous radar observations of the asteroid and continuous course corrections to improve the accuracy of the ballistic shot.*

8 Optimization

The next two chapters treat optimization and optimal control for the purpose of their application to guidance. Chapter 8 focuses on optimization as a stepping-stone toward optimal control, which is treated in Chapter 9. Optimization is concerned with finding the best option from among several to solve a problem.

Optimization is often necessary in aerospace engineering because of the merciless requirements that physics and chemistry impose on flying systems. For instance, it is well known that to lift 1 kg of payload from Earth's surface to orbit, using chemical rocket propulsion, it is required to use at least 80 kg of rocket structure, engine, fuel, and propellant [78]. This staggering 80/1 ratio is one of many stark reminders that, when it comes to flight, optimization is of the essence.

Throughout this chapter, we discuss necessary and sufficient conditions for optimality. Let us clarify what we mean by these. A necessary condition for optimality is a statement of the form: "If item x is optimal (i.e., is the best), then condition $NC(x)$ must be satisfied." Typically, condition $NC(x)$ provides enough information to determine x. A sufficient condition for optimality is a statement of the form: "If item x satisfies condition $SC(x)$, then item x must be optimal." Here also, typically condition $SC(x)$ provides enough information to determine x. In view of these statements, the following caution is in order: *necessary conditions guarantee neither optimality of* x*, nor even existence of a solution to the optimization problem.* Sufficient conditions, however, are more powerful: if item x satisfies the sufficient condition, then not only can we guarantee that the problem does have a solution, we can also guarantee that x is a solution. Note that if the necessary condition for optimality has no solution, then we can conclude that the optimization problem has no solution either. However, if the sufficient condition has no solution, we cannot necessarily conclude that the optimization problem has no solution.

Section 8.1 covers unconstrained optimization of real functions of a vector variable. Section 8.2 treats optimization under equality constraints, and Section 8.3 extends the results to the case of inequality constraints. Section 8.4 treats optimal control of discrete-time systems, which is to segue into optimal control in Chapter 9. Sections 8.5, 8.6, and 8.7 present a summary of the key results in the chapter, bibliographic notes for further reading, and homework problems, respectively.

REMARK 8.1 *Before proceeding with the study of optimization, another word of caution is in order. Optimization in engineering should always be approached with prudence. This is because engineering systems must often meet many competing performance specifications. As a consequence, optimizing an engineering system with respect to one performance metric is typically detrimental to other performance metrics that matter – as the adage says: the best is often the enemy of the good. Hence, optimization should be viewed as a convenient method for engineering synthesis, that is, to obtain a configuration (e.g., a trajectory) of the engineering system as a function of specified parameters (e.g., a cost function). In that context, it is often practical to adjust the cost function until optimization synthesizes a satisfactory configuration.*

8.1 Unconstrained Optimization on $\mathbb{R}^n$

This section presents a hierarchy of results on optimization of real functions of a vector variable. The hierarchy is based on the smoothness of the function, proceeding from continuous functions to functions whose first derivative is continuous and then to functions whose second derivative is continuous.

Consider the function f defined as

$$f : \mathbb{R}^n \to \mathbb{R} : x \mapsto f(x), \tag{8.1}$$

and let $X \subseteq \mathbb{R}^n$.

DEFINITION 8.1 *$x^* \in X$ is a* ***global minimum*** *of f on X if*

$$\forall x \in X,\ f(x^*) \leq f(x). \tag{8.2}$$

DEFINITION 8.2 *$x^* \in X$ is a* ***local minimum*** *of f on X if there exists an open neighborhood $\mathring{N}$ of x^* such that x^* is a global minimum of f on $\mathring{N} \cap X$.*

REMARK 8.2 *Note that if x^* is a global minimum of f on X, then x^* is also a local minimum of f on X.*

EXAMPLE 8.1 *Examples of global and local minima are shown in Figure 8.1 for f: $\mathbb{R} \to \mathbb{R}$.*

The following basic result guarantees existence and achievement of minimum for continuous functions.

PROPOSITION 8.1 *If X is closed and bounded, and if f: $X \to \mathbb{R}$ is continuous, then f achieves a minimum on X.*

EXAMPLE 8.2 *In the example shown in Figure 8.2, $X = (a, b]$ and the problem has no minimum. This illustrates the importance of the set X.*

DEFINITION 8.3 *We say that f is of* ***class*** *$\mathcal{C}^k$, written $f \in \mathcal{C}^k$, if all partial derivatives of f exist and are continuous up to order k, that is, the*

$$\frac{\partial^l f}{\partial x_{i_1} \partial x_{i_2} \cdots \partial x_{i_l}} \tag{8.3}$$

exist and are continuous for all l such that $0 \leq l \leq k$.

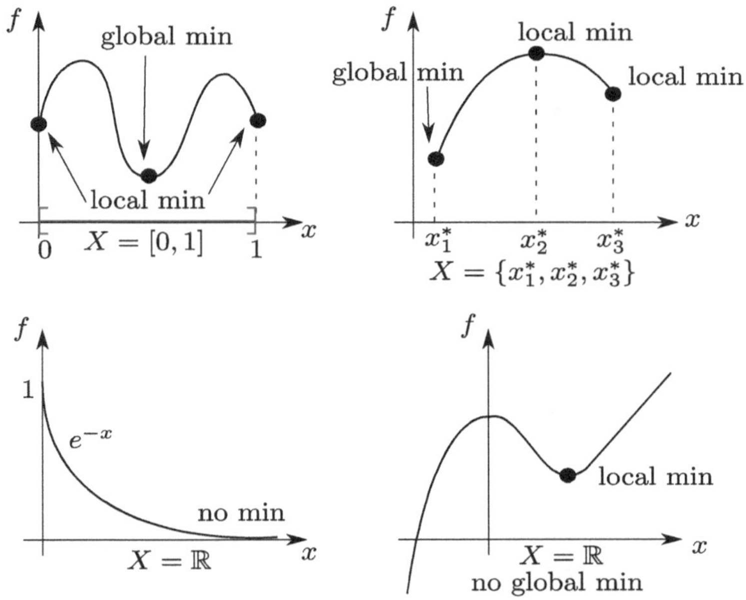

Figure 8.1. Examples of global and local minima. For all examples, $f : \mathbb{R} \to \mathbb{R}$.

DEFINITION 8.4 *We say that f is of **class** C^{*k}, written $f \in C^{*k}$, if $f \in C^{k-1}$ and all partial derivatives of f of order k exist (but are not necessarily continuous).*

The following result is the basis for necessary and sufficient conditions for local optimization of differentiable functions. It is stated in Appendix A.3 and repeated here for clarity.

PROPOSITION 8.2 (Taylor's Theorem) *Let $f : \mathbb{R} \to \mathbb{R}$, $f \in C^{*n+1}$ on $[a, x]$. Then,*

$$f(x) = f(a) + f'(a)(x - a) + \frac{f''(a)}{2!}(x - a)^2 + \cdots + \frac{f^{(n)}(a)}{n!}(x - a)^n + R_n(x, a), \tag{8.4}$$

where

$$R_n(x, a) = \int_a^x \frac{(x - \tau)^n}{n!} f^{(n+1)}(\tau)\, d\tau. \tag{8.5}$$

Moreover, if $f \in C^{n+1}$ on $[a, x]$, then there exists $c \in [a, x]$ such that

$$R_n(x, a) = \frac{f^{(n+1)}(c)}{(n + 1)!}(x - a)^{n+1}. \tag{8.6}$$

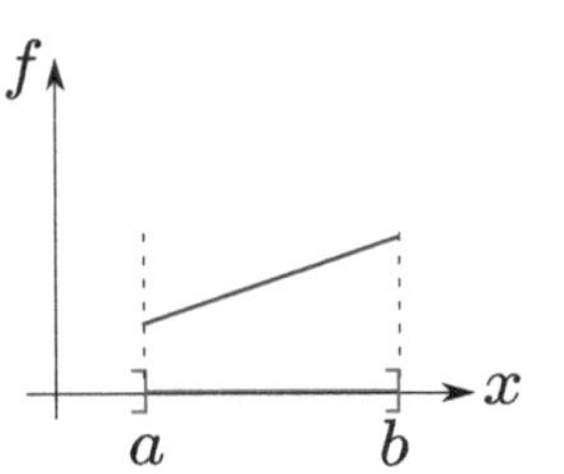

Figure 8.2. Importance of X for optimization problems: Example 8.2.

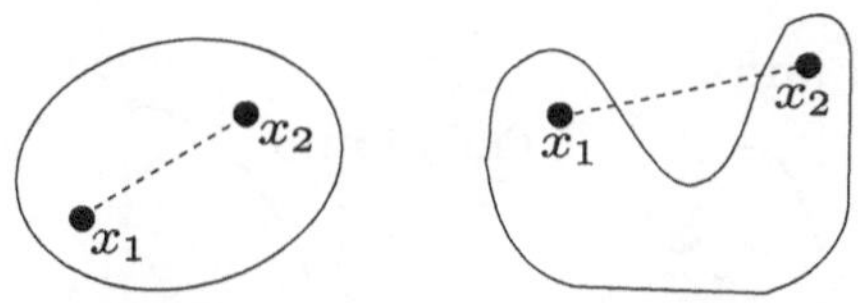

Figure 8.3. Examples of convex and nonconvex sets.

REMARK 8.3 (Interpretation of Taylor's Theorem) *Taylor's theorem states that*

$$\Delta f = f(x) - f(a) = \delta^1 f(x,a) + \delta^2 f(x,a) + \delta^3 f(x,a) + \dots, \tag{8.7}$$

where

$$\begin{aligned} \delta^1 f(x,a) &= f'(a)(x-a) \text{ is of order 1 in (x-a)}, \\ \delta^2 f(x,a) &= f''(a)(x-a)^2 \text{ is of order 2 in (x-a)}, \\ \delta^3 f(x,a) &= f'''(a)(x-a)^3 \text{ is of order 3 in (x-a)}, \\ &\dots; \end{aligned} \tag{8.8}$$

that is, the following limit is finite:

$$\lim_{|x-a|\to 0} \frac{\delta^k f(x,a)}{|x-a|^k}. \tag{8.9}$$

Hence, for $|x-a|$ small, $\delta^1 f$ dominates, followed by $\delta^2 f$, followed by $\delta^3 f$, and so on. In other words, Δf is decomposed into contributions that can be ranked.

These contributions, the $\delta^k f(x,a)$, are called the ***variations of** f **at** a. They lead to the necessary or sufficient conditions for optimality. Their computation in functional spaces leads to the* ***calculus of variations****.*

DEFINITION 8.5 *A set $X \subset \mathbb{R}^n$ is* ***convex*** *if*

$$\forall x_1, x_2 \in X, \forall \alpha \in [0,1], x = \alpha x_2 + (1-\alpha)x_1 \in X. \tag{8.10}$$

EXAMPLE 8.3 (Convex and Nonconvex Sets) *Examples of convex and nonconvex sets are shown in Figure 8.3.*

DEFINITION 8.6 *A function $f : X \to \mathbb{R}$ is* ***convex*** *if*

$$\forall x_1, x_2 \in X, \forall \alpha \in [0,1], f(\alpha x_2 + (1-\alpha)x_1) \le \alpha f(x_2) + (1-\alpha) f(x_1). \tag{8.11}$$

The function is called ***strictly convex*** *if*

$$\forall x_1, x_2 \in X, x_1 \ne x_2, \forall \alpha, 0 < \alpha < 1, f(\alpha x_2 + (1-\alpha)x_1) < \alpha f(x_2) + (1-\alpha) f(x_1). \tag{8.12}$$

EXAMPLE 8.4 (Convex Function) *An example of a convex function is shown in Figure 8.4.*

Taylor's theorem is generalized as follows. Let $X \in \mathbb{R}^n$ be convex, closed, and bounded. Let $f : X \to \mathbb{R}$, $f \in C^{m+1}$ on X. Assume that we want to calculate the

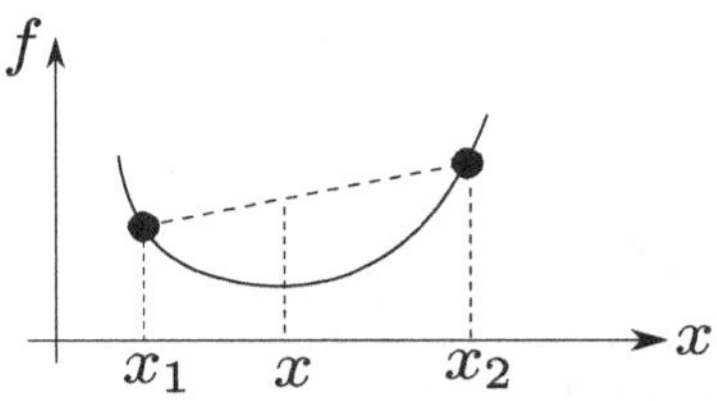

Figure 8.4. Convex function.

quantity $f(x+h) - f(x)$. Let $g(\alpha) = f(x + \alpha h)$, where $\alpha \in [0, 1]$ is a scalar. We apply the scalar version of Taylor's theorem to g to yield

$$g(1) = g(0) + g'(0) + \frac{g''(0)}{2!} + \cdots + R_m(1, 0). \tag{8.13}$$

Developing (8.13) yields

$$\begin{aligned} f(x+h) - f(x) = {} & \left(\frac{\partial f}{\partial x}\right)_x^T h + \frac{1}{2!} h^T \left(\frac{\partial^2 f}{\partial x^2}\right)_x h \\ & + \sum_{i_1}\sum_{i_2}\sum_{i_3} \frac{1}{3!} \left(\frac{\partial^3 f}{\partial x_{i_1} \partial x_{i_2} \partial x_{i_3}}\right)_x h_{i_1} h_{i_2} h_{i_3} \\ & + \dots \\ & + \sum_{i_1}\sum_{i_2} \cdots \sum_{i_m} \frac{1}{m!} \left(\frac{\partial^m f}{\partial x_{i_1} \partial x_{i_2} \dots \partial x_{i_m}}\right)_x h_{i_1} h_{i_2} \dots h_{i_m} \\ & + \sum_{i_1}\sum_{i_2} \cdots \sum_{i_{m+1}} \frac{1}{(m+1)!} \left(\frac{\partial^{m+1} f}{\partial x_{i_1} \partial x_{i_2} \dots \partial x_{i_{m+1}}}\right)_{x+\alpha_0 h, \alpha_0 \in [0,1]} h_{i_1} h_{i_2} \dots h_{i_{m+1}}, \end{aligned} \tag{8.14}$$

where

$$\frac{\partial f}{\partial x} = \begin{bmatrix} \frac{\partial f}{\partial x_1} \\ \frac{\partial f}{\partial x_2} \\ \cdots \\ \frac{\partial f}{\partial x_n} \end{bmatrix} \tag{8.15}$$

is the gradient of f, and

$$\frac{\partial^2 f}{\partial x^2} = \begin{bmatrix} \frac{\partial^2 f}{\partial x_1^2} & \frac{\partial^2 f}{\partial x_2 \partial x_1} & \cdots & \frac{\partial^2 f}{\partial x_n \partial x_1} \\ \frac{\partial^2 f}{\partial x_1 \partial x_2} & \frac{\partial^2 f}{\partial x_2^2} & \cdots & \frac{\partial^2 f}{\partial x_n \partial x_2} \\ \cdots & & & \\ \frac{\partial^2 f}{\partial x_1 \partial x_n} & \cdots & \cdots & \frac{\partial^2 f}{\partial x_n^2} \end{bmatrix} \tag{8.16}$$

is the (symmetric) Hessian matrix of f. Here also, defining

$$\Delta f(x; h) = f(x+h) - f(x), \tag{8.17}$$

we have

$$\Delta f(x; h) = \delta f(x; h) + \delta^2 f(x; h) + \delta^3 f(x; h) + \cdots + \delta^m f(x; h) + R_m(x; h), \tag{8.18}$$

where

$$\delta^k f(x; h) \tag{8.19}$$

is called the kth differential of f or the kth variation of f, and is homogeneous in h with degree k, that is,

$$\delta^k f(x;\alpha h) = \alpha^k \delta^k f(x;h). \tag{8.20}$$

Hence, the difference of a smooth function is a sum of variations and remainder. Each variation is a homogeneous function of the increment, the degree of homogeneity being the index of the variation.

Based on Taylor's theorem and its generalization, we can now state the following;

PROPOSITION 8.3 (Necessary Conditions for Optimality of Twice-differentiable Functions on Open Sets) *If $X \in \mathbb{R}^n$, $X = \mathring{X}$ (X is open), and $x^* \in \mathring{X}$, then*

- *$f \in C^1$ has a local minimum at x^* implies that*

$$\forall h, \delta^1 f(x^*;h) = 0 \tag{8.21}$$

- *$f \in C^2$ has a local minimum at x^* implies that*

$$\forall h, \delta^1 f(x^*;h) = 0 \tag{8.22}$$

$$\forall h, \delta^2 f(x^*;h) \geq 0. \tag{8.23}$$

REMARK 8.4 *Note that* (8.21) *is equivalent to requiring*

$$\frac{\partial f}{\partial x} = 0, \tag{8.24}$$

that is, to solving n equations for x^*. The solution yields what are called ***stationary*** *or* ***singular points****. Requirement* (8.23) *is equivalent to checking that*

$$\frac{\partial^2 f}{\partial x^2} \geq 0, \tag{8.25}$$

that is, that the Hessian matrix is positive semidefinite.

REMARK 8.5 *Note that the first-order necessary conditions for optimality have the form*

$$\frac{\partial f}{\partial x} = 0, \tag{8.26}$$

which is that of a system of n equations in n unknowns. Such systems can be solved systematically by using Newton's method as discussed in Appendix A.4.

REMARK 8.6 *To apply Newton's method to solve $\frac{\partial f}{\partial x} = 0$, the iteration becomes*

$$x^{k+1} = x^k - \left(\frac{\partial^2 f}{\partial x^2}\right)^{-1}_{x^k} \left(\frac{\partial f}{\partial x}\right)_{x^k}, \tag{8.27}$$

which can be interpreted as follows: approximate f locally by a ***quadratic*** *function and optimize it to get the next iterate.*

PROPOSITION 8.4 (Sufficient Conditions for Optimality of Twice-differentiable Functions on Open Sets) *If $X \in \mathbb{R}^n$, $X = \mathring{X}$ (X is open), $x^* \in \mathring{X}$, $f \in \mathcal{C}^2$ on X and*

$$\left(\frac{\partial f}{\partial x}\right)_{x^*} = 0 \tag{8.28}$$

$$\left(\frac{\partial^2 f}{\partial x^2}\right)_{x^*} > 0, \tag{8.29}$$

then f has a local minimum at x^*.

PROPOSITION 8.5 (Uniqueness of (Global) Minimum) *If X is convex, closed, and bounded, if $f \in \mathcal{C}^2$ on X, and if f is strictly convex on X, then there exists a unique $x^* \in X$ such that f has a global minimum at x^*.*

REMARK 8.7 *If $f \in \mathcal{C}^2$ on X and if f is strictly convex on X, then*

$$\frac{\partial^2 f}{\partial x^2} > 0 \text{ on } X. \tag{8.30}$$

8.2 Constrained Optimization on $\mathbb{R}^n$

Constrained optimization problems on $\mathbb{R}^n$ are of the following form: given $f : \mathbb{R}^n \to \mathbb{R} \in \mathcal{C}^2$, $g : \mathbb{R}^n \to \mathbb{R}^m \in \mathcal{C}^2$ with $m < n$, and given $X = \{x \in \mathbb{R}^n | g(x) = 0\}$, find

$$\min_{x \in X} J = f(x), \tag{8.31}$$

or

$$\min_{x \in X} J = f(x) \tag{8.32}$$
$$\text{subject to } g(x) = 0.$$

The key idea is to "eliminate" part of x using the constraint, and then to apply the results of Section 8.1.

DEFINITION 8.7 *$x^* \in X$ is a **regular point** if $\left(\frac{\partial g}{\partial x}\right)_{x^*}$ has full rank. Recall that*

$$\left(\frac{\partial g}{\partial x}\right)_{ij} = \frac{\partial g_j}{\partial x_i} \in \mathbb{R}^{n \times m} \tag{8.33}$$

$$\frac{\partial g}{\partial x} = \begin{bmatrix} \frac{\partial g_1}{\partial x} & \frac{\partial g_2}{\partial x} & \cdots & \frac{\partial g_m}{\partial x} \end{bmatrix}.$$

REMARK 8.8 *If x is a regular point, then it is always possible to reorder the entries of x so that $x = \begin{bmatrix} x_1 \\ x_2 \end{bmatrix}$, where $x_1 \in \mathbb{R}^m$, $x_2 \in \mathbb{R}^{n-m}$, and the Jacobian $\frac{\partial g}{\partial x_1}$ is nonsingular. Using the implicit function theorem of Appendix A.5, x_1 can be eliminated from the implicit constraint $g(x_1, x_2) = 0$ using an explicit function $x_1 = \phi(x_2)$. Moreover, to evaluate the Jacobian of the explicitation, we do not need a closed-form expression; all we need are the Jacobians of the implicit function.*

8.2.1 Lagrange Multipliers

We can now state necessary conditions for the optimization problem (8.32) when f and g are of class $\mathcal{C}^1$.

PROPOSITION 8.6 (Lagrange's Theorem) *If x^* is a solution to the minimization under constraints problem* (8.31) *or* (8.32), *where f and g are of class $\mathcal{C}^1$, and if x^* is regular, then there exist real numbers $p_1, p_2, \ldots, p_m$ such that*

$$\frac{\partial f}{\partial x} + \sum_{i=1}^{m} p_i \frac{\partial g_i}{\partial x} = 0. \tag{8.34}$$

REMARK 8.9 *The p_i are called **Lagrange multipliers**. There are as many of them as there are constraints. The necessary conditions for optimality of dynamic optimization problems, presented in Chapter 9, feature similar quantities called the co-states.*

PROOF. Here we present a sketch of the derivation of Proposition 8.6. Recall that x can be decomposed as $x = \begin{bmatrix} x_1 \\ x_2 \end{bmatrix}$, where $x_1 \in \mathbb{R}^m$ and $x_2 \in \mathbb{R}^{n-m}$. We then have

$$\begin{aligned} \delta J &= \left(\frac{\partial f}{\partial x_1}\right)^T \delta x_1 + \left(\frac{\partial f}{\partial x_2}\right)^T \delta x_2 \\ \delta g &= \left(\frac{\partial g}{\partial x_1}\right)^T \delta x_1 + \left(\frac{\partial g}{\partial x_2}\right)^T \delta x_2 = 0, \end{aligned} \tag{8.35}$$

where $\left(\frac{\partial g}{\partial x_1}\right)^T \in \mathbb{R}^{m \times m}$ is assumed nonsingular, without loss of generality. Then,

$$\delta x_1 = -\left(\frac{\partial g}{\partial x_1}\right)^{-T} \left(\frac{\partial g}{\partial x_2}\right)^T \delta x_2, \tag{8.36}$$

hence,

$$\delta J = \left[-\left(\frac{\partial g}{\partial x_2}\right)\left(\frac{\partial g}{\partial x_1}\right)^{-1} \left(\frac{\partial f}{\partial x_1}\right) + \left(\frac{\partial f}{\partial x_2}\right)\right]^T \delta x_2, \tag{8.37}$$

where J is now a function of x_2 alone, by use of the implicit function theorem. Define

$$p = -\left(\frac{\partial g}{\partial x_1}\right)^{-1} \left(\frac{\partial f}{\partial x_1}\right). \tag{8.38}$$

Then,

$$\left(\frac{\partial g}{\partial x_1}\right) p + \left(\frac{\partial f}{\partial x_1}\right) = 0, \tag{8.39}$$

or equivalently,

$$\frac{\partial}{\partial x_1}(f + p^T g) = 0. \tag{8.40}$$

Setting $\delta J = 0$ for all δx_2 in (8.37) yields

$$\frac{\partial}{\partial x_2}(f + p^T g) = 0. \tag{8.41}$$

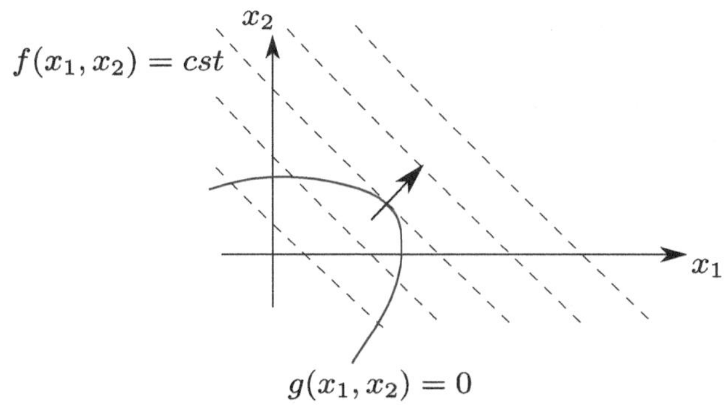

Figure 8.5. Geometric interpretation of minimization under constraints.

Combining (8.40) and (8.41) yields the desired result:

$$\frac{\partial}{\partial x}(f + p^T g) = 0. \tag{8.42}$$

REMARK 8.10 (Interpretation of p) *In (8.35), let $\delta x_2 = 0$, to obtain*

$$\delta x_1 = \left(\frac{\partial g}{\partial x_1}\right)^{-T} \delta g, \tag{8.43}$$

which implies that

$$\delta J = \left[\left(\frac{\partial g}{\partial x_1}\right)^{-1}\left(\frac{\partial f}{\partial x_1}\right)\right]^T \delta g = -p^T \delta g. \tag{8.44}$$

Thus, a possible interpretation of p is given by

$$p = -\left(\frac{\partial J}{\partial g}\right)_{x_2 \text{ fixed}}, \tag{8.45}$$

that is, p can be interpreted as the sensitivity of the cost with respect to the constraint.

REMARK 8.11 (Geometric Interpretation of p) *A geometric interpretation is as shown in Figure 8.5. At optimum, the gradients of f and g are collinear. Recalling that the gradient is a vector that is perpendicular to the local level curve, we obtain that the level curves of f and g have the same perpendicular, that is, they are tangent.*

For the problem of minimization under constraints as stated in (8.31) or (8.32), define

$$L(x, p) = p^T g(x) + f(x) \tag{8.46}$$

to be the **Lagrangian**. Then the first-order necessary conditions for optimality can be expressed as

$$\frac{\partial L}{\partial p} = g(x) = 0 \tag{8.47}$$

$$\frac{\partial L}{\partial x} = \frac{\partial f}{\partial x} + \left(\frac{\partial g}{\partial x}\right) p = 0,$$

which is a system of $n + m$ equations in $n + m$ unknowns.

REMARK 8.12 *We can generalize the contents of Remarks 8.5 and 8.6 to optimization under constraints as follows. Assume we use Newton's iteration to solve the system of $n+m$ equations with $n+m$ unknowns (8.47). Then, each iteration amounts to locally approximating the objective function f by a* ***quadratic*** *function, the constraint g by a* ***linear*** *function, and choosing as the next iterate the solution of this subsidiary optimization problem.*

8.2.2 Second-Order Conditions

DEFINITION 8.8 *For the constraints $g(x) = 0$, the* ***tangent plane*** *$\mathcal{T}$ is defined as*

$$\mathcal{T} = \left\{ \delta x \middle| \left(\frac{\partial g}{\partial x} \right)^T \delta x = 0 \right\}. \tag{8.48}$$

PROPOSITION 8.7 *For the problem*

$$\min_x J = f(x) \tag{8.49}$$

$$\text{subject to } g(x) = 0,$$

where f and g are of class C^2, if x^* is regular, a necessary condition for optimality of x^* is

$$\frac{\partial^2 L}{\partial x^2} \geq 0 \tag{8.50}$$

in the tangent plane $\mathcal{T}$ at x^*, where $L = p^T g(x) + f(x)$.

PROPOSITION 8.8 *For the problem*

$$\min_x J = f(x)$$

$$\text{subject to } g(x) = 0, \tag{8.51}$$

where f and g are of class C^2, if x^* is regular, a sufficient condition for optimality of x^* is

$$\frac{\partial L}{\partial x} = 0,$$

$$\frac{\partial L}{\partial p} = 0,$$

$$\frac{\partial^2 L}{\partial x^2} > 0 \text{ in } \mathcal{T} \text{ at } x^*. \tag{8.52}$$

REMARK 8.13 *Propositions 8.7 and 8.8 require ascertaining the sign of the Hessian of the Lagrangian in the tangent plane. Note that if a symmetric matrix is of definite sign (i.e., positive definite or negative definite), restricting its quadratic form to a plane preserves the sign of the quadratic form. Hence, if the Hessian of the Lagrangian is positive definite, we can conclude that the candidate optimum is a minimum without scrutinizing the tangent plane.*

EXAMPLE 8.5 (Least Energetic Ballistic Shot Revisited) *To illustrate the use of Lagrange multipliers, we revisit the optimization problem of Example 6.1: from among all ballistic trajectories that originate at polar coordinates $(r, \theta) = (r_0, \pi/2)$ and hit the target located at $(r, \theta) = (r_0, 0)$, we seek the one that does so with minimum energy. Here r_0 is the radius of the planet. Formally, the optimization problem is*

$$\min_{v_{r_1}, v_{\theta_1}} T(v_{r_1}, v_{\theta_1}) = \frac{1}{2}\left(v_{r_1}^2 + v_{\theta_1}^2\right) \tag{8.53}$$

subject to

$$v_{r_1} - v_{\theta_1} + \frac{k}{r_0 v_{\theta_1}} = 0, \tag{8.54}$$

where v_{r_1} and v_{θ_1} are the radial and tangential components of velocity at cutoff, respectively, k is the gravitational constant of the planet, the objective function is the specific kinetic energy at cutoff, and the constraint is the hit equation.

The Lagrangian is

$$L(v_{r_1}, v_{\theta_1}, p) = \frac{1}{2}(v_{r_1}^2 + v_{\theta_1}^2) + p\left(v_{r_1} - v_{\theta_1} + \frac{k}{r_0 v_{\theta_1}}\right), \tag{8.55}$$

where p is the Lagrange multiplier. The first-order necessary conditions for optimality are

$$\frac{\partial L}{\partial v_{r_1}} = v_{r_1} + p = 0, \tag{8.56}$$

$$\frac{\partial L}{\partial v_{\theta_1}} = v_{\theta_1} - p - \frac{pk}{r_0 v_{\theta_1}^2} = 0, \tag{8.57}$$

$$\frac{\partial L}{\partial p} = v_{r_1} - v_{\theta_1} + \frac{k}{r_0 v_{\theta_1}} = 0, \tag{8.58}$$

which constitute a system of three equations for the three unknowns v_{r_1}, v_{θ_1}, and p. This system is easily solved through a sequence of eliminations as follows. Equation (8.56) *yields*

$$p = -v_{r_1}. \tag{8.59}$$

Equations (8.57) *and* (8.59) *yield*

$$v_{\theta_1} + v_{r_1} + \frac{v_{r_1} k}{r_0 v_{\theta_1}^2} = 0. \tag{8.60}$$

Now, (8.58) *can be rewritten as*

$$v_{r_1} = v_{\theta_1} - \frac{k}{r_0 v_{\theta_1}}. \tag{8.61}$$

Equations (8.60) *and* (8.61) *yield, after grouping terms,*

$$2v_{\theta_1} - \frac{k^2}{r_0^2 v_{\theta_1}^3} = 0, \tag{8.62}$$

which can be solved for v_{θ_1} *as*

$$v_{\theta_1} = -\sqrt{\frac{k}{r_0\sqrt{2}}}, \tag{8.63}$$

where we have chosen the negative sign to ensure that $v_{r_1} > 0$*, that is, that the ballistic launch clears the surface of the planet. Then,* (8.61) *and* (8.63) *yield*

$$v_{r_1} = (\sqrt{2}-1)\sqrt{\frac{k}{r_0\sqrt{2}}}, \tag{8.64}$$

and, finally, (8.64) *and* (8.59) *yield*

$$p = (1-\sqrt{2})\sqrt{\frac{k}{r_0\sqrt{2}}}. \tag{8.65}$$

Equations (8.63), (8.64), *and* (8.65) *are the solution of the first-order necessary conditions for optimality.*

For the second-order analysis, we evaluate the Hessian of the Lagrangian as follows:

$$\frac{\partial^2 L}{\partial(v_{r_1}, v_{\theta_1})^2} = \begin{bmatrix} 1 & 0 \\ 0 & 1 + \frac{2pk}{r_0 v_{\theta_1}^3} \end{bmatrix}. \tag{8.66}$$

At the candidate optimum, (8.63)–(8.65) *imply that this Hessian is*

$$\frac{\partial^2 L}{\partial(v_{r_1}, v_{\theta_1})^2} = \begin{bmatrix} 1 & 0 \\ 0 & 1 + 2\sqrt{2}(\sqrt{2}-1) \end{bmatrix}. \tag{8.67}$$

At this point, the attentive reader may recognize that this matrix is positive definite and hence we could expediently conclude, as per Remark 8.13, that (8.63)–(8.65) *provide a minimum. However, for the sake of instruction, we carry out the second-order analysis to a dutiful end, as follows. The tangent plane* (8.48) *is defined by the linear homogenous equation*

$$\begin{bmatrix} 1 & -1 - \frac{k}{r_0 v_{\theta_1}^2} \end{bmatrix} \begin{bmatrix} \delta v_{r_1} \\ \delta v_{\theta_1} \end{bmatrix} = 0, \tag{8.68}$$

for which (8.63)–(8.65) *yield*

$$\begin{bmatrix} 1 & -(1+\sqrt{2}) \end{bmatrix} \begin{bmatrix} \delta v_{r_1} \\ \delta v_{\theta_1} \end{bmatrix} = 0. \tag{8.69}$$

A basis for this tangent plane is

$$\begin{bmatrix} 1+\sqrt{2} \\ 1 \end{bmatrix}. \tag{8.70}$$

Hence, we evaluate the sign of the 1×1 *matrix:*

$$\begin{bmatrix} 1+\sqrt{2} & 1 \end{bmatrix} \begin{bmatrix} 1 & 0 \\ 0 & 1 + 2\sqrt{2}(\sqrt{2}-1) \end{bmatrix} \begin{bmatrix} 1+\sqrt{2} \\ 1 \end{bmatrix} = (1+\sqrt{2})^2 + 1 + 2\sqrt{2}(\sqrt{2}-1). \tag{8.71}$$

Because this last expression is positive, we conclude, as per Proposition 8.8, that (8.63)–(8.65) provide the least energetic ballistic shot, which confirms the results of Example 6.1.

EXAMPLE 8.6 (Maximum Lift-to-drag Ratio) *Consider the computation of the maximum lift-to-drag ratio achievable by a conventional aircraft in steady flight, assuming standard models for lift and drag. Mathematically, the problem is*

$$\max_{C_L, C_D} \frac{C_L}{C_D}, \tag{8.72}$$

subject to

$$C_D = C_{D0} + KC_L^2, \tag{8.73}$$

where (8.73) is the **drag polar equation**, *which describes the aerodynamics of the aircraft, C_L and C_D are the nondimensional lift and drag coefficients, respectively, and C_{D0} and K are constants.*

The Lagrangian L is

$$L(C_L, C_D, p) = \frac{C_L}{C_D} + p\left(C_{D0} + KC_L^2 - C_D\right), \tag{8.74}$$

where p is the Lagrange multiplier. The first-order conditions are

$$\begin{aligned} \frac{\partial L}{\partial C_L} &= \frac{1}{C_D} + 2pKC_L = 0, \\ \frac{\partial L}{\partial C_D} &= -\frac{C_L}{C_D^2} - p = 0, \\ \frac{\partial L}{\partial p} &= C_{D0} + KC_L^2 - C_D = 0. \end{aligned} \tag{8.75}$$

Solving this system of algebraic equations yields

$$C_L = \sqrt{\frac{C_{D0}}{K}} \tag{8.76}$$

and

$$C_D = 2C_{D0}, \tag{8.77}$$

or equivalently,

$$\left(\frac{C_L}{C_D}\right)_{\max} = \frac{1}{2\sqrt{KC_{D0}}}. \tag{8.78}$$

The second-order analysis is left as an exercise (see Problem 8.4).

EXAMPLE 8.7 (Rocket Staging) *As mentioned in the introduction to this chapter, it takes 80 kg of rocket structure, engine, fuel, and propellant to lift 1 kg of payload from Earth's surface into orbit. Just before burnout, rockets use propellant to lift an empty propellant tank. This is clearly wasteful. The idea behind* **rocket staging** *is to jettison part of the tank "as we go."*

Let us start by consider the ***single-stage case****. Initially, the mass of the rocket is* $m_0 = m_p + m_s + m_l$*, where* m_p *is the mass of propellant,* m_s *is the mass of the structure and engine, and* m_l *is the payload mass. The final mass is* $m_s + m_l$*.*

The ***Tsiolkovsky rocket equation****, or* ***ideal rocket equation****, describes the motion of rockets. It relates the* Δv *(the maximum change of speed of the rocket if no other external forces act) with the effective exhaust velocity and the initial and final mass of the rocket:*

$$\begin{aligned} ||\Delta\vec{v}|| &= c \log \frac{m_0}{m_s + m_l} \\ &= c \log \left(1 + \frac{m_p}{m_s + m_l}\right), \end{aligned} \tag{8.79}$$

where c is the effective exhaust velocity and is dictated by the chemistry. Let

$$\epsilon = \frac{m_s}{m_p + m_s} \tag{8.80}$$

be the structural coefficient and

$$m = m_p + m_s \tag{8.81}$$

be the mass of the single stage. Then,

$$\Delta v = c \log \frac{m + m_l}{\epsilon m + m_l}. \tag{8.82}$$

In the ***two-stage case****, let the initial mass be* $m_0 = m_{s1} + m_{p1} + m_{l1} + m_{s2} + m_{p2} + m_{l2}$*. The mass right after the first burnout is* $m_{s1} + m_{s2} + m_{p2} + m_l$*, and* Δv_1 *is given by*

$$\Delta v_1 = c_1 \log \frac{m_{s1} + m_{p1} + m_{s2} + m_{p2} + m_l}{m_{s1} + m_{s2} + m_{p2} + m_l}. \tag{8.83}$$

The mass just before the second stage kicks in is $m_{s2} + m_{p2} + m_l$*, and the mass after the second burnout is* $m_{s2} + m_l$*. The second velocity increment,* Δv_2*, is given by*

$$\Delta v_2 = c_2 \log \frac{m_{s2} + m_{p2} + m_l}{m_{s2} + m_l}. \tag{8.84}$$

Overall,

$$\Delta v = \Delta v_1 + \Delta v_2. \tag{8.85}$$

Define $m_1 = m_{s1} + m_{p1}$ *and* $m_2 = m_{s2} + m_{p2}$ *to be the mass of the first and second stages, respectively. Let* $\epsilon_1 = \frac{m_{s1}}{m_1}$ *and* $\epsilon_2 = \frac{m_{s2}}{m_2}$ *be the first and second structural parameters, respectively. Then, the* ***two-stage rocket equation*** *is*

$$\Delta v = c_1 \log \frac{m_1 + m_2 + m_l}{\epsilon_1 m_1 + m_2 + m_l} + c_2 \log \frac{m_2 + m_l}{\epsilon_2 m_2 + m_l}. \tag{8.86}$$

We can extend this to the ***n-stage case****, where we have*

$$\Delta v = \sum_{i=1}^{n} \Delta v_i, \tag{8.87}$$

where

$$\Delta v_i = c_i \log \frac{m_i + m_{i+1} + \cdots + m_n + m_l}{\epsilon_i m_i + m_{i+1} + \cdots + m_n + m_l}. \tag{8.88}$$

Note that typically ϵ_i is known and depends on materials used in rockets. For example, for the Ariane IV rocket, $\epsilon_1 = 0.7$, $\epsilon_2 = 0.01$, and $\epsilon_3 = 0.1$.

Optimal staging can be used to decide how to distribute mass to obtain the largest Δv, or alternatively to minimize mass for a given Δv. Mathematically, the problem we wish to solve is

$$\min f(m_1, \dots, m_n) = m_1 + m_2 + \cdots + m_n \tag{8.89}$$

subject to

$$\sum_{i=1}^{n} c_i \log \frac{m_i + m_{i+1} + \cdots + m_n + m_l}{\epsilon_i m_i + m_{i+1} + \cdots + m_n + m_l} - \Delta v_d = g(m_1, \dots, m_n) = 0, \tag{8.90}$$

where Δv_d is the desired Δv.

Let the Lagrangian be

$$L(m_1, m_2, \dots, m_n, p) = f(m_1, \dots, m_n) + p g(m_1, \dots, m_n), \tag{8.91}$$

where p is a scalar. The first-order conditions are

$$\begin{aligned} \frac{\partial L}{\partial m_1} &= 0, \\ &\dots, \\ \frac{\partial L}{\partial m_n} &= 0, \\ \frac{\partial L}{\partial p} &= 0, \end{aligned} \tag{8.92}$$

yielding $n+1$ algebraic equations in the unknowns $m_1, \dots, m_n$ and p.

For example, in the two-stage case, the problem under consideration is

$$\min f(m_1, m_2) = m_1 + m_2 \tag{8.93}$$

subject to

$$c_1 \log \frac{m_1 + m_2 + m_l}{\epsilon_1 m_1 + m_2 + m_l} + c_2 \log \frac{m_2 + m_l}{\epsilon_2 m_2 + m_l} - \Delta v = 0. \tag{8.94}$$

The Lagrangian is

$$L(m_1, m_2, p) = m_1 + m_2 + p \left(c_1 \log \frac{m_1 + m_2 + m_l}{\epsilon_1 m_1 + m_2 + m_l} + c_2 \log \frac{m_2 + m_l}{\epsilon_2 m_2 + m_l} - \Delta v \right). \tag{8.95}$$

The first-order conditions are

$$\begin{aligned} \frac{\partial L}{\partial m_1} &= 1 + p c_1 \frac{\epsilon_1 m_1 + m_2 + m_l}{m_1 + m_2 + m_l} \frac{\epsilon_1 m_1 + m_2 + m_l - \epsilon_1 (m_1 + m_2 + m_l)}{(\epsilon_1 m_1 + m_2 + m_l)^2} = 0, \\ \frac{\partial L}{\partial m_2} &= 1 + p c_1 \frac{\epsilon_1 m_1 + m_2 + m_l}{m_1 + m_2 + m_l} \frac{\epsilon_1 m_1 + m_2 + m_l - \epsilon_1 (m_1 + m_2 + m_l)}{(\epsilon_1 m_1 + m_2 + m_l)^2} \\ &\quad p c_2 \frac{\epsilon_2 m_2 + m_l}{m_2 + m_l} \frac{\epsilon_2 m_2 + m_l - \epsilon_2 (m_2 + m_l)}{(\epsilon_2 m_2 + m_l)^2} = 0, \\ \frac{\partial L}{\partial p} &= c_1 \log \frac{m_1 + m_2 + m_l}{\epsilon_1 m_1 + m_2 + m_l} + c_2 \log \frac{m_2 + m_l}{\epsilon_2 m_2 + m_l} - \Delta v = 0, \end{aligned} \tag{8.96}$$

yielding three equations that can be solved for m_1, m_2, and p.

8.3 Inequality Constraints on $\mathbb{R}^n$

We use the following notation. Let $g : \mathbb{R}^n \to \mathbb{R}^m$. Then, $g(x) \leq 0$ if and only if $g_i(x) \leq 0, 1 \leq i \leq m$.

Let $X = \{x \in \mathbb{R}^n | g(x) \leq 0\}$. We seek

$$\min_{x \in X} f(x). \tag{8.97}$$

Note that this generalizes the results of the previous section dealing with equality constraints, because equality constraints of the form $g(x) = 0$ can be rewritten as inequality constraints of the form $g(x) \leq 0 \wedge -g(x) \leq 0$.

PROPOSITION 8.9 (Karush–Kuhn–Tucker (KKT) Conditions) *Assume that f and g are functions of class C^2, that x^* is an optimum, and that the active constraints (i.e., those for which $g_i(x^*) = 0$) are linearly independent at x^*. Then, there exist $p_1, p_2, \ldots, p_m \in \mathbb{R}$ such that*

$$\begin{aligned} \left(\frac{\partial f}{\partial x}\right)_{x^*} + \sum_{i=1}^{m} p_i \left(\frac{\partial g}{\partial x_i}\right)_{x^*} &= 0, \\ g_i(x^*) &\leq 0, \ 1 \leq i \leq m, \\ p_i g_i(x^*) &= 0, \ 1 \leq i \leq m, \\ p_i &\geq 0, \ 1 \leq i \leq m. \end{aligned} \tag{8.98}$$

REMARK 8.14 *Note that* (8.98) *is a system of $n + m$ equations in $n + m$ unknowns.*

EXAMPLE 8.8 *Let us consider the following example:*

$$\min_{x,y} J = (5 - x - y)^2 \tag{8.99}$$

$$\textit{subject to } x^2 + y^2 = 1,$$

that is, the problem of finding the point on the unit disk that is closest to the line $x + y = 5$, as shown in Figure 8.6.

The Lagrangian is given by

$$L = (5 - x - y)^2 + p(x^2 + y^2 - 1). \tag{8.100}$$

The necessary conditions for optimality are

$$\begin{aligned} \frac{\partial L}{\partial x} &= -2(5 - x - y) + 2px = 0, \\ \frac{\partial L}{\partial y} &= -2(5 - x - y) + 2py = 0, \\ p(x^2 + y^2 - 1) &= 0, \\ p &\geq 0, \\ x^2 - y^2 &\leq 1. \end{aligned} \tag{8.101}$$

We consider first the case of $p = 0$, and second the case of $p > 0$.

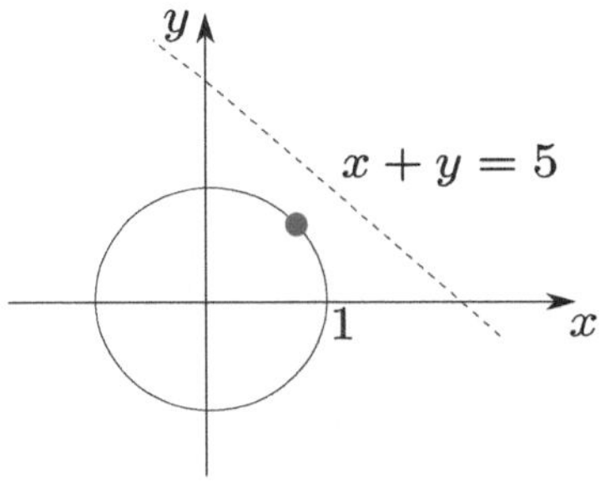

Figure 8.6. Setup for example illustrating use of the KKT conditions.

Case 1: *$p = 0$: If $p = 0$, then $x + y = 5$, and $y = 5 - x$. Then,*

$$x^2 + (5 - x)^2 \leq 1, \tag{8.102}$$

or, equivalently,

$$2x^2 - 10x + 24 \leq 0. \tag{8.103}$$

The left-hand side achieves a minimum for $4x - 10 = 0$, that is, for $x = 5/2$. For that value of x, the minimum is $23/2 > 0$, which is impossible.

Case 2: *$p > 0$: If $p > 0$, then $x = y = \frac{5}{p+2}$. In that case,*

$$x^2 + y^2 - 1 = 0, \tag{8.104}$$

so

$$p = -2 + \sqrt{50}, \tag{8.105}$$

and

$$x = y = \frac{\sqrt{2}}{2}, \tag{8.106}$$

which is illustrated in Figure 8.6.

8.4 Optimal Control of Discrete-Time Systems

Let us consider the following dynamic system, where the state $x(k) \in \mathbb{R}^n$, the control $u(k) \in \mathbb{R}^m$, and the dynamics are given by

$$\begin{aligned} x(k+1) &= x(k) + f(x(k), u(k), k), \;\; k_0 \leq k \leq k_f, \\ x(k_0) &= x_0 \text{ given}, \\ x(k_f + 1) & \quad \text{free}, \end{aligned} \tag{8.107}$$

and where the unknowns are $u(k_0), u(k_1), \ldots, u(k_f)$. We seek to solve the minimization problem:

$$\min_{u(k_0),\ldots,u(k_f)} J = K(x(k_f + 1)) + \sum L(x(k), u(k), k). \tag{8.108}$$

Let us define x as

$$x = \begin{bmatrix} x_1 \\ x_2 \end{bmatrix} = \begin{bmatrix} x(k_1) \\ x(k_2) \\ \dots \\ x(k_f+1) \\ u(k_0) \\ \dots \\ u(k_f) \end{bmatrix}, \tag{8.109}$$

where x_1 is of order $(k_f - k_0 + 1)n$ and x_2 is of order $(k_f - k_0 + 1)m$. Then, we seek

$$\min_x J(x) \tag{8.110}$$
$$\text{subject to } F(x) = 0,$$

where $F(x)$ has components

$$F(x) = \begin{bmatrix} x(k_1) - x(k_0) - f(x(k_0), u(k_0), k_0) \\ x(k_2) - x(k_1) - f(x(k_1), u(k_1), k_1) \\ \dots \\ x(k_f+1) - x(k_f) - f(x(k_f), u(k_f), k_f) \end{bmatrix} \tag{8.111}$$

and $F(x)$ is of order $(k_f - k_0 + 1)n$. To check regularity, we show that $\frac{\partial F}{\partial x_1}$ is nonsingular, as follows:

$$\frac{\partial F}{\partial x_1} = \begin{bmatrix} I & -I - \frac{\partial f(x(k_1),u(k_1),k_1)}{\partial x(k_1)} & 0 & \dots & \dots & 0 \\ 0 & I & -I - \frac{\partial f(x(k_2),u(k_2),k_2)}{\partial x(k_2)} & 0 & \dots & 0 \\ 0 & 0 & I & \dots & \dots & 0 \\ 0 & 0 & \dots & \dots & 0 & I \end{bmatrix}. \tag{8.112}$$

This matrix is square, of order $(k_f - k_0 + 1)n$, upper triangular, and with ones on the diagonal. Therefore, it is nonsingular. Note that this nonsingularity guarantees uniqueness of the solution to the difference equation (8.107).

Consider the sequence of Lagrange multipliers $p(k) \in \mathbb{R}^n$, $p(k_1)$, $p(k_2)$, ..., $p(k_f + 1)$. Then, the Lagrangian is

$$\begin{aligned} \mathcal{L} = {} & K(x(k_f+1)) + \sum_{k=k_0}^{k_f} L(x(k_f), u(k_f), k) \\ & + \sum_{k=k_0}^{k_f} p^T(k+1)\,(x(k+1) - x(k) - f(x(k), u(k), k)). \end{aligned} \tag{8.113}$$

The necessary conditions for optimality are

$$\frac{\partial \mathcal{L}}{\partial x(k)} = p(k) - p(k+1) - \frac{\partial}{\partial x(k)} f(x(k), u(k), k) p(k+1)$$
$$+ \frac{\partial}{\partial x(k)} L(x(k), u(k), k) = 0, \; k_1 \le k \le k_f, \tag{8.114}$$
$$\frac{\partial \mathcal{L}}{\partial x(k_f+1)} = \frac{\partial K(x(k_f+1))}{\partial x(k_f+1)} + p(k_f+1) = 0,$$
$$\frac{\partial \mathcal{L}}{\partial u(k)} = \frac{\partial}{\partial u(k)} L(x(k), u(k), k) - \frac{\partial}{\partial u(k)} f(x(k), u(k), k) p(k+1) = 0, k_1 \le k \le k_f.$$

Summarizing our progress, we have four types of equations so far: the **state dynamics**, which are propagated forward in time, are given by (8.107) (line 1) and repeated here for convenience:

$$x(k+1) = x(k) + f(x(k), u(k), k), \; k_0 \le k \le k_f. \tag{8.115}$$

The **co-state dynamics**, propagated backward in time, are obtained from (8.114) (line 1) and also repeated for convenience:

$$p(k) = p(k+1) + \frac{\partial}{\partial x(k)} f(x(k), u(k), k) p(k+1)$$
$$- \frac{\partial}{\partial x(k)} L(x(k), u(k), k), \; k_1 \le k \le k_f. \tag{8.116}$$

The **optimality condition**, given by (8.114) (line 3), is

$$\frac{\partial \mathcal{L}}{\partial u(k)} = \frac{\partial}{\partial u(k)} L(x(k), u(k), k) - \frac{\partial}{\partial u(k)} f(x(k), u(k), k) p(k+1) \tag{8.117}$$
$$= 0, k_1 \le k \le k_f.$$

The **boundary conditions**, given by (8.107) (line 2) and (8.114) (line 2), are

$$x(k_0) = x_0 \tag{8.118}$$
$$p(k_f+1) = -\frac{\partial K(x(k_f+1))}{\partial x(k_f+1)}.$$

8.5 Summary of Key Results

The key results in Chapter 8 are as follows:

1. Proposition 8.1, which guarantees existence of a minimum
2. Proposition 8.2, which provides the basis for all necessary or sufficient conditions for local optimization
3. Proposition 8.3, which provides necessary conditions for unconstrained optimization
4. Proposition 8.4, which provides sufficient conditions for unconstrained optimization
5. Proposition 8.5, which guarantees existence and uniqueness of an optimizer
6. Proposition 8.6, which provides first-order necessary conditions for constrained optimization

7. Proposition 8.7, which provides second-order necessary conditions for constrained optimization
8. Proposition 8.8, which provides sufficient conditions for constrained optimization
9. Proposition 8.9, which provides necessary conditions for optimization under inequality constraints

8.6 Bibliographic Notes for Further Reading

The material in Chapter 8 is standard and is well covered in many texts, including [12] and [63].

One of the earliest formal applications of optimization in aeronautics was the optimization of range for a conventional airplane [2]. Since then, the theory has been applied in many aerospace instances, including optimization of wings [2], airline cruise altitude [75], rocket staging [36], ballistic reentry [75], and many more.

8.7 Homework Problems

PROBLEM 8.1 *Consider the constraint given by*

$$x_1^2 + x_2^2 - 1 = 0. \tag{8.119}$$

When can x_1 *be expressed as a function of* x_2*? Evaluate* $\frac{dx_1}{dx_2}$ *using the implicit function theorem.*

PROBLEM 8.2 *Prove that*

$$\underset{|u|\leq 1}{\text{argmax}}(au) = \text{sign}(a), \tag{8.120}$$

where the function sign(.) *is defined as follows:*

$$\text{sign}[\alpha] = \begin{cases} 1 & \text{if } \alpha > 0, \\ -1 & \text{if } \alpha < 0, \\ \textit{undefined} & \text{if } \alpha = 0. \end{cases} \tag{8.121}$$

PROBLEM 8.3 *Let* $(x, y) \in \mathbb{R}^2$*. Use the method of Lagrange multipliers to optimize* $J(x, y) = x^2 - y^2$ *subject to the constraint* $2x - y + 1 = 0$*. Carry out the analysis to second order to ascertain whether the optimum is a minimum or a maximum.*

PROBLEM 8.4 *Carry out the second-order analysis in Example 8.6 to ascertain that* (8.76) *and* (8.77) *do indeed provide the* ***maximum*** *lift-to-drag ratio.*

PROBLEM 8.5 *Perform numerically the optimization of Example 8.7 for a two-stage rocket, using the numerical values* $\epsilon_1 = 0.7$, $\epsilon_2 = 0.01$, $m_l = 1$, *and* Δv *appropriate for low-Earth orbit. Assume an exhaust speed of 3,000 m/s. Carry out the analysis to second order to ascertain the nature of the optimum.*

PROBLEM 8.6 *Let $x \in \mathbb{R}^n$, $A \in \mathbb{R}^{n\times n}$ be symmetric positive definite, $b \in \mathbb{R}^n$ and $c \in \mathbb{R}$.*

1. *Obtain in closed form the solution of the unconstrained optimization problem:*

$$\min_x f(x) = \frac{1}{2}x^T Ax + x^T b + c.$$

2. *Let $m < n$, $D \in \mathbb{R}^{m\times n}$ have full rank, and $e \in \mathbb{R}^m$. Obtain in closed form the solution of the constrained optimization problem:*

$$\min_x f(x) = \frac{1}{2}x^T Ax + x^T b + c,$$

subject to

$$Dx = e.$$

PROBLEM 8.7 *Let M and K be real $n \times n$ symmetric positive definite matrices. Obtain in closed form the solution of the constrained optimization problem:*

$$\min_x f(x) = \frac{1}{2}x^T Kx,$$

subject to

$$\frac{1}{2}x^T Mx = 1.$$

Give a geometric interpretation to the results.

PROBLEM 8.8 *Answer both parts.*

1. *For a fixed perimeter, what is the rectangle with maximum area?*
2. *For a fixed area, what is the rectangular parallelepiped with maximum volume?*

PROBLEM 8.9 *In designing a cylindrical canister, we want to maximize the volume for a fixed area. What is the optimal shape (e.g., radius to height ratio)?*

PROBLEM 8.10 *What is the parallelepiped with maximum volume inscribed in the ellipsoid:*

$$\frac{x^2}{a^2} + \frac{y^2}{b^2} + \frac{z^2}{c^2} = 1?$$

Assume that the edges are parallel to the axes.

PROBLEM 8.11 *Let $p > 1$ and $q > 1$ be such that*

$$\frac{1}{p} + \frac{1}{q} = 1.$$

Show that for all $a > 0$ and $b > 0$, we have

$$\frac{a^p}{p} + \frac{b^q}{q} \geq ab.$$

Hint: *Solve the constrained minimization problem:*

$$\min_{x,y} f(x,y) = \frac{x^p}{p} + \frac{y^q}{q},$$

subject to

$$xy = 1.$$

PROBLEM 8.12 *Let m and n be integers satisfying $m < n$, $Q \in \mathbb{R}^{n\times n}$ be symmetric, $b \in \mathbb{R}^n$, $A \in \mathbb{R}^{n\times m}$, and $c \in \mathbb{R}^m$. Show that if $x \in \mathbb{R}^n$ is a regular local optimum for the problem*

$$\min_x \frac{1}{2}x^T Qx - b^T x,$$

subject to

$$A^T x = c,$$

then x is a global optimum.

PROBLEM 8.13 *Evaluate the sign of the matrix*

$$M = \begin{bmatrix} 4 & 3 & 2 \\ 3 & 1 & 1 \\ 2 & 1 & 1 \end{bmatrix},$$

restricted to the subspace

$$T = \left\{x \in \mathbb{R}^3 : x_1 + x_2 + x_3 = 0\right\}.$$

PROBLEM 8.14 (Least Squares Polynomial Interpolation) *An experiment has produced n measurements $(x_1, y_1), (x_2, y_2), \ldots, (x_n, y_n)$ where, for $1 \le i, j \le n$, we have $x_i \in \mathbb{R}$, $y_i \in \mathbb{R}$, and $x_i \neq x_j$. We want to model the variable y as a polynomial function of the variable x with degree $m < n - 1$. In other words, given the polynomial*

$$p(\xi) = \alpha_m \xi^m + \alpha_{m-1}\xi^{m-1} + \cdots + \alpha_1 \xi + \alpha_0,$$

and defining

$$v_i = p(x_i) - y_i,$$

we want to find the values of the coefficients $\alpha_0, \alpha_1, \ldots, \alpha_m$ that minimize the cost function

$$f(\alpha) = \frac{1}{2}\sum_{i=1}^{n} v_i^2.$$

1. *What is the optimal value of the coefficients?*
2. *Show that this is indeed a minimum.*

PROBLEM 8.15 *Answer both parts.*

1. *Find, if possible, a minimizer for the function*

$$f(x,y,z) = 2x^2 + xy + y^2 + yz + z^2 - 6x - 7y - 8z + 9.$$

2. *For the same function, find, if possible, a minimizer under the constraint*

$$x + y + z = 1.$$

PROBLEM 8.16 *Let $x \in \mathbb{R}^n$, $y \in \mathbb{R}^n$, $a \in \mathbb{R}^n$, $b \in \mathbb{R}^n$, and $\alpha \in \mathbb{R}$. Solve the constrained minimization problem*

$$\min_{x,y} f(x, y) = x^T x + x^T y + y^T y - 2b^T x - 2a^T y,$$

subject to

$$a^T x + b^T y = \alpha.$$

Show that you obtain a minimum indeed.

PROBLEM 8.17 *Solve the minimization problem*

$$\min_{x,y} f(x, y) = -xy,$$

subject to

$$g(x, y) = (x - 3)^2 + y^2 - 5 = 0.$$

PROBLEM 8.18 *For the problem*

$$\max_{x,y} xy,$$

subject to

$$(x + y - 2)^2 = 0,$$

an (almost) obvious solution is

$$(x, y) = (1, 1).$$

However, writing the first-order necessary conditions using a Lagrange multiplier leads to a difficulty. Explain why.

9 Optimal Guidance

This chapter treats optimal control for the purpose of application to optimal guidance. Optimal control can be used to find, from among several candidates, the best trajectory to accomplish a mission.

Note that Chapter 8 provides the tools to perform optimization in finite dimensional spaces. In practice, trajectory optimization can be, through appropriate discretization, transformed into a finite dimensional optimization problem; see, for instance, Section 8.4. So, the reader may ask: Why study optimal control beyond finite dimensional optimization? There are at least two compelling pragmatic answers to this question. First, after discretization, evaluating the gradients of the objective function and the constraints is generally much more cumbersome than writing the differential equations stemming from optimal control. Second, optimal control reveals the special structure of the state – co-state dynamics, the Hamiltonian or symplectic structure, which can be exploited in both analysis and computation.

Section 9.1 introduces and formulates the optimal control problem. Section 9.2 gives examples of optimal guidance problems fitting the formulation. Section 9.3 extends the results of Section 8.4 to obtain necessary conditions for optimal control without control constraints. Section 9.4 treats the case of control constraints, yielding the maximum principle. Section 9.5 is devoted to dynamic programming, which provides sufficient conditions. Section 9.6 elucidates the relationship between the results on optimal control and dynamic programming. Sections 9.7, 9.8, and 9.9 present a summary of the key results in the chapter, bibliographic notes for further reading and homework problems, respectively.

9.1 Problem Formulation

Optimal control is concerned with optimizing a functional, defined by a dynamic system, with respect to the time history of the control input.

DEFINITION 9.1 *A **functional** is a mapping that assigns a real number to each function belonging to some class.*

EXAMPLE 9.1 *Let* $\mathcal{C}^1 = \{f : \mathbb{R} \to \mathbb{R} : x \mapsto f(x) | f \text{ is continuous}\}$. *Then,*

$$J_0 : \mathcal{C}^1 \to \mathbb{R} : f \mapsto J_0(f) = f(x_0) \tag{9.1}$$

is called a ***sampling functional;***

$$J_1 : \mathcal{C}^1 \to \mathbb{R} : f \mapsto J_1(f) = \int_0^1 f(x)\,\mathrm{d}x \tag{9.2}$$

is an ***integral functional.***

REMARK 9.1 *Note that in optimal control, terminology is not unique, and that the following expressions all relate to functionals: cost functional, penalty function, objective function, performance index, and performance criterion.*

For the formulation of the optimal control problem, we need four basic ingredients: the system dynamics, the boundary conditions, a performance index, and constraints. We review these in turn.

The *system dynamic equations* are of the form

$$\begin{aligned} \dot{x} &= f(x, u, t) \\ y &= g(x, t), \end{aligned} \tag{9.3}$$

where $x \in \mathbb{R}^n$ is the state vector, $u \in \mathbb{R}^m$ is the input vector, and $y \in \mathbb{R}^p$ is the output vector.

The *boundary conditions* include the starting time t_0, the initial state $x_0 = x(t_0)$, the final time t_f, and the final state $x_f = x(t_f)$. Some of the boundary conditions may be specified (fixed), and some may be free. Different combinations are possible, including the specification of target sets.

The *performance index* is a functional of the form

$$J = K(x_f, t_f) + \int_{t_0}^{t_f} L(x(t), u(t), t)\,\mathrm{d}t. \tag{9.4}$$

Constraints can be on the control input, for example, $|u_i| \leq 1$, which is a very common type of constraint. Constraints can also be imposed on the state, for example, $G(x_f, t_f) = 0$, known as a target set constraint, or $|x_i(t)| \leq X_i$, which is a very challenging form of constraint. The constraints determine $\mathcal{U}$, the set of admissible control histories, and $\mathcal{X}$, the set of admissible state trajectories.

With these ingredients in place, we can now state the **optimal control problem** as follows: Find $u(.) \in \mathcal{U}$ that drives the system from x_0 at t_0 to x_f at t_f, subject to the dynamics (9.3), so as to minimize (9.4). In other words,

$$\min_{u \in \mathcal{U}, x_f, t_f} J = K(x_f, t_f) + \int L(x(t), u(t), t)\,\mathrm{d}t, \tag{9.5}$$

subject to

$$\begin{aligned} \dot{x} &= f(x, u, t), \\ x(t_0) &= x_0, \\ x(t_f) &= x_f. \end{aligned} \tag{9.6}$$

Our goals are to characterize solutions through necessary conditions, sufficient conditions, and/or necessary and sufficient conditions.

Particular cases of performance indices include

$$J = K(x_f, t_f), \tag{9.7}$$

which leads to the **Mayer Problem**;

$$J = \int_{t_0}^{t_f} L(x(t), u(t), t)\,\mathrm{d}t, \tag{9.8}$$

which leads to the **Lagrange Problem**; and

$$J = K(x_f, t_f) + \int_{t_0}^{t_f} L(x(t), u(t), t)\,\mathrm{d}t, \tag{9.9}$$

which is known as the **Bolza Problem**.

Control taxonomies can be organized around the form of control input that is sought. For the general system description (9.3), different types of control inputs may be derived:

1. If $u = u(t)$, we are searching for an **open-loop control input**.
2. If $u = u(x(t), t)$, we are searching for a **closed-loop control input**, in the form of a **state feedback policy**.
3. If $u = u(y(.), t)$, we are searching for a **closed-loop control input**, based on **output feedback**, where $y(.)$ denotes the whole time history of $y(t)$.

Generally, optimal control can be used to yield open-loop control inputs relatively inexpensively and closed-loop state feedback control inputs with much effort. Occasionally, closed-loop output feedback laws can be obtained relatively inexpensively, as in the Linear Quadratic Gaussian (LQG) control case.

Another possible organization of control strategies pertains to the space $\mathcal{U}$ in which the control u is to be found. Traditionally, two different control input development techniques are used, depending on the nature of $\mathcal{U}$:

1. $\mathcal{U}$ is a continuum. Here control inputs can be obtained relatively easily using methods of differential calculus.
2. $\mathcal{U}$ is discrete. Here control inputs are relatively difficult to obtain, requiring efficient search methods.

In this chapter, we focus on continuous optimal control problems.

An optimal control problem (9.5), (9.6) is called **scleronomic** or time invariant if it has the following three properties:

$$\frac{\partial}{\partial t} f(x, u, t) \equiv 0,$$
$$\frac{\partial}{\partial t} L(x, u, t) \equiv 0,$$
$$\frac{\partial}{\partial t_f} K(x_f, t_f) \equiv 0. \tag{9.10}$$

Otherwise, it is called **rheonomic** or time varying. In a scleronomic problem, shifting the initial and final time by a common value just causes the solution to be time shifted by the same amount. Hence, in a scleronomic problem, closed-loop control laws are stationary, that is, of the form $u = u(x(t))$ rather than $u = u(x(t), t)$. In other words, in a scleronomic problem, closed-loop control laws do not depend explicitly on time. In a rheonomic problem, closed-loop control laws typically depend explicitly on time.

9.2 Examples

In this section, we illustrate the problem formulation (9.5), (9.6) by showing how several optimal control problems fit this formalism. Specifically, we show in each case what the dynamic equations, boundary conditions, performance index, and constraints are.

EXAMPLE 9.2 (Time-optimal Control of Double Integrator) *Consider the large class of double integrator systems introduced and motivated in Example 2.3. In such systems, it is sometimes desirable to drive the state from an arbitrary initial condition to a specified final position with zero velocity as quickly as possible, using a bounded input.*

The system dynamic equations have the form (2.18) *and are repeated here for convenience:*

$$\begin{bmatrix} \dot{x}_1 \\ \dot{x}_2 \end{bmatrix} = \begin{bmatrix} 0 & 1 \\ 0 & 0 \end{bmatrix} \begin{bmatrix} x_1 \\ x_2 \end{bmatrix} + \begin{bmatrix} 0 \\ 1 \end{bmatrix} u$$
$$y = \begin{bmatrix} 1 & 0 \end{bmatrix} \begin{bmatrix} x_1 \\ x_2 \end{bmatrix}, \tag{9.11}$$

where x_1 is the linear or angular position, x_2 is the linear or angular velocity, the input u is a specific force or moment, and the output y is the linear or angular position. The boundary conditions are

$$\begin{bmatrix} x_1(0) \\ x_2(0) \end{bmatrix} \quad \textit{given},$$
$$\begin{bmatrix} x_1(t_f) \\ x_2(t_f) \end{bmatrix} = \begin{bmatrix} 0 \\ 0 \end{bmatrix}, \tag{9.12}$$

where, without loss of generality, the initial time is 0 and the final condition is at the origin of the state space.

The performance index is

$$J(u) = \int_0^{t_f} \mathrm{d}t, \tag{9.13}$$

which quantifies the elapsed time.

Boundedness of the input can be expressed as the constraint

$$|u| \leq 1. \tag{9.14}$$

Note that there is no loss of generality in assuming that the amplitude constraint for the input is 1: any constant amplitude constraint can be reformulated as (9.14) *through appropriate time scaling of* (9.11) *(see Problem 9.2).*

Equations (9.11)–(9.14) *are the four ingredients of an optimal control problem.*

EXAMPLE 9.3 (Path Planning with Obstacle Avoidance) *Here the problem is to steer a vehicle in planar motion with constant speed and bounded turn rate, between specified locations, in minimum time, while avoiding specified obstacles (see Figure 9.1).*

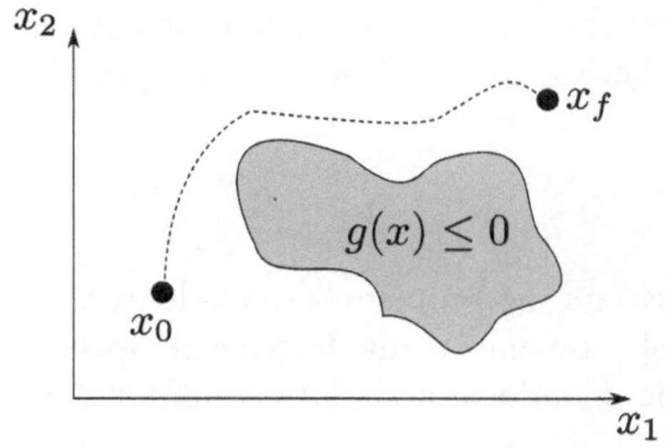

Figure 9.1. Path planning with obstacle avoidance: An example of optimal control problem.

The system dynamic equations have the form

$$\dot{x}_1 = \cos\theta,$$
$$\dot{x}_2 = \sin\theta,$$
$$\dot{\theta} = u, \tag{9.15}$$

where x_1 and x_2 are the Cartesian coordinates of the vehicle, θ is its heading angle, the control input u is the turn rate, and, without loss of generality, the constant speed is 1.

The boundary conditions are

$$\begin{bmatrix} x_1(0) \\ x_2(0) \end{bmatrix} = x_0 \text{ given} \tag{9.16}$$

$$\begin{bmatrix} x_1(t_f) \\ x_2(t_f) \end{bmatrix} = x_f \text{ given.}$$

The performance index is

$$J(u) = \int_0^{t_f} \mathrm{d}t. \tag{9.17}$$

The constraints have the form

$$|u| \leq U$$
$$g(x) \geq 0, \tag{9.18}$$

which reflect boundedness of the turn rate and obstacle avoidance, respectively.

Equations (9.15)–(9.18) *are the four ingredients of an optimal control problem.*

EXAMPLE 9.4 (Path Planning for Optimal Information Collection) *Here the problem is to steer a vehicle in planar motion with constant speed, between specified locations, while collecting the maximum amount of information about an object of interest. We make the following standard assumptions: (1) that the rate of information collection is a logarithmic function of the signal-to-noise ratio of the sensors, following Shannon's equation [65]; and (2) that the signal-to-noise ratio of the sensors decays as the fourth power of the range, as typical of radars [67]. Further details are given in [39].*

The system dynamic equations have the form

$$\dot{x}_1 = \cos\theta,$$
$$\dot{x}_2 = \sin\theta,$$
$$\dot{x}_3 = a\log\left(1 + \frac{b}{\left(x_1^2 + x_2^2\right)^2}\right), \tag{9.19}$$

where x_1 and x_2 are the Cartesian coordinates of the vehicle, x_3 is the amount of information collected, θ is the heading angle and control input, a and b are positive constants that characterize the quality of the sensor and visibility of the object, respectively, and we assume, without loss of generality, that the constant speed is 1.

The boundary conditions are

$$\begin{aligned} x_1(0), x_2(0) &\quad \textit{given}, \\ x_3(0) &= 0, \\ x_1(t_f), x_2(t_f) &\quad \textit{given}, \\ x_3(t_f) &\quad \textit{free}, \\ t_f &\quad \textit{given}. \end{aligned} \tag{9.20}$$

The objective function is

$$J(\theta(.)) = -x_3(t_f), \tag{9.21}$$

where the negative sign reflects the fact that we want to ***maximize*** *the final amount of information. Equations* (9.19)–(9.21) *specify an optimal control problem.*

EXAMPLE 9.5 (Orbital Maneuvers using Low-thrust Propulsion) *Consider the motion of a spacecraft in a long-term, deep-space mission, subject to a low-thrust propulsion system such as electric propulsion or solar sails. We write the equations of motion in terms of the six classical orbital elements: the semimajor axis a, the eccentricity e, the argument at periapse ω, the inclination i, the longitude of the ascending node Ω, and the mean anomaly M. It is convenient to use these elements as states because, in the absence of thrust, these elements are constant, except the last one, which is a linear function of time. Hence, the equations of motion take a very simple form. Assume that the thrusters provide a specific force with components u_x, u_y, u_z. Then Gauss's equations of motion for the orbital elements are [19]*

$$\begin{aligned}
\frac{da}{dt} &= \frac{2}{n\sqrt{1-e^2}}\left(u_z e \sin f + \frac{a(1-e^2)}{r}u_y\right), \\
\frac{de}{dt} &= \frac{\sqrt{1-e^2}}{na}\left(u_z \sin f + \left(\frac{e+\cos f}{1+e\cos f} + \cos f\right)u_y\right), \\
\frac{d\omega}{dt} &= \frac{\sqrt{1-e^2}}{nea}\left(-u_z \cos f + \left(1+\frac{r}{a(1-e^2)}\right)u_y \sin f\right) - \dot{\Omega}\cos i, \\
\frac{di}{dt} &= \frac{1}{na\sqrt{1-e^2}}\frac{r}{a}\cos(\omega+f)u_x, \\
\frac{d\Omega}{dt} &= \frac{1}{na\sqrt{1-e^2}}\frac{r}{a}\frac{\sin(\omega+f)}{\sin i}u_x, \\
\frac{dM}{dt} &= n + \frac{1}{na}\left(\frac{2r}{a} - \frac{(1-e^2)}{e}\cos f\right)u_z - \frac{(1-e^2)}{nae}\left(1+\frac{r}{a(1-e^2)}\right)u_y \sin f,
\end{aligned} \tag{9.22}$$

where n is the mean motion, r is the radius, and f is the true anomaly. In this framework, an orbital maneuver consists of changing the orbital elements according to (9.22).

For given initial and final orbits, the boundary conditions are

$$\begin{bmatrix} a(0) \\ e(0) \\ \omega(0) \\ i(0) \\ \Omega(0) \\ M(0) \end{bmatrix} \text{given,} \quad \begin{bmatrix} a(t_f) \\ e(t_f) \\ \omega(t_f) \\ i(t_f) \\ \Omega(t_f) \\ M(t_f) \end{bmatrix} \text{given.} \tag{9.23}$$

Consider the following performance index:

$$J(u) = \int_0^{t_f} \sqrt{u_x^2 + u_y^2 + u_z^2}\,\mathrm{d}t, \tag{9.24}$$

which quantifies the consumption of propellant.

Finally, consider the constraint

$$\sqrt{u_x^2 + u_y^2 + u_z^2} \le U, \tag{9.25}$$

which models the limitation of the thruster.

Then, (9.22)–(9.25) *specify the optimal control problem of changing orbit with minimal propellant consumption.*

9.3 Optimal Control without Control Constraints

Consider the simple optimal control problem

$$\min_{u(.)} J = K(x(t_f)) + \int_{t_0}^{t_f} L(x, u, t)\,\mathrm{d}t, \tag{9.26}$$

subject to

$$\begin{aligned} \dot{x} &= f(x, u, t) \\ x(t_0) &= x_0, \end{aligned} \tag{9.27}$$

where the initial condition is specified, the final condition is free, the final time is specified, and there are no other constraints on the state or control.

To derive necessary conditions of optimality for the optimal control problem specified by (9.26), (9.27), we revisit the example of Section 8.4, modifying it by multiplying $f(., ., .)$ and $L(., ., .)$ by a parameter Δt. We then take the limits $\Delta t \to 0$ and $k_f \to \infty$, causing derivatives and integrals to appear.

The state dynamics, as given by (8.115), become

$$\begin{aligned} x(k+1) &= x(k) + f(x(k), u(k), k)\Delta t, \ \ k_0 \le k \le k_f \\ x(k_0) &= x_0, \end{aligned} \tag{9.28}$$

which in the limit $\Delta t \to 0$ become

$$\begin{aligned} \dot{x} &= f(x, u, t) \\ x(t_0) &= x_0, \end{aligned} \tag{9.29}$$

which matches (9.27).

The minimization problem (8.108) is now given by

$$\min_{u(k_0),\ldots,u(k_f)} J = K(x(k_f+1)) + \sum_{k=k_0}^{k_f} L(x(k), u(k), k)\Delta t, \tag{9.30}$$

and, in the limit $\Delta t \to 0$,

$$\min_{u(.)} J = K(x(t_f)) + \int_{t_0}^{t_f} L(x, u, t)\,\mathrm{d}t, \tag{9.31}$$

which matches (9.26).

The co-state dynamics (8.116) become

$$\begin{aligned} p(k) &= p(k+1) + \frac{\partial}{\partial x(k)} f(x(k), u(k), k) p(k+1)\Delta t \\ &\quad - \frac{\partial}{\partial x(k)} L(x(k), u(k), k)\Delta t, \;\; k_1 \le k \le k_f \\ p(k_f+1) &= -\frac{\partial K(x(k_f+1))}{\partial x(k_f+1)}, \end{aligned} \tag{9.32}$$

or, in the limit $\Delta t \to 0$,

$$\begin{aligned} \dot{p} &= -\left(\frac{\partial f}{\partial x}\right) p + \frac{\partial L}{\partial x} \\ p(t_f) &= -\frac{\partial K(x(t_f))}{\partial x(t_f)}. \end{aligned} \tag{9.33}$$

The optimality condition (8.117) becomes

$$\frac{\partial L}{\partial u} - \left(\frac{\partial f}{\partial u}\right) p = 0. \tag{9.34}$$

For this type of system, it is customary to introduce the **Hamiltonian**

$$H(x, p, u, t) = p^T f(x, u, t) - L(x, u, t). \tag{9.35}$$

Then, the necessary conditions for optimality are given by

The state dynamics:

$$\dot{x} = \frac{\partial H}{\partial p}, \tag{9.36}$$

The co-state dynamics:

$$\dot{p} = -\frac{\partial H}{\partial x}, \tag{9.37}$$

The optimality condition:

$$\frac{\partial}{\partial u} H(x, p, u, t) = 0, \tag{9.38}$$

The boundary conditions:

$$x(t_0) = x_0,$$
$$p(t_f) = -\frac{\partial K(x_f)}{\partial x_f}. \tag{9.39}$$

In the optimality condition (9.38), H must be maximized.

REMARK 9.2 *Note that the necessary conditions for optimality for problem (9.26), (9.27) consist of the same four ingredients anticipated in Section 8.4, namely: state equations (9.36), co-state equations (9.37), optimality condition (9.38), and boundary conditions (9.39). These four ingredients typically yield arcs of curve in the state space, the* ***extremal arcs****, that the solution of the optimal guidance problem must follow. The determination of these extremal arcs constitutes a solution "in the small" of the optimal guidance problem. Typically, the solution "in the large" is obtained by piecing together extremal arcs through a switching logic to be determined. This is illustrated in Section 9.4.4.*

EXAMPLE 9.6 (Controllability Gramian) *In this example, we show how the controllability Gramian (2.138) arises naturally as a quantitative measure of controllability. Recall that, in a linear time varying system, the state x_0 is controllable at time t_0 if there exists a time $t_1 > t_0$ and a control defined on the interval $[t_0, t_1]$ that drives the state of the system from initial condition x_0 at time t_0 to final condition 0 at time t_1. Accordingly, consider the unconstrained optimal control problem defined by the following ingredients:*

Controllable linear time varying dynamics:

$$\dot{x}(t) = A(t)x(t) + B(t)u(t), \tag{9.40}$$

Boundary conditions:

$$x(t_0) = x_0,$$
$$x(t_1) = 0,$$
$$t_0,\ x_0,\ t_1 \quad \textit{all given}, \tag{9.41}$$

Performance index:

$$J(u(.), x_0) = \frac{1}{2}\int_{t_0}^{t_1} u^T(t)u(t)\,\mathrm{d}t. \tag{9.42}$$

Clearly, if the optimal cost for the preceding problem is large (resp. small), then the state x_0 is poorly controllable (resp. very controllable). The Hamiltonian (9.35) is

$$H(x, p, u, t) = p^T(Ax + Bu) - \frac{1}{2}u^T u. \tag{9.43}$$

The state dynamics (9.36) are

$$\dot{x} = \frac{\partial H}{\partial p} = Ax + Bu. \tag{9.44}$$

The co-state dynamics (9.37) *are*

$$\dot{p} = -\frac{\partial H}{\partial x} = -A^T p. \tag{9.45}$$

Let $\Phi(t, \tau)$ *be the state transition matrix associated with* $A(t)$*. In Problem 2.5, we evaluate the state transition matrix associated with* $-A^T(t)$ *as* $\Phi^T(\tau, t)$*. Hence,* (9.45) *can be integrated as*

$$p(t) = \Phi^T(t_0, t)p_0, \tag{9.46}$$

where $p_0 = p(t_0)$ *is the initial condition of the co-state vector. The optimality condition* (9.38) *is*

$$\frac{\partial H}{\partial u} = B^T p - u = 0, \tag{9.47}$$

which yields

$$u(t) = B^T(t)p(t). \tag{9.48}$$

Therefore, (9.44), (9.46), *and* (9.48) *yield*

$$\dot{x}(t) = A(t)x(t) + B(t)B^T(t)\Phi^T(t_0, t)p_0. \tag{9.49}$$

Applying the variation of constants formula (2.34)*, we evaluate* x *at time* t_1 *as*

$$\begin{aligned} x(t_1) &= \Phi(t_1, t_0)x_0 + \int_{t_0}^{t_1} \Phi(t_1, \tau)B(\tau)B^T(\tau)\Phi^T(t_0, \tau)\,\mathrm{d}\tau p_0 \\ &= \Phi(t_1, t_0)\left(x_0 + \int_{t_0}^{t_1} \Phi(t_0, \tau)B(\tau)B^T(\tau)\Phi^T(t_0, \tau)\,\mathrm{d}\tau p_0\right), \end{aligned} \tag{9.50}$$

where we recognize the matrix integral as the controllability Gramian $W(t_0, t_1)$ *defined in* (2.138)*. Therefore, the boundary condition* (9.41) *(line 2), together with* (9.50)*, the nonsingularity of* $\Phi(t_1, t_0)$*, and the controllability assumption yield*

$$p_0 = -W^{-1}(t_0, t_1)x_0. \tag{9.51}$$

Hence, (9.46), (9.48), *and* (9.51) *imply that the optimal control must have the form*

$$u(t) = -B^T(t)\Phi^T(t_0, t)W^{-1}(t_0, t_1)x_0, \tag{9.52}$$

that is, (2.141)*. The control* (9.52) *is indeed optimal, i.e., the performance index it achieves cannot be improved – proving this is left as an exercise for the reader. Finally,* (9.42) *and* (9.52) *imply that the optimal cost has the form*

$$\begin{aligned} J^*(x_0) &= \frac{1}{2}\int_{t_0}^{t_1} u^T(t)u(t)\,\mathrm{d}t, \\ &= \frac{1}{2}\int_{t_0}^{t_1} x_0^T W^{-1}(t_0, t_1)\Phi(t_0, t)B(t)B^T(t)\Phi^T(t_0, t)W^{-1}(t_0, t_1)x_0\,\mathrm{d}t, \\ &= \frac{1}{2}x_0^T W^{-1}(t_0, t_1)\int_{t_0}^{t_1} \Phi(t_0, t)B(t)B^T(t)\Phi^T(t_0, t)\,\mathrm{d}t W^{-1}(t_0, t_1)x_0, \end{aligned} \tag{9.53}$$

where, again, we recognize the matrix integral as $W(t_0, t_1)$. *Therefore, the optimal cost is, after canceling factors,*

$$J^*(x_0) = \frac{1}{2} x_0^T W^{-1}(t_0, t_1) x_0. \tag{9.54}$$

The practical consequence of the preceding is that, when (9.54) *is large (resp. small), the state* x_0 *is poorly controllable (resp. very controllable) at time* t_0 *in the sense that it takes a large (resp. small) amount of energy, as measured by* (9.42), *to drive the state of the system from* x_0 *at time* t_0 *to the origin at time* t_1. *This happens, for instance, if* x_0 *is an eigenvector of* $W(t_0, t_1)$ *associated with a small (resp. large) eigenvalue.*

EXAMPLE 9.7 (Kalman Filtering) *In this example, we follow up on the idea, suggested in Proposition 4.5, of optimizing the gain of an asymptotic observer to obtain the Kalman filter. Consider the unconstrained optimal control problem defined by the following ingredients:*

Dynamics given by the Lyapunov equation (4.86) *and repeated here for convenience;*

$$\begin{aligned} \dot{P}(t) = {} & [A(t) + G(t)C(t)]P(t) + P(t)[A(t) + G(t)C(t)]^T \\ & + G(t)R_v(t)G^T(t) + R_w(t), \end{aligned} \tag{9.55}$$

where $P(t)$ *is the covariance matrix of the estimation error,* $A(t)$ *and* $C(t)$ *are the state and output matrices of the system, respectively,* $G(t)$ *is the gain of the observer, and* $R_v(t) > 0$ *and* $R_w(t) \geq 0$ *are covariance matrices of noise processes.*

Boundary conditions:

$$\begin{aligned} P(t_0) &= P_0, \\ P(t_f) & \quad \textit{free}, \\ t_0, P_0, t_f & \quad \textit{all given}, \end{aligned} \tag{9.56}$$

Performance index:

$$J = \mathrm{tr}(P(t_f)), \tag{9.57}$$

where tr *denotes the trace of a square matrix, that is, the sum if its diagonal elements. Note that in this optimal control problem, the "state vector" is the square symmetric matrix* $P(t)$, *the "control vector" is the matrix* $G(t)$, *and the performance index is the mean square estimation error at the final time.*

Define, as co-state, the symmetric matrix $\Lambda(t)$, *of same order as* $P(t)$, *so that the Hamiltonian* (9.35) *is then*

$$H(P, \Lambda, G, t) = \mathrm{tr}\left\{\Lambda\left[(A + GC)P + P(A + GC)^T + GR_vG^T + R_w\right]\right\}. \tag{9.58}$$

Henceforth, we use the identity

$$\frac{\partial}{\partial M_2}\mathrm{tr}(M_1 M_2 M_3) = M_1^T M_3^T,$$

where M_1, M_2, M_3 *are matrices of compatible dimensions such that* $M_1 M_2 M_3$ *is square.*

The state dynamics (9.36) *are*

$$\dot{P} = \frac{\partial H}{\partial \Lambda} = (A + GC)P + P(A + GC)^T + GR_v G^T + R_w. \tag{9.59}$$

The co-state dynamics (9.37) *are*

$$\dot{\Lambda} = -\frac{\partial H}{\partial P} = -(A + GC)^T \Lambda - \Lambda(A + GC) \tag{9.60}$$

with boundary condition (9.39) *(line 2)*

$$\Lambda(t_f) = -\frac{\partial}{\partial P(t_f)} \operatorname{tr}(P(t_f)) = -I. \tag{9.61}$$

Let $\Psi(t, \tau)$ *be the state transition matrix associated with* $-(A(t) + G(t)C(t))^T$. *Then* (9.60) *and* (9.61) *yield*

$$\Lambda(t) = -\Psi(t, t_f)\Psi^T(t, t_f), \tag{9.62}$$

which implies that $\Lambda(t)$ *is nonsingular at all times.*

The optimality condition (9.38) *yields*

$$\frac{\partial H}{\partial G} = 2\Lambda PC^T + 2\Lambda GR_v = 0. \tag{9.63}$$

Since Λ *is nonsingular,* (9.63) *implies*

$$G = -PC^T R_v^{-1}, \tag{9.64}$$

which matches (4.88). *Finally, substituting* (9.64) *into* (9.59) *and canceling terms yields the differential Riccati equation*

$$\dot{P} = AP + PA^T - PC^T R_v^{-1} CP + R_w, \tag{9.65}$$

which matches (4.89).

9.4 The Maximum Principle

We now consider the following generalization of the problem solved in the previous section:

$$\min_{u(.), x_0, t_0, x_f, t_f} \quad J = K(x_0, t_0, x_f, t_f) + \int_{t_0}^{t_f} L(x, u, t)\, \mathrm{d}t, \tag{9.66}$$

subject to

$$\begin{aligned} \dot{x} &= f(x, u, t), \\ x(t_0) &= x_0, \\ x(t_f) &= x_f, \\ u(t) &\in U, \end{aligned} \tag{9.67}$$

where the boundary conditions may be fixed, free, or on target sets, and the compact set U describes magnitude constraints on the control input.

PROPOSITION 9.1 (Pontryagin's Maximum Principle) *A necessary condition for optimality in the problem* (9.66), (9.67) *is that, defining the* ***Hamiltonian*** *as*

$$H(x, p, u, t) = p^T f(x, u, t) - L(x, u, t), \tag{9.68}$$

we satisfy

$$\begin{aligned} \dot{x} &= \frac{\partial H}{\partial p}(x, p, u, t), \\ \dot{p} &= -\frac{\partial H}{\partial x}(x, p, u, t), \\ u &= \operatorname*{argmax}_{v \in U} H(x, p, v, t), \\ \delta K + \left[p^T \delta x - H \delta t\right]_{t_0}^{t_f} &= 0 \textit{ for all admissible variations}. \end{aligned} \tag{9.69}$$

REMARK 9.3 *In the four lines of* (9.69), *we recognize the four ingredients of the necessary conditions for optimal guidance alluded to in Remark 9.2: state equations, co-state equations, optimality condition, and boundary conditions. Equation* (9.69) *(line 4) takes the general form of* ***transversality conditions*** *and yields boundary conditions, as explained in Section 9.4.2.*

Optimal control problems sometimes yield two situations that are beyond the coverage of this book: **singular controls** and **conjugate points**. Singular controls happen when (9.69) (line 3) does not provide enough information to determine u, for example, if the Hamiltonian does not depend explicitly on u. Conjugate points are locations, on an extremal arc, where the bundle of neighboring extremals becomes singular, for example, in seeking the shortest path between two points on a sphere, when the two points are polar opposites. Both singular controls and conjugate points are symptoms of an ill-posed problem in the sense that a change in a control or boundary condition candidate yields, to first order, no change in the cost function or the extremal, respectively. In practice, rather than devoting much effort to solve an ill-posed problem, it is more expedient to modify the problem data, for example, the cost function, to make the problem well posed. This procedure is justified in the light of Remark 8.1.

EXAMPLE 9.8 (Calculus of Variations) *To illustrate the maximum principle, we use it to derive a standard result in the calculus of variations. Consider the minimization of a scalar functional of the form*

$$J(x(.)) = \int_{t_0}^{t_f} L(x(t), \dot{x}(t), t)\, \mathrm{d}t \tag{9.70}$$

with respect to the time history of the differentiable vector function of time $x(t)$, *subject to given boundary conditions at the given times* t_0 *and* t_f. *We formulate this as a Lagrange optimal control problem* (9.8) *by defining the control* u *such that*

$$\dot{x}(t) = u(t). \tag{9.71}$$

The Hamiltonian (9.68) *is then*

$$H(x, p, u, t) = p^T u - L(x, u, t). \tag{9.72}$$

The co-state dynamics (9.69) *(line 2) are*

$$\dot{p} = \frac{\partial}{\partial x} L(x, u, t). \tag{9.73}$$

The optimality condition (9.69) *(line 3) implies*

$$\frac{\partial}{\partial u} H(x, p, u, t) = p - \frac{\partial}{\partial u} L(x, u, t) = 0, \tag{9.74}$$

yielding

$$p = \frac{\partial}{\partial u} L(x, u, t). \tag{9.75}$$

Therefore, (9.71), (9.73), *and* (9.75) *imply that extremals must satisfy*

$$\frac{\mathrm{d}}{\mathrm{d}t}\left(\frac{\partial L}{\partial \dot{x}}\right) - \left(\frac{\partial L}{\partial x}\right) = 0, \tag{9.76}$$

which is known as the ***Euler–Lagrange equation****.*

9.4.1 Greed

Consider the distinction between *static* decision problems (i.e., problems where taking a decision does not change the state) and *dynamic* decision problems (i.e., problems where taking a decision u changes the state according to $\dot{x} = f(x, u, t)$). Assume that we want

$$\min_{u \in U} J(u) = \int_{t_0}^{t_f} L(x, u, t)\,\mathrm{d}t. \tag{9.77}$$

For a static decision problem, an obvious solution is

$$\begin{aligned} u &= \operatorname*{argmin}_v L(x, v, t) \\ &= \operatorname*{argmax}_v (-L(x, v, t)). \end{aligned} \tag{9.78}$$

However, for a dynamic decision problem, the solution is

$$u = \operatorname*{argmax}_v \left(p^T f(x, v, t) - L(x, v, t)\right). \tag{9.79}$$

In (9.79), the term $p^T f(x, v, t)$ accounts for the fact that taking a decision changes the state. Note that the co-state dynamics are specified backward. This means that, in dynamic situations, taking good decisions implies accounting for the effect of those decisions *in the future*.

For a dynamic decision problem, the strategy

$$u = \operatorname*{argmax}_v (-L(x, v, t)) \tag{9.80}$$

is not optimal and is called **greedy**. Obviously, if the dynamics are negligible, a greedy strategy can be expected to be close to optimal.

9.4.2 The Transversality Conditions

The transversality conditions are of the form

$$\delta K + \left[p^T \delta x - H\delta t\right]_{t_0}^{t_f} = 0 \text{ for all admissible variations.} \tag{9.81}$$

They provide the boundary conditions for x, p, t_0, and t_f. Recall that $x \in \mathbb{R}^n$, $p \in \mathbb{R}^n$, $t_0 \in \mathbb{R}$, and $t_f \in \mathbb{R}$. Let us consider some particular cases.

Case 1: x_0, t_0, x_f, and t_f are all given. Then, the transversality conditions are of the form

$$\begin{aligned} x(t_0) &= x_0 \\ x(t_f) &= x_f. \end{aligned} \tag{9.82}$$

Case 2: x_0, t_0, and t_f are given, and x_f is free. Then, the transversality conditions are of the form

$$\begin{aligned} x(t_0) &= x_0 \\ p(t_f) &= -\frac{\partial K}{\partial x_f}. \end{aligned} \tag{9.83}$$

Case 3: x_0 and t_0 are given, and x_f and t_f are free. Then, the transversality conditions are of the form

$$\begin{aligned} x(t_0) &= x_0, \\ p(t_f) &= -\frac{\partial K}{\partial x_f}, \\ \frac{\partial K}{\partial t_f} - H(x, u, p, t)\Big|_{t_f} &= 0. \end{aligned} \tag{9.84}$$

Case 4: t_0 and t_f are given, and x_0 and x_f are free. Then, the transversality conditions are of the form

$$\begin{aligned} p(t_0) &= -\frac{\partial K}{\partial x_0} \\ p(t_f) &= -\frac{\partial K}{\partial x_f}. \end{aligned} \tag{9.85}$$

Case 5: x_0, t_0, x_f, and t_f are all free. Then, the transversality conditions are of the form

$$\begin{aligned} \frac{\partial K}{\partial t_0} + H\Big|_{t_0} &= 0, \\ \frac{\partial K}{\partial t_f} + H\Big|_{t_f} &= 0, \\ p(t_0) &= \frac{\partial K}{\partial x_0}, \\ p(t_f) &= -\frac{\partial K}{\partial x_f}. \end{aligned} \tag{9.86}$$

9.4.3 Target Sets

Next we consider how to handle boundary conditions of the form $g(x_f, t_f) = 0$.

PROPOSITION 9.2 *Let $M \in \mathbb{R}^{m\times n}$, and $N \in \mathbb{R}^{1\times n}$. If*

$$\forall x \in \mathbb{R}^n, Mx = 0 \text{ implies } Nx = 0, \tag{9.87}$$

then

$$\exists \mu \in \mathbb{R}^m \text{ such that } N^T = M^T \mu. \tag{9.88}$$

For boundary conditions of the form $g(x_f, t_f) = 0$, admissible variations satisfy

$$\left(\frac{\partial g}{\partial x_f}\right)^T \delta x_f + \left(\frac{\partial g}{\partial t_f}\right) \delta t_f = 0, \tag{9.89}$$

or, equivalently,

$$\begin{bmatrix} \left(\frac{\partial g}{\partial x_f}\right)^T & \left(\frac{\partial g}{\partial t_f}\right) \end{bmatrix} \begin{bmatrix} \delta x_f \\ \delta t_f \end{bmatrix} = M \begin{bmatrix} \delta x_f \\ \delta t_f \end{bmatrix} = 0. \tag{9.90}$$

The transversality condition specifies that

$$\begin{bmatrix} \left(\frac{\partial K}{\partial x_f}\right)^T + p^T(t_f) & \frac{\partial K}{\partial t_f} - H\Big|_{t_f} \end{bmatrix} \begin{bmatrix} \delta x_f \\ \delta t_f \end{bmatrix} = N \begin{bmatrix} \delta x_f \\ \delta t_f \end{bmatrix} = 0 \tag{9.91}$$

whenever

$$M \begin{bmatrix} \delta x_f \\ \delta t_f \end{bmatrix} = 0. \tag{9.92}$$

From Proposition 9.2, there exists $\mu \in \mathbb{R}^n$ such that

$$\begin{bmatrix} \frac{\partial K}{\partial x_f} + p(t_f) \\ \frac{\partial K}{\partial t_f} - H\Big|_{t_f} \end{bmatrix} = \begin{bmatrix} \frac{\partial g}{\partial x_f} \\ \left(\frac{\partial g}{\partial t_f}\right)^T \end{bmatrix} \mu. \tag{9.93}$$

Therefore, we obtain the following system of $n + m + 1$ equations in the $n + m + 1$ unknowns x_f, μ, and t_f:

$$\begin{aligned} \frac{\partial K}{\partial x_f} + p(t_f) &= \frac{\partial g}{\partial x}\mu, \\ \frac{\partial K}{\partial t_f} - H\Big|_{t_f} &= \left(\frac{\partial g}{\partial t}\right)^T \mu, \\ g(x_f, t_f) &= 0. \end{aligned} \tag{9.94}$$

REMARK 9.4 *Note that $p(t_f) = \left(\frac{\partial g}{\partial x}\right)\mu$ implies that $p(t_f)$ is normal to the target set, hence the word **transversality**.*

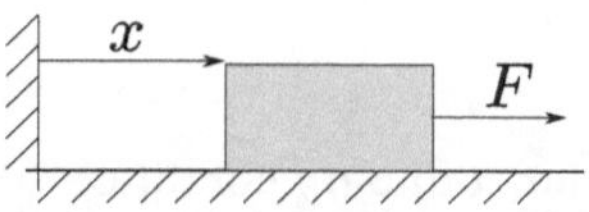

Figure 9.2. Double integrator.

We turn our attention to **Rendez–Vous**, which is a particular form of target set useful in aerospace guidance. In these problems, we want

$$x_f = g(t_f), \tag{9.95}$$

which describes a target trajectory. We need

$$\left(\frac{\partial K}{\partial x_f}\right)^T \delta x_f + \left(\frac{\partial K}{\partial t_f}\right)^T \delta t_f + p^T(t_f)\delta x_f - H\Big|_{t_f} \delta t_f = 0 \tag{9.96}$$

for all admissible variations, defined by

$$\delta x_f = \left(\frac{\mathrm{d}g}{\mathrm{d}t_f}\right)\delta t_f. \tag{9.97}$$

This implies

$$\forall \delta t_f, \left[\left(\frac{\partial K}{\partial x_f} + p(t_f)\right)^T \left(\frac{\mathrm{d}g}{\mathrm{d}t_f}\right) + \left(\frac{\partial K}{\partial t_f}\right) - H\Big|_{t_f}\right]\delta t_f = 0, \tag{9.98}$$

or

$$\left(\frac{\partial K}{\partial x_f} + p(t_f)\right)^T \left(\frac{\mathrm{d}g}{\mathrm{d}t_f}\right) + \left(\frac{\partial K}{\partial t_f}\right) - H\Big|_{t_f} = 0. \tag{9.99}$$

Note that (9.99) is a single equation in the unknown t_f.

REMARK 9.5 *Note that we could have obtained the same result by using result* (9.94) *on target sets. Indeed, define the target set as*

$$G(x_f, t_f) = x_f - g(t_f) = 0, \tag{9.100}$$

so that

$$\begin{aligned}\frac{\partial G}{\partial x_f} &= I\\ \frac{\partial G}{\partial t_f} &= -\frac{\mathrm{d}g}{\mathrm{d}t_f}.\end{aligned} \tag{9.101}$$

Then, (9.94)*, with the target set* $G(x_f, t_f) = 0$*, yields* (9.99).

9.4.4 Time-Optimal Control of Double Integrator

Consider again the double integrator, introduced and motivated in Example 2.3, and shown in Figure 9.2. The equations of motion have the form

$$\begin{aligned}\dot{x}_1 &= x_2\\ \dot{x}_2 &= u.\end{aligned} \tag{9.102}$$

In addition, assume that we have an amplitude constraint on the input of the form

$$|u| \leq 1. \tag{9.103}$$

Let the boundary conditions be of the form

$$\begin{aligned} x_1(0) &= x_{10}, x_2(0) = x_{20}, t_0 = 0 \text{ given} \\ x_1(t_f) &= 0, x_2(t_f) = 0 \text{ given.} \end{aligned} \tag{9.104}$$

We minimize the duration of transfer to the origin, that is,

$$J = t_f = \int_0^{t_f} \mathrm{d}t. \tag{9.105}$$

For this system, the Hamiltonian is given by

$$\begin{aligned} H(x, p, u, t) &= p^T f(x, u, t) - L(x, u, t) \\ &= p_1 x_2 + p_2 u - 1. \end{aligned} \tag{9.106}$$

The necessary conditions for optimality yield

$$\begin{aligned} \dot{x} &= \frac{\partial H}{\partial p} \\ \dot{p} &= -\frac{\partial H}{\partial x}, \end{aligned} \tag{9.107}$$

or equivalently,

$$\begin{aligned} \dot{x}_1 &= x_2, \\ \dot{x}_2 &= u, \\ \dot{p}_1 &= 0, \\ \dot{p}_2 &= -p_1. \end{aligned} \tag{9.108}$$

Then, u is given by

$$u = \underset{|v| \leq 1}{\operatorname{argmax}}(p_1 x_2 + p_2 v - 1), \tag{9.109}$$

with

$$\begin{aligned} x_1(0) &= x_{10}, \\ x_2(0) &= x_{20}, \\ H\Big|_{t_f} &= (p_1 x_2 + p_2 u - 1)\Big|_{t_f} = 0. \end{aligned} \tag{9.110}$$

Define

$$\operatorname{sign}[\alpha] = \begin{cases} 1 & \text{if } \alpha > 0, \\ -1 & \text{if } \alpha < 0, \\ \text{undefined} & \text{if } \alpha = 0. \end{cases} \tag{9.111}$$

To maximize H, we have (see Problem 8.2)

$$u = \operatorname{sign}[p_2(t)]. \tag{9.112}$$

If $p_2(t) \equiv 0$ on an interval, u is undetermined and we have a *singular* control. We analyze whether this is possible.

Recall, from (9.108), that

$$\dot{p}_1 = 0$$
$$\dot{p}_2 = -p_1; \tag{9.113}$$

that is,

$$p_1 = \text{constant} = \pi_1$$
$$p_2 = -\pi_1 t + \pi_2. \tag{9.114}$$

If $p_2(t) \equiv 0$ on an interval, then $\pi_1 = \pi_2 = 0$. Then, $p_1(t) = 0$, and $H\big|_{t_f} = -1$, which contradicts the condition $H\big|_{t_f} = 0$ specified in (9.110) (line 3). This implies that singular control is impossible.

The optimal trajectory is given by

$$\begin{aligned}
\dot{x}_1 &= x_2, \\
\dot{x}_2 &= \text{sign}[p_2], \\
\dot{p}_1 &= 0, \\
\dot{p}_2 &= -p_1, \\
x(0) & \quad \text{given}, \\
x(t_f) &= 0.
\end{aligned} \tag{9.115}$$

Note that $p_2(t)$ is a first-degree polynomial and changes sign at most once. Therefore, there exist four possible control sequences, $(+1)$, (-1), $(+1, -1)$, and $(-1, +1)$. Let $\sigma = \pm 1$. Then, x_2 and x_1 are given by

$$x_2 = x_{20} + \sigma t$$
$$x_1 = x_{10} + x_{20} t + \frac{1}{2}\sigma t^2. \tag{9.116}$$

We can express x_1 as a function of x_2 by eliminating t, using

$$t = \frac{1}{\sigma}(x_2 - x_{20}) = \sigma(x_2 - x_{20}), \tag{9.117}$$

and substituting in the expression for x_1 as follows:

$$x_1 = x_{10} + \sigma x_{20}(x_2 - x_{20}) + \frac{1}{2}\sigma(x_2 - x_{20})^2, \tag{9.118}$$

or equivalently,

$$x_1 - \left(x_{10} - \frac{1}{2}\sigma x_{20}\right)^2 = \frac{1}{2}\sigma x_2^2. \tag{9.119}$$

Equation (9.119) describes the extremals, which are parabolas in the x_1, x_2 plane, as shown in Figure 9.3. Figure 9.4 shows the optimal switching curve. The equation for the switching curve is

$$x_1 = -\frac{1}{2}x_2|x_2|. \tag{9.120}$$

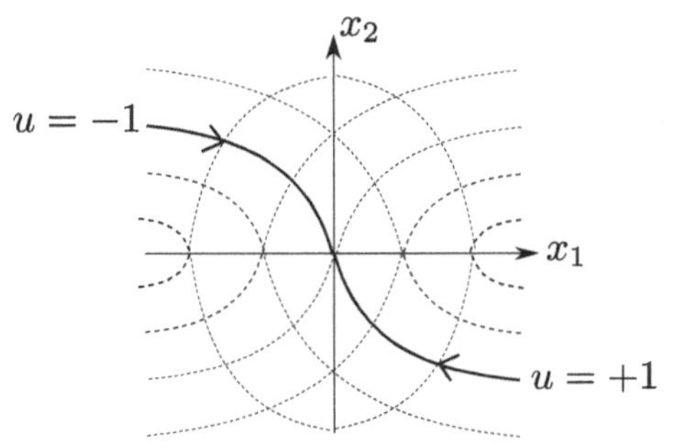

Figure 9.3. Extremals for time-optimal control of double integrator.

A feedback control implementation is shown in Figure 9.5. Define $z = x_1 + \frac{1}{2}x_2|x_2|$. Then, the switching logic is as follows: if $z > 0$, use $u = -1$, and if $z < 0$, use $u = +1$.

REMARK 9.6 *Note that for an* nth-*order system with nonpositive real poles only and* u *scalar, the optimal control switches at most* $n - 1$ *times. This result is established in [61].*

9.4.5 Optimal Evasion through Jinking

Here we revisit the idea, suggested in Section 5.6.2, of using target lateral acceleration to evade the attack of a missile that is guided by proportional navigation. We formulate evasion as a Mayer optimal control problem and obtain necessary conditions for optimality. Through these conditions, we demonstrate that the optimal evasive maneuver must jink, that is, oscillate between perfectly timed hard turns to the left and to the right.

A state space model for the engagement is given by (5.94) and repeated here for convenience:

$$\begin{bmatrix} \dot{x}_1 \\ \dot{x}_2 \end{bmatrix} = \begin{bmatrix} 0 & 1 \\ -\frac{\Lambda}{(t_f^*-t)T} & -\frac{1}{T} \end{bmatrix} \begin{bmatrix} x_1 \\ x_2 \end{bmatrix} + \begin{bmatrix} 0 \\ \frac{\Lambda}{(t_f^*-t)T} \end{bmatrix} y_T$$

$$y_M = \begin{bmatrix} 1 & 0 \end{bmatrix} \begin{bmatrix} x_1 \\ x_2 \end{bmatrix}, \tag{9.121}$$

where the input y_T and the output y_M are the displacement of the target and missile in the direction perpendicular to the line of sight, respectively, T is the time constant of the first-order autopilot, Λ is the effective navigation constant, and t_f^* is the nominal time of intercept.

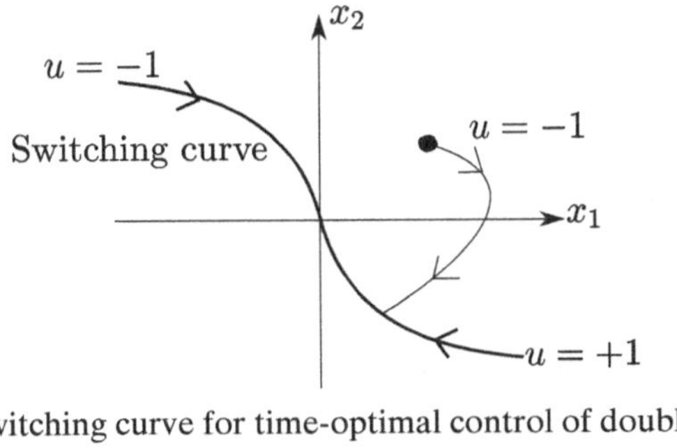

Figure 9.4. Switching curve for time-optimal control of double integrator.

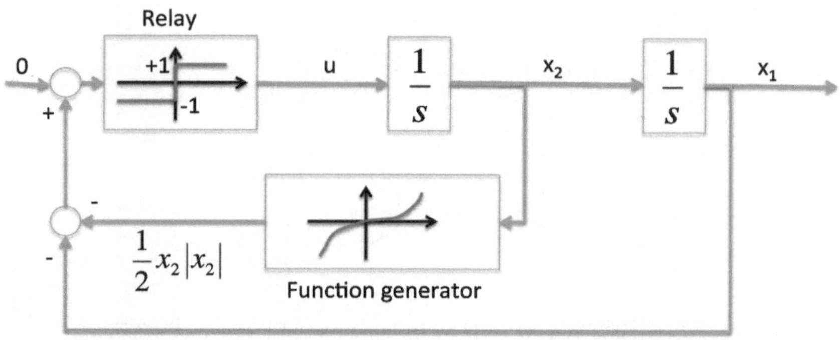

Figure 9.5. Implementation of optimal control of double integrator.

The miss distance is approximated in (5.89) as the difference of lateral displacements at the nominal time of intercept, that is,

$$M = y_T(t_f^*) - y_M(t_f^*). \tag{9.122}$$

To use target lateral acceleration as an input, we define two additional states as

$$\begin{aligned} x_3 &= y_T \\ x_4 &= \dot{y}_T. \end{aligned} \tag{9.123}$$

Then, the system dynamic equations become

$$\begin{bmatrix} \dot{x}_1 \\ \dot{x}_2 \\ \dot{x}_3 \\ \dot{x}_4 \end{bmatrix} = \begin{bmatrix} 0 & 1 & 0 & 0 \\ -\frac{\Lambda}{(t_f^*-t)T} & -\frac{1}{T} & \frac{\Lambda}{(t_f^*-t)T} & 0 \\ 0 & 0 & 0 & 1 \\ 0 & 0 & 0 & 0 \end{bmatrix} \begin{bmatrix} x_1 \\ x_2 \\ x_3 \\ x_4 \end{bmatrix} + \begin{bmatrix} 0 \\ 0 \\ 0 \\ 1 \end{bmatrix} u, \tag{9.124}$$

or equivalently,

$$\dot{x} = A(t)x + bu, \tag{9.125}$$

where the definitions of A and b are obvious, and the input u is the lateral target acceleration. We assume that this input has bounded magnitude, and without loss of generality:

$$|u| \leq 1. \tag{9.126}$$

To maximize the miss distance (9.122), we use the cost functional

$$\begin{aligned} J &= -y_T(t_f^*) + y_M(t_f^*) \\ &= \begin{bmatrix} 1 & 0 & -1 & 0 \end{bmatrix} x(t_f^*), \end{aligned} \tag{9.127}$$

which is in Mayer form.

Equations (9.125)–(9.127) specify an optimal control problem where we assume that the initial time t_0 and initial state are given; this is the **Optimal Evasion Problem.** The Hamiltonian (9.68) is

$$H(x, p, u, t) = p^T(A(t)x + bu). \tag{9.128}$$

The necessary conditions for optimality, (9.69), are

$$\begin{aligned}
\dot{x} &= \frac{\partial H}{\partial p} = Ax + bu, \\
\dot{p} &= -\frac{\partial H}{\partial x} = -A^T(t)p, \\
u &= \operatorname{argmax}(H) = \operatorname{sign}(p_4), \\
x(t_0) & \quad \text{given}, \\
p(t_f^{*-}) &= \begin{bmatrix} -1 \\ 0 \\ 1 \\ 0 \end{bmatrix}.
\end{aligned} \tag{9.129}$$

We eliminate the possibility of singular control as follows. First, explicitly write the co-state dynamics (9.129), line 2, as:

$$\begin{bmatrix} \dot{p}_1 \\ \dot{p}_2 \\ \dot{p}_3 \\ \dot{p}_4 \end{bmatrix} = \begin{bmatrix} 0 & \frac{\Lambda}{(t_f^*-t)T} & 0 & 0 \\ -1 & \frac{1}{T} & 0 & 0 \\ 0 & -\frac{\Lambda}{(t_f^*-t)T} & 0 & 0 \\ 0 & 0 & -1 & 0 \end{bmatrix} \begin{bmatrix} p_1 \\ p_2 \\ p_3 \\ p_4 \end{bmatrix}. \tag{9.130}$$

Note that (9.130) is a linear time varying ordinary differential equation with a right-hand side that is analytic, with a singularity at t_f^*. Hence, the solution is analytic except perhaps at t_f^*. Now assume $p_4 \equiv 0$ on an open interval. Analyticity implies that $p_4 \equiv 0$ everywhere except perhaps at t_f^*. Then the last row of (9.130) yields $p_3 \equiv 0$ almost everywhere. However, as per (9.129) (line 5), the boundary condition for p_3 is $p_3(t_f^{*-}) = 1$. This contradiction guarantees that there is no singular control in the optimal evasion problem.

The optimal evasive maneuver must therefore satisfy

$$u = \operatorname{sign}(p_4), \tag{9.131}$$

which is **jinking**, as illustrated in Figures 9.6 and 9.7. For all simulations, $\Lambda = 4$, $T = 1$, and $t_f^* = 8$.

REMARK 9.7 *The preceding analysis assumes that the target can change instantaneously its lateral acceleration; hence the double integrator dynamics* (9.123). *For a highly maneuverable target, that is, one that can change its velocity instantaneously, the analysis can be modified by using a simple integrator, as follows. The dynamic equations are now of order three, with* $x_1 = y_M$, $x_2 = \dot{y}_M$, *and* $x_3 = y_T$:

$$\begin{bmatrix} \dot{x}_1 \\ \dot{x}_2 \\ \dot{x}_3 \end{bmatrix} = \begin{bmatrix} 0 & 1 & 0 \\ -\frac{\lambda}{(t_f^*-t)T} & -\frac{1}{T} & \frac{\lambda}{(t_f^*-t)T} \\ 0 & 0 & 0 \end{bmatrix} \begin{bmatrix} x_1 \\ x_2 \\ x_3 \end{bmatrix} + \begin{bmatrix} 0 \\ 0 \\ 1 \end{bmatrix} u = Ax + bu, \tag{9.132}$$

and the cost functional is

$$J = -y_T(t_f^*) + y_M(t_f^*) = \begin{bmatrix} 1 & 0 & -1 \end{bmatrix} x(t_f^*). \tag{9.133}$$

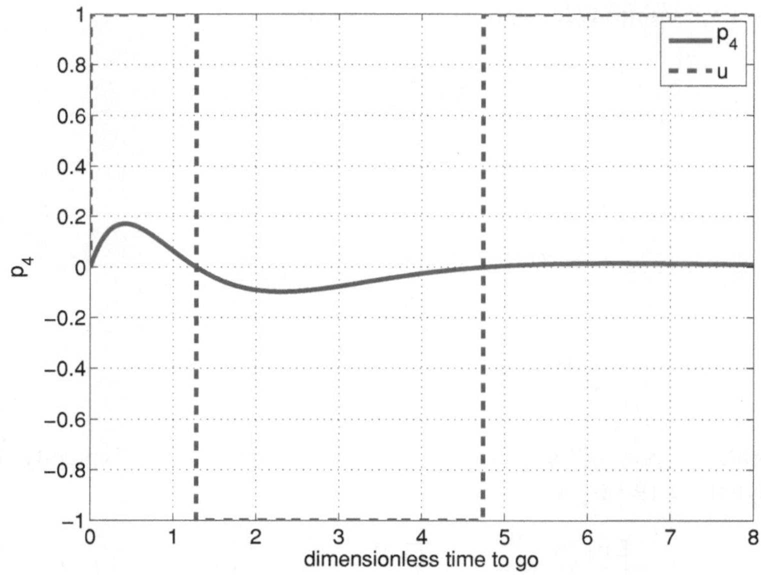

Figure 9.6. Co-state variable p_4 and control u in the optimal evasion problem (jinking maneuver).

The necessary conditions for optimality, (9.69), *are*

$$\dot{x} = \frac{\partial H}{\partial p} = Ax + bu,$$

$$\dot{p} = -\frac{\partial H}{\partial x} = -A^T(t)p,$$

$$u = \operatorname{argmax}(H) = \operatorname{sign}(p_3),$$

$$x(t_0) \quad \textit{given},$$

$$p(t_f^{*-}) = \begin{bmatrix} -1 \\ 0 \\ 1 \end{bmatrix}. \tag{9.134}$$

We eliminate the possibility of indeterminacy in (9.134), *(line 3), as follows. First, explicitly write the co-state dynamics as*

$$\begin{bmatrix} \dot{p}_1 \\ \dot{p}_2 \\ \dot{p}_3 \end{bmatrix} = \begin{bmatrix} 0 & \frac{\lambda}{(t_f^*-t)T} & 0 \\ -1 & \frac{1}{T} & 0 \\ 0 & -\frac{\lambda}{(t_f^*-t)T} & 0 \end{bmatrix} \begin{bmatrix} p_1 \\ p_2 \\ p_3 \end{bmatrix}. \tag{9.135}$$

Using the same argument as for (9.130), *now assume* $p_3 \equiv 0$. *Then the last row of* (9.135) *yields* $p_2 \equiv 0$, *which, combined with* (9.135) *(line 2), yields* $p_1 \equiv 0$. *However, as per* (9.134) *(line 5), the boundary condition for* p_1 *is* $p_1(t_f^{*-}) = -1$. *This contradiction guarantees that there is no indeterminacy in this optimal evasion problem either.*

The optimal evasive maneuver must therefore satisfy

$$u = \operatorname{sign}(p_3), \tag{9.136}$$

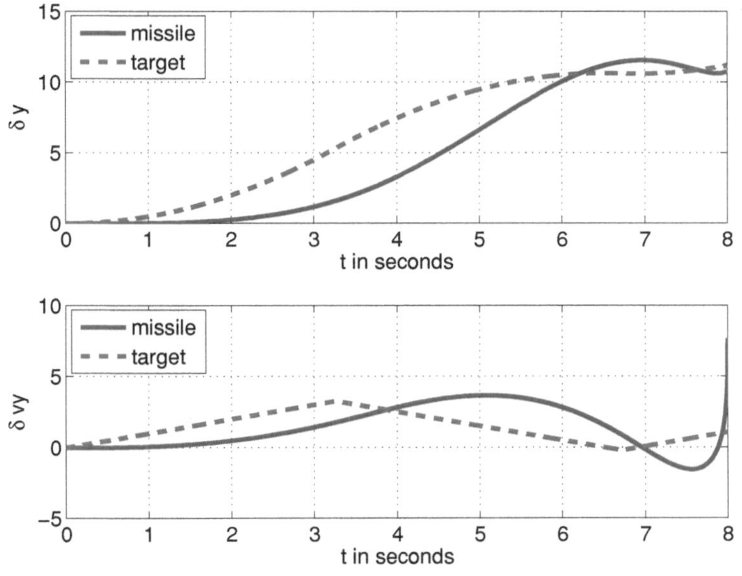

Figure 9.7. Lateral displacements and velocities in the optimal evasion problem (jinking maneuver).

*which in this case we call **juking**, as illustrated in Figures 9.8 and 9.9.*

Note that juking is a well-known tactic in interception sports such as football and soccer [51]. As Figure 9.8 indicates, it uses perfectly timed hard turns to the left and to the right. Moreover, it assumes a target that is capable of changing instantaneously its velocity. This ability to "turn on a dime" is also a well-known requirement for a successful juke.

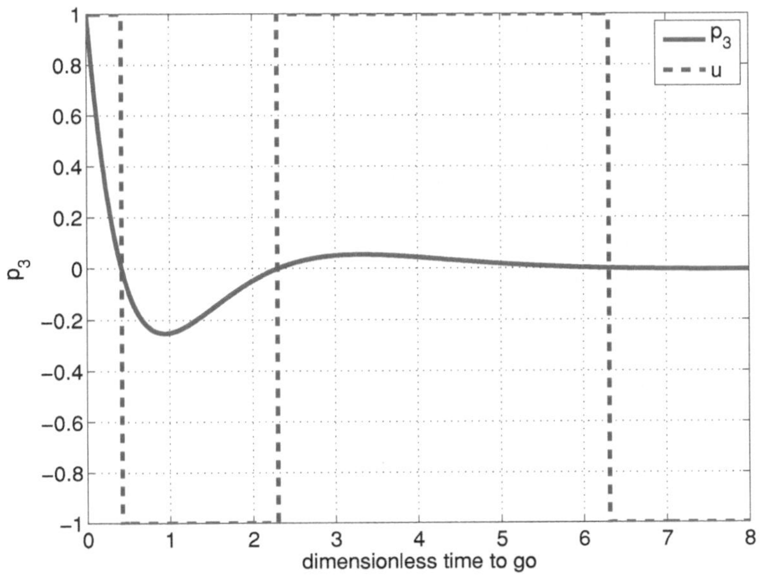

Figure 9.8. Co-state variable p_3 and control u in the optimal evasion problem (juking maneuver).

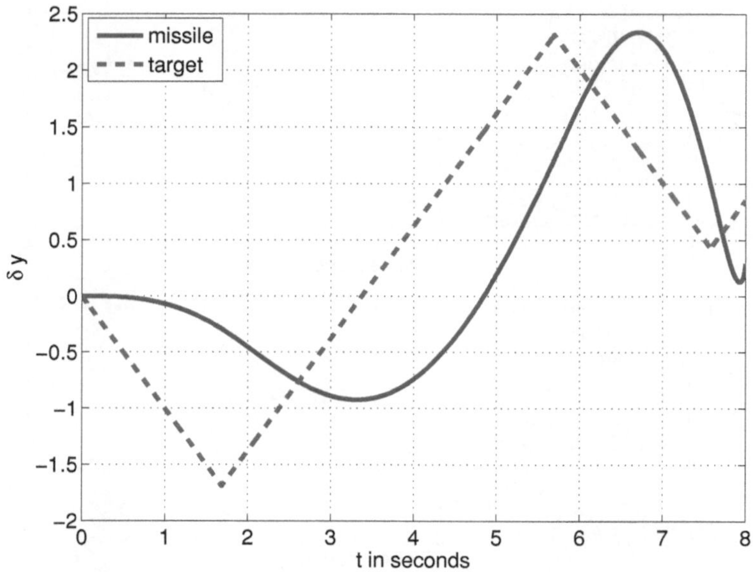

Figure 9.9. Lateral displacements for the pursuer and target in the optimal evasion problem (juking maneuver).

9.5 Dynamic Programming

While Section 9.4 presents the maximum principle as a necessary condition for optimal control, this section presents dynamic programming as a sufficient condition. We start the section with a motivational example, and then we discuss the principle of optimality, backward dynamic programming, continuous-time dynamic programming, and present an example. We then discuss the relationship between Pontryagin's maximum principle and Bellman's equation and discuss the Hamilton–Jacobi–Bellman equation. We end the section with a summary.

9.5.1 Motivational Example: Dynamic Programming

Referring to the example shown in Figure 9.10, we consider the problem of moving, in the t, x space, from (t_0, x_0) to (t_f, x_f) with minimal cumulative cost. In this problem, let x_f be free. We face a sequence of four decisions, each binary (up or down). Hence, there are $2^4 = 16$ possible paths.

The system can be described by the state equations $x_{k+1} = f(x_k, u_k)$, the input is given by $u_k = \pm 1$, there is a total of five steps, given by t_0, t_1, t_2, t_3, and $t_4 = t_f$, and the problem is to minimize the cost function:

$$J = \sum_{k=0}^{3} L(x_k, u_k, t_k). \tag{9.137}$$

The solution to the problem is to proceed *backward in time*, and store, as a function of location and time, the optimal control input and cost achievable from that location and time, $\begin{bmatrix} u_{opt} & J_{TG} \end{bmatrix}^T$, where J_{TG} is called the **cost-to-go**.

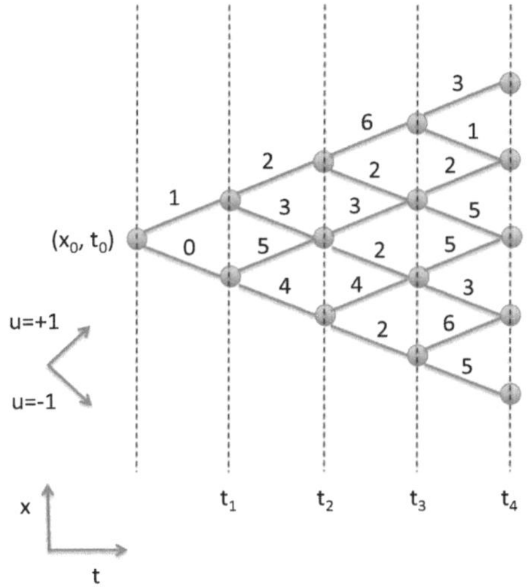

Figure 9.10. Motivational example for dynamic programming.

The result of this procedure is shown in Figure 9.11. In this particular example, the **optimal cost-to-go** from location (t_0, x_0) is 7, and the optimal sequence of decisions is $+1, +1, -1, +1$.

REMARK 9.8 *We have obtained the optimal control as a function of location and time, that is, a feedback control or policy, $u = u(x, t)$.*

REMARK 9.9 *We have had to solve for and store $\begin{bmatrix} u_{opt} & J_{TG} \end{bmatrix}^T$ **for all** x and t. This leads to what is called the **curse of dimensionality**, which is aggravated if the state space or the time line is a continuum.*

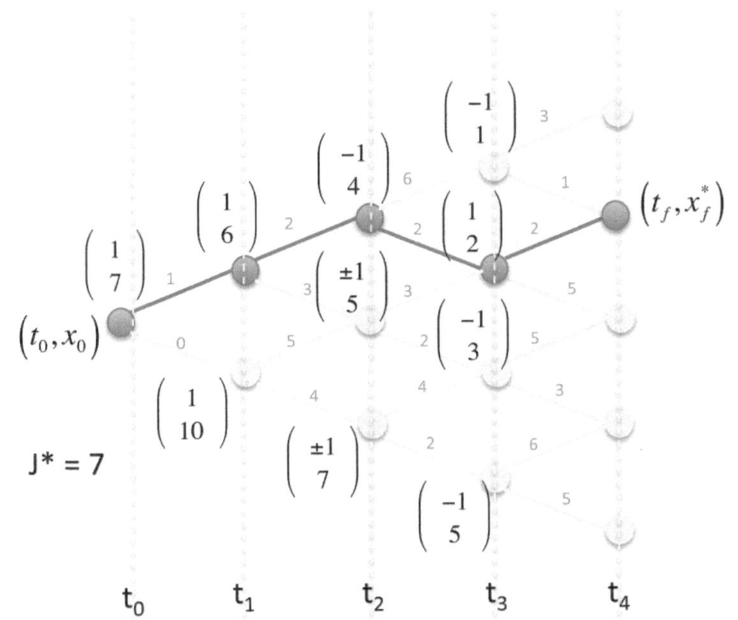

Figure 9.11. Optimal decision and cost-to-go for motivational example for dynamic programming.

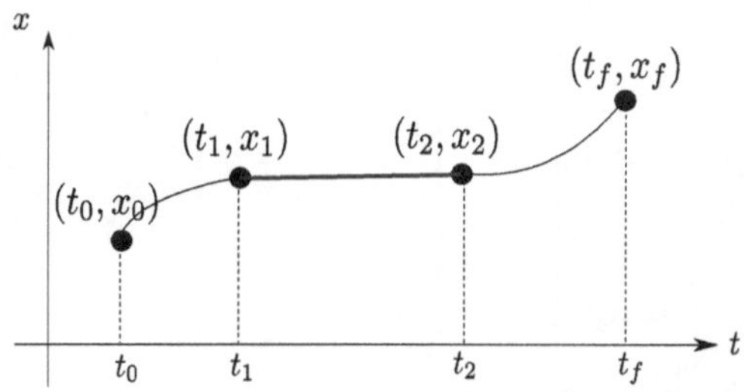

Figure 9.12. Illustration of the principle of optimality.

REMARK 9.10 *The optimal solution is not greedy (see Section 9.4.1). The greedy control sequence, $-1, -1, -1, -1$, yields $J_{greedy} = 11$.*

REMARK 9.11 *Sometimes optimality allows several alternative decisions (in our example, at $(t, x) = (t_2, x_1)$ and $(t, x) = (t_2, x_2)$, both $u = +1$ and $u = -1$ are optimal).*

9.5.2 The Principle of Optimality

This section presents a simple yet powerful principle that is the foundation of dynamic programming. It stems from the realization that *restriction must preserve optimality*.

Let us consider the system

$$\begin{aligned} \dot{x} &= f(x, u, t), \\ u &\in U, \\ x(t_0) &= x_0, \\ x(t_f) &= x_f. \end{aligned} \tag{9.138}$$

In Figure 9.12, assume that a trajectory from (t_0, x_0) to (t_f, x_f) optimizes the functional

$$J = \int_{t_0}^{t_f} L(x, u, t)\, \mathrm{d}t \tag{9.139}$$

subject to (9.138), first two lines. Let $[t_1, t_2]$ be a subinterval of $[t_0, t_f]$, and consider the subtrajectory from (t_1, x_1) to (t_2, x_2) obtained by restricting the original trajectory to the interval $[t_1, t_2]$. The question at hand is whether the restricted trajectory optimizes the restricted functional

$$J' = \int_{t_1}^{t_2} L(x, u, t)\, \mathrm{d}t \tag{9.140}$$

subject to (9.138), first two lines. The answer is as follows.

PROPOSITION 9.3 (Bellman's Principle of Optimality) *The optimal trajectory going from x_0 at t_0 to x_f at t_f is such that for all t_1, t_2 belonging to $[t_0, t_f]$, the portion of the optimal trajectory from $x(t_1)$ at t_1 to $x(t_2)$ at t_2 is also an optimal trajectory relative to the same cost functional restricted to the interval $[t_1, t_2]$.*

PROOF. Let $u(t), t \in [t_0, t_f]$ be an optimal control. Rewrite the cost functional as

$$J = \int_{t_0}^{t_1} L(x, u, t)\,\mathrm{d}t + \int_{t_1}^{t_2} L(x, u, t)\,\mathrm{d}t + \int_{t_2}^{t_f} L(x, u, t)\,\mathrm{d}t. \tag{9.141}$$

Assume that $\bar{u}(t)$ exists such that, with end points (t_1, x_1) and (t_2, x_2),

$$\int_{t_1}^{t_2} L(x, \bar{u}, t)\,\mathrm{d}t < \int_{t_1}^{t_2} L(x, u, t)\,\mathrm{d}t. \tag{9.142}$$

Then, define $u^*(t)$ as

$$u^*(t) = \begin{cases} u(t) & \text{on } [t_0, t_1], \\ \bar{u}(t) & \text{on } [t_1, t_2], \\ u(t) & \text{on } [t_2, t_f]. \end{cases} \tag{9.143}$$

In that case, $J(u^*) < J(u)$, which implies that $u(t)$ is not optimal. This contradiction completes the proof.

REMARK 9.12 *The principle of optimality is a* ***necessary*** *condition for optimality,* ***not a sufficient*** *one. Piecing together two optimal trajectories does not necessarily yield an optimal trajectory. For instance, it is well known (see Problem 9.1) that, in $\mathbb{R}^2$, the shortest path between two points is a straight line. However, piecing together two straight lines generally does not yield a straight line but rather a broken line.*

9.5.3 Backward Dynamic Programming

Let us revisit the same kind of system as in the motivational example of Section 9.5.1, that is, systems with discrete state space, discrete input space, and discrete time. The dynamics are

$$\begin{aligned} x(k+1) &= f(x(k), u(k), k) \\ x(0) &\quad \text{given}, \end{aligned} \tag{9.144}$$

and the cost function is

$$J = \sum_{k=0}^{N} L(x(k), u(k), k). \tag{9.145}$$

Let $J^*(x, k)$ be the optimal cost-to-go from location x in the state space and step k on the time line. Then, applying the principle of optimality yields the **fundamental equation of dynamic programming**:

$$J^*(x(k), k) = \min_{u(k) \in U} \left\{ L(x(k), u(k), k) + J^*(x(k+1), k+1) \right\}. \tag{9.146}$$

REMARK 9.13 *Note that in* (9.146), $J^*(x(k+1), k+1)$ *is a function of* $u(k)$ *because* $x(k+1) = f(x(k), u(k), k)$.

EXAMPLE 9.9 *Referring back to the example of Section 9.5.1, apply this equation sequentially for* $k = 3, 2, 1, 0$. *Then, we obtain* $J^*(x_0, 0) = 7$, *as expected.*

REMARK 9.14 *The minimization (9.146) yields the optimal policy $u^*(x(k), k)$ as*

$$u^*(x(k), k) = \operatorname*{argmin}_{v \in U} \{L(x(k), v, k) + J^*(x(k+1), k+1)\}$$
$$= \operatorname*{argmin}_{v \in U} \{L(x(k), v, k) + J^*(f(x(k), v, k), k+1)\}. \quad (9.147)$$

Hence, the optimal policy chooses u^* along the direction of maximum descent of the function

$$L(x(k), v, k) + J^*(f(x(k), v, k), k+1). \quad (9.148)$$

REMARK 9.15 (Imbedding Principle) *The dynamic programming equations (9.146), (9.147) reflect an imbedding principle, as follows. Given the problem of finding a control sequence that optimizes (9.145) subject to (9.144), we have imbedded this problem into the larger class of finding the optimal control* ***at every location in the state space****, including locations that are not on the optimal trajectory. The solution of this larger class of problems is provided by (9.146), (9.147). This imbedding is the root cause of the curse of dimensionality.*

REMARK 9.16 *Dynamic programming (9.147) yields the optimal decision as a function of state and time, that is, as a policy. Hence, it assumes feedback control.*

REMARK 9.17 *Equations (9.146), (9.147) reflect backward dynamic programming. Forward dynamic programming can be used as an alternative.*

9.5.4 Continuous-Time Dynamic Programming

Let us now consider the continuous-time system and optimization problem:

$$\begin{aligned}
\dot{x} &= f(x, u, t), \\
x(t_0) &= x_0, \\
u &\in U, \\
t &\in [t_0, t_f], \\
J &= \int_{t_0}^{t_f} L(x, u, t)\,\mathrm{d}t, \\
x_f &\quad \text{free}, \\
t_f &\quad \text{free}.
\end{aligned} \quad (9.149)$$

Here $J(x_0, u(.), t_0)$ for t_f fixed is a functional in u. We seek to minimize it to obtain $J^*(x_0, t_0)$, the optimal cost, that is,

$$J^*(x_0, t_0) = \min_{u \in U, t_0 \le t \le t_f} \int_{t_0}^{t_f} L(x(u(.), t), u(t), t)\,\mathrm{d}t$$
$$= \min_{u \in U, t_0 \le t \le t_f} \left\{ \int_{t_0}^{t_1} L(x, u, t)\,\mathrm{d}t + \int_{t_1}^{t_f} L(x, u, t)\,\mathrm{d}t \right\}, \quad (9.150)$$

where $t_1 \in [t_0, t_f]$ is arbitrary. Applying the fundamental equation of dynamic programming (9.146) yields

$$J^*(x_0, t_0) = \min_{u \in U, t_0 \leq t \leq t_1} \left\{ \int_{t_0}^{t_1} L(x, u, t)\,dt + J^*(x(t_1), t_1) \right\}, \tag{9.151}$$

where $x(t_1)$ is the state resulting from applying control $u(t)$ on the interval $[t_0, t_1]$. Assume that the optimal control u^* is used. Then, (9.151) becomes

$$J^*(x_0, t_0) = \int_{t_0}^{t_1} L(x(u^*(.), t), u^*, t)\,dt + J^*(x(t_1), t_1), \tag{9.152}$$

or equivalently,

$$-\frac{J^*(x_0, t_0) - J^*(x(t_1), t_1)}{t_1 - t_0} = \frac{1}{t_1 - t_0} \int_{t_0}^{t_1} L(x(u^*(.), t), u^*, t)\,dt. \tag{9.153}$$

Taking the limit as $t_1 \to t_0$ and assuming that the limit of the left-hand side exists, we obtain

$$\left(\frac{dJ^*}{dt}\right)_{u^*, t_0} + L(x_0, u^*(t_0), t_0) = 0. \tag{9.154}$$

If we had used a nonoptimal control $u \neq u^*$ instead of (9.152), we would have obtained

$$J^*(x_0, t_0) \leq \int_{t_0}^{t_1} L(x, u, t)\,dt + J^*(x(t_1), t_1), \tag{9.155}$$

and the same limit process that led to (9.154) would have yielded

$$\left(\frac{dJ^*}{dt}\right)_{u, t_0} + L(x_0, u(t_0), t_0) \geq 0. \tag{9.156}$$

Because using the optimal u^* yields (9.154), we have

$$\min_u \left\{ \left(\frac{dJ^*}{dt}\right)_{u,t} + L(x, u, t) \right\} = 0, \tag{9.157}$$

which is called **Bellman's functional equation**, in general form, in continuous time. Note that (9.157) assumes that the time derivative in the left-hand side exists.

If, for a given state x at time t, $\frac{\partial J^*}{\partial x}$, $\frac{\partial J^*}{\partial t}$ and $f(x, u, t)$ are all continuous, then $\left(\frac{dJ^*}{dt}\right)_{u,t}$ is given by the chain rule

$$\left(\frac{dJ^*}{dt}\right)_{u,t} = \left(\frac{\partial J^*}{\partial x}\right)^T f(x, u, t) + \frac{\partial J^*}{\partial t}. \tag{9.158}$$

Note that $\frac{\partial J^*}{\partial t}$ depends on u^*, but not on u. Then, (9.157) can be rewritten as

$$-\frac{\partial J^*}{\partial t} = \min_{u \in U} \left[L(x, u, t) + \left(\frac{\partial J^*}{\partial x}\right)^T f(x, u, t) \right]. \tag{9.159}$$

This is a more common form of **Bellman's equation**, but, as noted previously, it requires $J^* \in \mathcal{C}^1$.

REMARK 9.18 *When one performs the minimization* (9.159), *obtains the optimal u as a function of* $\frac{\partial J^*}{\partial x}$, *and substitutes the result into* (9.159), *one obtains a first-order partial* *differential equation, which is usually nonlinear. One then solves it for J^* and uses the resulting $\frac{\partial J^*}{\partial x}$ to obtain the optimal u^*. The boundary condition for that partial differential equation is*

$$\lim_{t \to t_f} J^*(x(t), t) = \lim_{t \to t_f} \int_t^{t_f} L(x^*, u^*, \tau)\, d\tau = 0. \tag{9.160}$$

REMARK 9.19 *Note that in (9.159), the quantity $L(x, u, t) + \left(\frac{\partial J^*}{\partial x}\right)^T f(x, u, t)$ looks suspiciously like a Hamiltonian. This resemblance is investigated in Section 9.5.7.*

REMARK 9.20 *The optimal control u^* must satisfy*

$$-\frac{\partial J^*}{\partial t} = L(x^*, u^*, t) + \left(\frac{\partial J^*}{\partial x}\right)^T f(x^*, u^*, t). \tag{9.161}$$

For many problems, this is not only a necessary condition but is also sufficient. In any case, (9.161) can be used to test the optimality of a candidate u^*.

9.5.5 The Linear Quadratic Regulator

MOTIVATION. In Remark 8.12, we establish that, to optimize a twice-differentiable function of a vector variable subject to differentiable constraints, it is possible to set up an iteration wherein the function is approximated locally by a quadratic function, the constraint by a linear function, and choose as the next iterate the solution of this subsidiary optimization problem. It turns out that a similar process is possible in optimal control. Indeed, consider the minimization problem

$$\min_{u(.)} J = \int_{t_0}^{t_f} L(x, u, t)\, dt, \tag{9.162}$$

subject to

$$\begin{aligned} \dot{x} &= f(x, u, t), \\ x(t_0) &= x_0, \\ x_f, t_f &\quad \text{free}. \end{aligned} \tag{9.163}$$

Assume that we are given a nominal trajectory, $u^0(t), x^0(t)$ such that $\dot{x}^0 = f(x^0, u^0, t)$. Also assume perturbations $\delta u(t)$ and $\delta x(t)$ of the nominal trajectory so that $u(t)$ and $x(t)$ satisfy

$$\begin{aligned} u(t) &= u^0(t) + \delta u(t) \\ x(t) &= x^0(t) + \delta x(t). \end{aligned} \tag{9.164}$$

We seek δu for an optimal decrease in J.

We start by linearizing the dynamic constraint

$$\begin{aligned} \dot{x}^0 + \delta\dot{x} &= f(x^0 + \delta x, u^0 + \delta u, t) \\ &= f(x^0, u^0, t) + \left(\frac{\partial f}{\partial x}\right)^T_{x^0, u^0} \delta x + \left(\frac{\partial f}{\partial u}\right)^T_{x^0, u^0} \delta u + \text{H.O.T.} \end{aligned} \tag{9.165}$$

We can then use the nominal condition $\dot{x}^0 = f(x^0, u^0, t)$ to eliminate terms on both sides, yielding to first order

$$\delta\dot{x} = \left(\frac{\partial f}{\partial x}\right)^T_{x^0,u^0} \delta x + \left(\frac{\partial f}{\partial u}\right)^T_{x^0,u^0} \delta u. \tag{9.166}$$

Let $A(t) = \left(\frac{\partial f}{\partial x}\right)^T_{x^0(t),u^0(t)}$ and $B(t) = \left(\frac{\partial f}{\partial u}\right)^T_{x^0(t),u^0(t)}$. Then, (9.166) becomes

$$\delta\dot{x} = A(t)\delta x(t) + B(t)\delta u(t). \tag{9.167}$$

In response to the perturbations $\delta u(t)$ and $\delta x(t)$, the increment in cost functional has the form

$$\Delta J = \int_{t_0}^{t_f} \left\{ \left(\frac{\partial L}{\partial x}\right)^T \delta x + \left(\frac{\partial L}{\partial u}\right)^T \delta u + \frac{1}{2}\begin{bmatrix}\delta x \\ \delta u\end{bmatrix}^T \left(\frac{\partial^2 L}{\partial (x,u)}\right) \begin{bmatrix}\delta x \\ \delta u\end{bmatrix} + \text{H.O.T.} \right\} \mathrm{d}t. \tag{9.168}$$

Consider the particular case where

$$\frac{\partial L}{\partial x} = 0,$$
$$\frac{\partial L}{\partial u} = 0,$$
$$\frac{\partial^2 L}{\partial x \partial u} = 0. \tag{9.169}$$

In this case, the second variation of the cost functional is

$$\delta^2 J = \frac{1}{2}\int_{t_0}^{t_f} \left\{ (\delta x)^T \left(\frac{\partial^2 L}{\partial x^2}\right)_{x^0(t),u^0(t)} \delta x + (\delta u)^T \left(\frac{\partial^2 L}{\partial u^2}\right)_{x^0(t),u^0(t)} \delta u \right\} \mathrm{d}t. \tag{9.170}$$

Let $Q(t) = \left(\frac{\partial^2 L}{\partial x^2}\right)_{x^0(t),u^0(t)}$ and $R(t) = \left(\frac{\partial^2 L}{\partial u^2}\right)_{x^0(t),u^0(t)}$. Then, (9.169) can be rewritten as

$$\delta^2 J = \frac{1}{2}\int_{t_0}^{t_f} \left\{ (\delta x)^T Q(t) \delta x + (\delta u)^T R(t) \delta u \right\} \mathrm{d}t. \tag{9.171}$$

Equations (9.167) and (9.171) define the **Linear Quadratic Regulator (LQR) Problem**. This problem can be solved in closed form, as shown next. Once the optimal time histories of δu and δx are computed, u and x are updated according to (9.164), and the iteration proceeds.

SOLUTION OF THE LQR PROBLEM. Let us consider as an example application of LQR the following dynamic system and functional:

$$\dot{x}(t) = A(t)x(t) + B(t)u(t),$$
$$x_0 = 0,$$
$$t_0 = 0,$$
$$t_f \quad \text{given},$$
$$J = \frac{1}{2}\int_{t_0}^{t_f} \left(x^T(t)Q(t)x(t) + u^T(t)R(t)u(t)\right) \mathrm{d}t. \tag{9.172}$$

Let us in addition assume that $Q(t) \geq 0$ and that $R(t) > 0$.

The Bellman equation applied to (9.172) yields

$$-\frac{\partial J^*}{\partial t} = \min_u \left(\frac{1}{2} x^T Q x + \frac{1}{2} u^T R u + \frac{\partial J^*}{\partial x^T}(Ax + Bu) \right). \tag{9.173}$$

The first-order necessary condition for optimality is

$$\frac{\partial}{\partial u} \left(\frac{1}{2} x^T Q x + \frac{1}{2} u^T R u + \frac{\partial J^*}{\partial x^T}(Ax + Bu) \right) = Ru + B^T \frac{\partial J^*}{\partial x} = 0, \tag{9.174}$$

that is,

$$u^* = -R^{-1} B^T \frac{\partial J^*}{\partial u}, \tag{9.175}$$

and the second-order condition yields

$$\frac{\partial^2}{\partial u^2} \left(\frac{1}{2} x^T Q x + \frac{1}{2} u^T R u + \frac{\partial J^*}{\partial x^T}(Ax + Bu) \right) = R > 0, \tag{9.176}$$

which, together with (9.175), guarantees a minimum.

Substituting (9.175) for u^* in Bellman's equation (9.173), we obtain

$$0 = \frac{\partial J^*}{\partial t} + \frac{1}{2} x^T Q x - \frac{1}{2} \left(\frac{\partial J^*}{\partial x} \right)^T B R^{-1} B^T \left(\frac{\partial J^*}{\partial x} \right) + \left(\frac{\partial J^*}{\partial x} \right)^T Ax. \tag{9.177}$$

We use separation of variables to solve this partial differential equation. Let us postulate that the optimal cost-to-go is quadratic, of the form

$$J^*(x, t) = \frac{1}{2} x^T K(t) x. \tag{9.178}$$

Then,

$$\frac{\partial J^*}{\partial t} = \frac{1}{2} x^T \dot{K} x, \tag{9.179}$$

and

$$\frac{\partial J^*}{\partial x} = Kx. \tag{9.180}$$

Equation (9.177) becomes

$$\frac{1}{2} x^T \dot{K} x + \frac{1}{2} x^T Q x - \frac{1}{2} x^T K B R^{-1} B^T K x + \frac{1}{2} x^T (KA + A^T K) x = 0. \tag{9.181}$$

Note that (9.181), together with boundary condition (9.160), is easily satisfied by setting

$$\begin{aligned} -\dot{K}(t) &= A^T(t) K(t) + K(t) A(t) - K(t) B(t) R^{-1}(t) B^T(t) K(t) + Q(t) \\ K(t_f) &= 0. \end{aligned} \tag{9.182}$$

Then, the optimal policy and cost-to-go are given by

$$\begin{aligned} u^*(x(t), t) &= -R^{-1}(t) B^T(t) K(t) x(t) \\ J^*(x_0, t_0) &= \frac{1}{2} x_0^T K(t_0) x_0. \end{aligned} \tag{9.183}$$

REMARK 9.21 *A comparison of* (9.182) *with* (4.89) *reveals that the former is a differential Ricatti equation, dual to the latter, in the sense of duality introduced in*

Proposition 2.23. Hence, for linear dynamic systems with quadratic performance and observation costs, the processes of control and observation are dual of one another. The problems of optimal control and optimal observation are both solved, in closed form, using differential Ricatti equations. Moreover, the Ricatti equations for optimal control and optimal observation are also dual of one another.

REMARK 9.22 *Note from* (9.183) *(line 1) that when $R(t)$ is close to being singular, $u^*(x(t), t)$ is generally large. This situation is called* ***cheap control.***

9.5.6 The Linear Quadratic Gaussian Regulator

In this section, we present the solution to a basic problem in stochastic optimal control that is applicable to navigation and guidance. The problem is, given a guidance system governed by a linear Gauss–Markov model, such as in Section 3.8, design a guidance law, based on output feedback, that minimizes the expected value of a quadratic cost functional, such as in Section 9.5.5. Remarkably, the solution to this stochastic optimal control problem consists of two *independent* parts: an optimal navigator, such as the Kalman filter of Section 4.7, followed by an optimal state-based guidance law, such as the linear quadratic regulator of Section 9.5.5.

Consider the equations of motion of a vehicle, linearized about a nominal trajectory and described by the linear Gauss–Markov model:

$$\begin{aligned} \dot{x}(t) &= A(t)x(t) + B(t)u(t) + w(t), \quad t \geq t_0 \\ y(t) &= C(t)x(t) + v(t), \end{aligned} \tag{9.184}$$

where $x(t) \in \mathbb{R}^n$ is the guidance error, $u(t) \in \mathbb{R}^m$ is the thrust acceleration, $y(t) \in \mathbb{R}^p$ is the reading of position and velocity sensors, t_0 is a given initial time, and $w(t)$ and $v(t)$ represent disturbance and noise processes, respectively. We make the same standard assumptions as in Section 3.8, namely,

$$\begin{aligned} x(t_0) &= \mathcal{N}\left(\bar{x}(t_0), P_0\right), \\ E[w(t)] &= 0, \\ E[w(t)w^T(\tau)] &= R_w(t)\delta(t-\tau), \\ E[w(t)(x(t_0) - \bar{x}(t_0))^T] &= 0, \\ E[v(t)] &= 0, \\ E[v(t)v^T(\tau)] &= R_v(t)\delta(t-\tau), \\ E[v(t)(x(t_0) - \bar{x}(t_0))^T] &= 0, \\ E[w(t)v^T(\tau)] &= 0. \end{aligned} \tag{9.185}$$

For the system (9.184) under assumptions (9.185), we are to design a guidance law, based on feedback of the measured output $y(t)$, that solves the following minimization problem:

$$\min_u J(u) = E\left(\frac{1}{2}x^T(t_f)K_f x(t_f) + \frac{1}{2}\int_{t_0}^{t_f} \left(x^T(t)Q(t)x(t) + u^T(t)R(t)u(t)\right)\right), \tag{9.186}$$

where $E(.)$ represents expected value, $K_f \geq 0$, $Q(t) \geq 0$, and $R(t) > 0$ are symmetric matrices of appropriate dimensions, and $t_f > t_0$ is a specified final time.

The solution to the optimization problem specified by (9.184)–(9.186) is as follows:

$$u^*(t) = -R^{-1}(t)B^T(t)K(t)\hat{x}(t), \tag{9.187}$$

where the symmetric matrix $K(t)$ satisfies

$$\begin{aligned} -\dot{K}(t) &= A^T(t)K(t) + K(t)A(t) - K(t)B(t)R^{-1}(t)B^T(t)K(t) + Q(t) \\ K(t_f) &= K_f, \end{aligned} \tag{9.188}$$

and the vector $\hat{x}(t)$ satisfies

$$\begin{aligned} \dot{\hat{x}}(t) &= A(t)\hat{x}(t) + B(t)u(t) + -P(t)C^T(t)R_v^{-1}(t)\left(C(t)\hat{x}(t) - y(t)\right) \\ \hat{x}(t_0) &= \bar{x}(t_0), \end{aligned} \tag{9.189}$$

where the symmetric matrix $P(t)$ satisfies

$$\begin{aligned} \dot{P}(t) &= A(t)P(t) + P(t)A^T(t) - P(t)C^T(t)R_v^{-1}(t)C(t)P(t) + R_w(t) \\ P(t_0) &= P_0. \end{aligned} \tag{9.190}$$

Comparing (9.188) and (9.189) with (9.182) and (9.183) reveals that the optimal guidance law is to feed back $\hat{x}(t)$ through the optimal state-based guidance law defined by the matrices $Q(t)$ and $R(t)$. Moreover, comparing (9.189) and (9.190) with (4.79), (4.88) and (4.89) reveals that $\hat{x}$ is itself the optimal estimate of $x(t)$ defined by the matrices $R_v(t)$ and $R_w(t)$. Most remarkably, the optimal state-based guidance law and the optimal estimator are designed separately. This is the **stochastic separation principle**, which strengthens the deterministic separation principle of Chapter 7: not only can navigation and guidance laws be designed separately, they can also be optimized separately under the assumptions of linear dynamics, quadratic cost, and Gaussian noise processes. Hence the name **linear quadratic Gaussian (LQG) regulator** for the guidance law (9.187)–(9.190).

9.5.7 Relationship between the Maximum Principle and Dynamic Programming

In this section, we follow up on a suspicion raised in Remark 9.19. Recall Bellman's equation (9.159), repeated here for convenience:

$$-\frac{\partial J^*}{\partial t} = \min_u \left[L(x,u,t) + \left(\frac{\partial J^*}{\partial x}\right)^T f(x,u,t) \right], \tag{9.191}$$

which can be rewritten as

$$\frac{\partial J^*}{\partial t} = \max_u \left[-\left(\frac{\partial J^*}{\partial x}\right)^T f(x,u,t) - L(x,u,t) \right]. \tag{9.192}$$

At the optimum, we have

$$u^* = \operatorname*{argmax}_u \left[-\left(\frac{\partial J^*}{\partial x}\right)^T f(x,u,t) - L(x,u,t) \right]. \tag{9.193}$$

Now, also recall the Hamiltonian (9.68), repeated here for convenience:

$$H(x, p, u, t) = p^T f(x, u, t) - L(x, u, t), \tag{9.194}$$

and for which the maximum principle stipulates that

$$u^* = \underset{v}{\operatorname{argmax}} \left(p^T f(x, v, t) - L(x, v, t)\right). \tag{9.195}$$

A comparison of (9.195) with (9.193) suggests that $p(t)$, the co-state vector in the maximum principle, and $\frac{\partial J^*}{\partial x}(x, t)$, the gradient of the optimal cost-to-go, are related by

$$p(t) = -\frac{\partial J^*}{\partial x}(x, t). \tag{9.196}$$

To show that (9.196) indeed holds, we show that $p(t)$ and $-\frac{\partial J^*}{\partial x}$ satisfy the same differential equation with boundary conditions. Then their equality follows from standard results on uniqueness of solutions of smooth ordinary differential equations.

The co-state vector satisfies (9.69) (line 2) and (9.83) (line 2), that is,

$$\begin{aligned} \dot{p}(t) &= -\frac{\partial H}{\partial x}(x, p, u, t) \\ p(t_f) &= 0. \end{aligned} \tag{9.197}$$

Now, at the optimum, (9.192) yields

$$\frac{\partial J^*}{\partial t} + L(x^*, u^*, t) + \left(\frac{\partial J^*}{\partial x}\right)^T f(x^*, u^*, t) = 0. \tag{9.198}$$

Assume that $J \in C^2$. Differentiate (9.198) with respect to x to obtain

$$\frac{\partial^2 J^*}{\partial x \partial t} + \frac{\partial L}{\partial x} + \left(\frac{\partial^2 J^*}{\partial x^2}\right) f + \left(\frac{\partial f}{\partial x}\right)^T \left(\frac{\partial J^*}{\partial x}\right) = 0. \tag{9.199}$$

Now, the chain rule also yields

$$\frac{\mathrm{d}}{\mathrm{d}t}\left(-\frac{\partial J^*}{\partial x}\right) = -\left(-\frac{\partial^2 J^*}{\partial x^2}\right) f - \left(-\frac{\partial^2 J^*}{\partial x \partial t}\right). \tag{9.200}$$

Combining (9.199) and (9.200) yields

$$\begin{aligned} \frac{\mathrm{d}}{\mathrm{d}t}\left(-\frac{\partial J^*}{\partial x}\right) &= \left(\frac{\partial f}{\partial x}\right)^T \left(\frac{\partial J^*}{\partial x}\right) + \frac{\partial L}{\partial x} \\ &= -\frac{\partial}{\partial x} H\left(x^*, -\frac{\partial J^*}{\partial x}, u^*, t\right). \end{aligned} \tag{9.201}$$

Moreover, the definition of J^* given in (9.150) implies that $J^*(x, t_f) \equiv 0$, yielding

$$\left.\left(\frac{\partial J^*}{\partial x}\right)\right|_{t_f} = 0. \tag{9.202}$$

Comparing (9.197) with (9.201)–(9.202), we conclude that the vectors $p(t)$ and $-\frac{\partial J^*}{\partial x}$ satisfy the same differential equation and boundary condition. Therefore they must agree, and (9.196) must hold.

REMARK 9.23 *Note the similarity between* (8.45) *and* (9.196). *Whereas in finite dimensional constrained optimization the vector of Lagrange multipliers quantifies the sensitivity of the objective function with respect to the constraints, in optimal control the co-state vector quantifies the sensitivity of the optimal cost-to-go with respect to the state vector.*

9.5.8 The Hamilton–Jacobi–Bellman Equation

If the optimal cost-to-go J^* is twice continuously differentiable and one uses an optimal control u^*, (9.191) and (9.196) imply

$$-\frac{\partial J^*}{\partial t} = L + \left(\frac{\partial J^*}{\partial x}\right)^T f(x^*, u^*, t)$$
$$= -H(x^*, \frac{\partial J^*}{\partial x}, u^*, t), \tag{9.203}$$

which can be rewritten as

$$\frac{\partial J^*}{\partial t} - H(x^*, \frac{\partial J^*}{\partial x}, u^*, t) = 0. \tag{9.204}$$

This partial differential equation for J^* is called the **Hamilton–Jacobi–Bellman equation**, which, as noted earlier, assumes that $J^* \in \mathcal{C}^2$.

9.5.9 Dynamic Programming Summary

In this section, we summarize the hierarchy of results on dynamic programming obtained so far. This hierarchy is based on assumptions on the differentiability of J^*. We consider dynamic systems of the form

$$\dot{x} = f(x, u, t),$$
$$x(t_0) = x_0,$$
$$u \in U. \tag{9.205}$$

We have a cost functional of the form

$$J(x_0, t_0, u(.)) = \int_{t_0}^{t_f} L(x, u, t)\, \mathrm{d}t. \tag{9.206}$$

Define the optimal cost-to-go as

$$J^*(x_0, t_0) = \min_{u(.)\in U} \int_{t_0}^{t_f} L(x, u, t)\, \mathrm{d}t \tag{9.207}$$

subject to (9.205). Then, assuming that $\frac{\mathrm{d}J^*}{\mathrm{d}t}$ exists, we must have

$$\min_{u\in U} \left[\frac{\mathrm{d}J^*}{\mathrm{d}t} + L(x, u, t)\right] = 0. \tag{9.208}$$

If we further assume that $J^* \in \mathcal{C}^1$, that is, that all partial derivatives of J^* exist and are continuous, then we must have

$$-\frac{\partial J^*}{\partial t} = \min_{u\in U} \left[L(x, u, t) + \left(\frac{\partial J^*}{\partial x}\right)^T f(x, u, t)\right]. \tag{9.209}$$

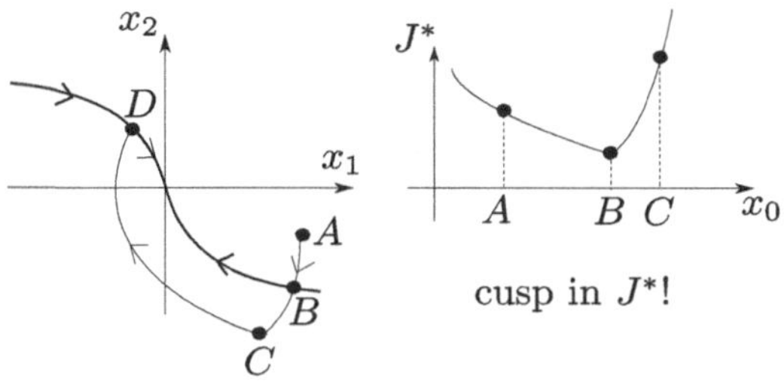

Figure 9.13. Switching curve and optimal cost-to-go for time-optimal control of double integrator example.

Finally, if we further assume that $J^* \in \mathcal{C}^2$, that is, that all second-order partial derivatives of J^* exist and are continuous, then we must have

$$\begin{aligned} p &= -\frac{\partial J^*}{\partial x} \\ \frac{\partial J^*}{\partial t} &- H\left(x^*, -\frac{\partial J^*}{\partial x}, u^*, t\right) = 0. \end{aligned} \tag{9.210}$$

REMARK 9.24 *Note that $J^* \in \mathcal{C}^2$ may not be true. For instance, in the example of Section 9.4.4 where we considered time-optimal control of a double integrator, $\frac{\partial J^*}{\partial x}$ is discontinuous at point B, as shown in Figure 9.13. This is because the optimal trajectory originating from A needs only to connect to the ascending switching line at B before moving to the origin. However, the optimal trajectory originating from C must first cross the x_1 axis, connect to the descending switching line at D, and then move to the origin. Even though in the x_1, x_2 plane B is equidistant from A and C, the optimal trajectory originating from C lasts substantially longer than that from A. As a consequence, on the plot of J^* as a function of the location on the arc ABC, the slope of J^* on the C-side of B is substantially steeper than the slope on the A-side of B. This implies that the plot of J^* has a cusp at B. The optimal cost-to-go is not continuously differentiable at points on the switching line.*

Note that, even when J^* is not continuously differentiable, we can still use the general form of Bellman's equation (9.157):

$$\min_u \left\{ \left(\frac{\mathrm{d}J^*}{\mathrm{d}t} \right) + L(x, u, t) \right\} = 0. \tag{9.211}$$

9.6 The Maximum Principle and Dynamic Programming

In this section, we summarize and interrelate the results obtained on both the maximum principle and dynamic programming. For convenience, we restate the problem of interest. We consider dynamic systems of the form

$$\begin{aligned} \dot{x} &= f(x, u, t), \\ x(t_0) &= x_0, \\ t_f &\quad \text{given}, \\ x(t_f) &\quad \text{free}, \\ u(t) &\in U. \end{aligned} \tag{9.212}$$

We are also given a cost functional of the form

$$J(x_0, t_0, u(.)) = K(x(t_f)) + \int_{t_0}^{t_f} L(x(t), u(t), t)\, \mathrm{d}t. \tag{9.213}$$

Our goal is to minimize $J(x_0, t_0, u(.))$ over the time history of the control inputs, $u(.)$, subject to the constraint that for all t, $u(t) \in U$. Assume that $f(., ., .)$, $L(., ., .)$ and $K(.)$ are all in $\mathcal{C}^1$. We define

$$J^*(x_0, t_0) = \min_{u(.):\forall t, u(t) \in U} J(x_0, t_0, u(.)) \tag{9.214}$$

to be the **cost-to-go**, and:

$$H(x, p, u, t) = p^T f(x, u, t) + L(x, u, t) \tag{9.215}$$

to be the **Hamiltonian**.

PROPOSITION 9.4 *Given the optimal control problem specified by* (9.212) *and* (9.213), *consider the following propositions:*

(a) $J^*(x_0, t_0)$ *exists, with optimal control* u^*.
(b) $J^*(x_0, t_0)$ *exists and is in* $\mathcal{C}^1$ *with optimal control* u^*.
(c) $J^*(x_0, t_0)$ *exists and is in* $\mathcal{C}^2$ *with optimal control* u^*.
(d) $\forall t_1 \in [t_0, t_f]$,

$$J^*(x_0, t_0) = \min_{u(t) \in U, t_0 \le t \le t_1} \left[\int_{t_0}^{t_1} L(x, u, t)\, \mathrm{d}t + J^*(x(t_1), t_1) \right]. \tag{9.216}$$

(e) We have

$$\min_{u \in U} \left\{ \frac{\mathrm{d}J^*}{\mathrm{d}t} + L(x, u, t) \right\} = 0,$$
$$u^* = \operatorname*{argmin}_{u \in U} \left\{ \frac{\mathrm{d}J^*}{\mathrm{d}t} + L(x, u, t) \right\}. \tag{9.217}$$

(f) We have

$$-\frac{\partial J^*}{\partial t} = \min_{u \in U} \left[L(x, u, t) + \left(\frac{\partial J^*}{\partial x} \right)^T f(x, u, t) \right],$$
$$u^* = \operatorname*{argmin}_{u \in U} \left[L(x, u, t) + \left(\frac{\partial J^*}{\partial x} \right)^T f(x, u, t) \right]. \tag{9.218}$$

(g) At the optimum, the following condition holds:

$$\frac{\partial J^*}{\partial t} + H\left(x^*, \frac{\partial J^*}{\partial x}, u^*, t \right) = 0. \tag{9.219}$$

(h) There exists a vector $p(t)$ such that $u^*(t)$ satisfies

$$\dot{p} = -\frac{\partial H}{\partial x}(x^*, p, u^*, t),$$

$$p(t_f) = \frac{\partial K}{\partial x},$$

$$u^* = \operatorname*{argmin}_{u \in U} H(x^*, p, u, t), \tag{9.220}$$

and u^* solves $\min_{u\in U} H(x^*, p, u, t)$.

(i) There exist $V(x,t) \in \mathcal{C}^1$ and $u^*(x,t) \in \mathcal{C}^{*0}$ satisfying

$$\frac{\partial V}{\partial t} + \min_{u \in U}\left[\left(\frac{\partial V}{\partial x}\right)^T f(x,u,t) + L(x,u,t)\right] = 0, \tag{9.221}$$

$$V(x_f, t_f) \equiv K(x_f),$$

$$\left(\frac{\partial V}{\partial x}\right)^T f(x,u^*,t) + L(x,u^*,t) = \min_{u \in U}\left[\left(\frac{\partial V}{\partial x}\right)^T f(x,u,t) + L(x,u,t)\right].$$

(j) $V(x,t) \equiv J^*(x,t)$.

Then the following implications hold:

(1) (c) ⇒ (b) ⇒ (a).
(2) (a) ⇒ (d).
(3) (a) ⇒ (e) (under the provision that $\frac{dJ^*}{dt}$ exist).
(4) (b) ⇒ (f).
(5) (b) ⇒ (g).
(6) (c) ⇒ (h).
(7) (i) ⇒ (b).
(8) (i) ⇒ (j).

REMARK 9.25 *In the preceding theorem, statements (2), (3), (4), (5), and (6) are necessary conditions for optimality, and statements (7) and (8) are sufficient conditions for optimality.*

9.7 Summary of Key Results

The key results in Chapter 9 are as follows:

1. Proposition 9.1 (Pontryagin's maximum principle), which provides a necessary condition for optimal control allowing control constraints
2. Proposition 9.3 (Bellman's principle of optimality), which provides the basis for dynamic programming
3. Equations (9.157) and (9.159), Bellman's equations, and (9.204), the Hamilton–Jacobi–Bellman equation, which provide necessary conditions for optimal control under a hierarchy of smoothness conditions on the optimal cost-to-go

4. Equations (9.182) and (9.183), which provide the solution to the linear quadratic regulator
5. Equations (9.187)–(9.190), which provide the solution to the linear quadratic Gaussian regulator
6. Equation (9.196), which clarifies the relationship between the co-state vector in the maximum principle and the optimal cost-to-go in dynamic programming
7. Proposition 9.4, which interrelates the results on the maximum principle and dynamic programming

9.8 Bibliographic Notes for Further Reading

The material in Chapter 9 is standard and is well covered in many texts, including [62], [12], [15], and [63].

Some of the earliest applications of optimal control to flight were the proof that Hohmann transfers in orbital mechanics are propellant optimal ([6]), the propellant optimization of Moon landing ([53]), and the optimization of aircraft cruise performance through periodic control ([28], [29], [70], [11]).

9.9 Homework Problems

PROBLEM 9.1 *Show that, in $\mathbb{R}^2$, the shortest path between two points is a straight line.*

PROBLEM 9.2 *For the double integrator system* (9.102), *show that any constant amplitude constraint can be reformulated as* (9.103) *through appropriate time scaling.*

PROBLEM 9.3 *In a two-dimensional plane with Cartesian coordinates, we seek the shortest path from a point A, with coordinates (a, b), to the curve specified by the equation*

$$f(x, y) = 0.$$

If s is the arc length, we have the equations:

$$\frac{dx}{ds} = \cos\alpha$$
$$\frac{dy}{ds} = \sin\alpha,$$

where α is the heading angle, used as the control.

1. *Show that the optimal path must be such that*

$$\alpha^* = constant.$$

2. *Write all the necessary conditions for the computation of the optimal path.*
3. *As a particular case, let $(a, b) = (0, 0)$ and $f(x, y) = mx + ny + p$, where m, n, and p are given constants. Calculate the final values of x and y in terms of m, n, and p.*

PROBLEM 9.4 (The Brachistochrone Problem) *Consider the system described by*

$$\dot{x} = v \sin\alpha,$$

$$\dot{y} = v \cos\alpha,$$

$$\dot{v} = g \cos\alpha,$$

where α *is the heading angle, used as the control. We want to find an optimal control* $\alpha^*(t)$ *to bring the system from the initial conditions:*

$$t = 0, \ x(0) = 1, \ y(0) = 0, \ v(0) = 0,$$

to the final conditions:

$$x(t_f) = \pi R, \ y(t_f) \ \textit{free}, \ v(t_f) \ \textit{free},$$

while minimizing t_f.

1. *Introduce the co-state variables* p_x, p_y, *and* p_v *and use the optimality conditions to show that*

$$p_x = c_1 \neq 0,$$

$$p_y = c_2 = 0,$$

$$c_1 v = \sin\alpha,$$

$$g p_v = \cos\alpha. \tag{9.222}$$

2. *Show that*

$$\dot{\alpha} = g c_1. \tag{9.223}$$

Use the boundary conditions to evaluate $\alpha(0)$ *and* $\alpha(t_f)$ *from* (9.222). *Deduce that*

$$\alpha = g c_1 t.$$

3. *Use* (9.223) *to rewrite the state equations as*

$$\frac{dx}{d\alpha} = \frac{1}{g c_1^2} \sin^2\alpha$$

$$\frac{dy}{d\alpha} = \frac{1}{g c_1^2} \sin\alpha \cos\alpha.$$

By integrating the system using the boundary conditions, deduce the complete solution

$$x = R(2\alpha - \sin 2\alpha),$$

$$y = R(1 - \cos 2\alpha),$$

$$\alpha = \frac{1}{2}\sqrt{\frac{g}{R}}\, t,$$

$$v^2 = 2gy,$$

$$t_f = \pi\sqrt{\frac{R}{g}}.$$

PROBLEM 9.5 *A rocket ascends in a fixed vertical plane subject to a constant gravitational acceleration and a preprogrammed thrust acceleration $f(t)$. The equations of motion are*

$$\ddot{x}(t) = f(t)\cos\theta(t)$$

$$\ddot{y}(t) = f(t)\sin\theta(t) - g,$$

where x and y are the Cartesian coordinates of the rocket, θ is the thrusting angle and control, and g is the acceleration of gravity. We want to find an optimal thrusting angle schedule $\theta^*(t)$ that brings the rocket from the initial conditions,

$$t_0 = 0,\ x(0) = y(0) = 0,\ \dot{x}(0) = \dot{y}(0) = 0,$$

to the final conditions;

$$x(t_f)\ \textit{free},\ y(t_f) = h\ \textit{given},\ \ \dot{x}(t_f) = V\ \textit{given},\ \dot{y}(t_f) = 0,$$

while minimizing the final time t_f.

1. *Show that the optimal control law must satisfy*

$$\tan\theta^*(t) = at + b,$$

   where a and b are constants.
2. *Show how to compute a and b by numerical integration.*

PROBLEM 9.6 *A rocket in planar flight in a vacuum is subjected to the gravitational acceleration g and a preprogrammed thrust acceleration $f(t)$. The thrust angle θ above the horizontal is used as control.*

1. *Write the equations of motion using the coordinates x, y and the projections u, v of the velocity vector as state variables.*
2. *We want to find the optimal thrust angle $\theta^*(t)$ to bring the rocket from the initial state,*

$$t = 0,\ x = y = u = v = 0,$$

   to the final state,

$$x_f\ \textit{free},\ y_f = h\ \textit{given},\ u_f = u_c\ \textit{given},\ v_f = 0,$$

   such that the final time t_f is minimized.
   Show that the optimal control law must satisfy

$$\tan\theta^*(t) = at + b,$$

   where a and b are constants.
3. *Discuss briefly how to evaluate a and b by numerical integration. (You are not required to integrate the equations of motion.)*

PROBLEM 9.7 *A dynamic system is given by*

$$\dot{x}_1 = x_2,$$

$$\dot{x}_2 = -x_2 + u,$$

$$\dot{x}_3 = (x_1 - u)^2.$$

We want to find an optimal control $u^*(t)$, assumed unbounded, to bring the system from the initial conditions

$$t = 0,\ x_1(0) = 1,\ x_2(0) = x_3(0) = 0,$$

to the final conditions

$$t = t_f \text{ given},\ x_1(t_f) = -1,\ x_2(t_f) = 0,$$

while minimizing $x_3(t_f)$.

1. *Show that the optimal control must satisfy*

$$u^*(t) = x_1(t) + a_1 e^{\lambda_1 t} + a_2 e^{\lambda_2 t},$$

where a_1 and a_2 are constants and

$$\lambda_{1,2} = \frac{1}{2}\left(1 \pm \sqrt{5}\right).$$

2. *Derive the equation for x_1 and show that its solution is*

$$x_1(t) = b_1 e^{-\lambda_1 t} + b_2 e^{-\lambda_2 t} + \frac{a_1}{2\lambda_1} e^{\lambda_1 t} + \frac{a_2}{2\lambda_2} e^{\lambda_2 t},$$

where b_1 and b_2 are constants.

3. *Derive a system of four linear equations for a_1, a_2, b_1, and b_2, and solve it numerically.*

PROBLEM 9.8 *A dynamic system with state (x_1, x_2) is given by*

$$\dot{x}_1 = x_2 + u^2$$

$$\dot{x}_2 = -x_2 + u,$$

where u is a scalar control. We want to find an optimal control $u^*(t)$, assumed unbounded, to bring the system from the initial conditions

$$t = 0,\ x_1(0) = d \textit{ given},\ x_2(0) = 0,$$

to the final conditions

$$t_f \textit{ given},\ x_1(t_f) = 0,\ x_2(t_f) = 0,$$

such that the area enclosed by the curve $x_2(t)$ and the t-axis is a maximum.

1. *Show that the optimal control must satisfy*

$$u^*(t) = \alpha + \beta e^t,$$

where α and β are constants to be determined.

2. *Find the solution $x_2 = x_2(t, \alpha, \beta)$. In particular, show that*

$$\alpha = -\frac{\beta}{2}\left(1 + e^{t_f}\right)$$

and that the curve $x_2(t)$ is symmetric with respect to the line $t = \frac{1}{2}t_f$.
3. *Derive another equation for evaluating α and β.*

PROBLEM 9.9 *A dynamic system with state (x_1, x_2) is given by*

$$\dot{x}_1 = x_2 + u^2$$
$$\dot{x}_2 = -x_2 + u,$$

where u is a scalar control. We want to find an optimal control $u^*(t)$, assumed unbounded, to bring the system from the initial conditions

$$t = 0,\ x_1(0) = 0,\ x_2(0) = 0$$

to the final conditions

$$t_f \text{ given},\ x_1(t_f) = 0,$$

while maximizing $x_2(t_f)$.

1. *Show that the optimal control must satisfy*

$$u^*(t) = -\left(\alpha e^t + \frac{1}{2}\right),$$

where α is a constant to be determined.
2. *Show that, for any prescribed final time t_f, α is obtained by solving a quadratic equation.*
3. *Show that the solution always exists for $t_f > 0$.*

PROBLEM 9.10 *Consider the following 2-inputs, 1-output, linear time invariant controllable system with control constraint:*

$$\begin{aligned}\dot{x}(t) &= Ax(t) + Bu(t),\\ y &= Cx(t),\\ x(0) &= 0,\\ u^T(t)u(t) &\le 1,\end{aligned} \tag{9.223}$$

where $x(t) \in \mathbb{R}^n$, $u(t) \in \mathbb{R}^2$, and $y(t) \in \mathbb{R}$. We want to find a control that maximizes $y^2(t_f)$, where $t_f > 0$ is given.

1. *Obtain necessary conditions of optimality for this problem. Discuss the possibility of singular controls.*
2. *Solve the necessary conditions in closed form in terms of $G(t)$, the impulse response of (9.223), first two lines.*
3. *Derive a necessary and sufficient condition on $G(t)$ so that the optimal objective remains bounded as $t_f \to \infty$.*

PROBLEM 9.11 *Consider the following system:*

$$\dot{x} = y$$
$$\dot{y} = u,$$

where u is the control, subject to

$$|u| \leq 1.$$

We want to bring the system from an arbitrary initial condition (x_0, y_0) at time $t_0 = 0$ to the final state (x_f, y_f) such that the transfer time t_f is minimized.

1. *Derive the switching function for the general case and show that there can be at most one switch at a certain time t_1.*
2. *In the (x, y) space, plot the trajectory. In particular, compute t_0 and t_f when*

$$x_0 = -1, \quad y_0 = 0, \quad x_f = 0, \quad y_f = 0.$$

PROBLEM 9.12 *A particle moves in a fixed plane, subject only to an acceleration with known constant magnitude f and inclined at an angle θ with respect to the x-axis. Hence, the equations of motion are*

$$\ddot{x}(t) = f \cos\theta(t)$$
$$\ddot{y}(t) = f \sin\theta(t),$$

where x and y are the Cartesian coordinates, and θ is the control input. Use the initial conditions

$$t_0 = 0, \quad x(0) = y(0) = 0, \quad \dot{x}(0) = \dot{y}(0) = 0.$$

In each of the following two parts, we want to find an optimal control $\theta^*(t)$ that brings the particle to a final condition in minimum time.

First, show that in general, the optimal control must satisfy

$$\tan\theta^*(t) = \frac{-c_2 t + c_4}{-c_1 t + c_3},$$

where the c_i, $1 \leq i \leq 4$ are constants.

1. *In the problem of interception of a fixed target located at given coordinates $(x, y) = (a, b)$, we require that at the final time t_f,*

   $$x(t_f) = a, \quad y(t_f) = b, \quad \dot{x}(t_f) \;\; \textit{free}, \quad \dot{y}(t_f) \;\; \textit{free}.$$

   Show that the optimal control must satisfy

   $$\theta^*(t) = \textit{constant}.$$

   Calculate this constant and the minimum time to reach the destination in terms of a, b, and f.
2. *In the problem of achieving a cruise condition, we require that at the final time,*

   $$x(t_f) \;\textit{free}, \quad y(t_f) = b \;\textit{given}, \quad \dot{x}(t_f) = V \;\textit{given}, \quad \dot{y}(t_f) = 0.$$

   Show that the optimal control must then satisfy

   $$\tan\theta^*(t) = \alpha t + \beta,$$

   where α and β are constants. Discuss how to evaluate these constants.

PROBLEM 9.13 *Consider the second-order linear time invariant system*

$$\dot{x} = 2y + 2u - 2$$
$$\dot{y} = -2x + u - 1,$$

where u is the control, subject to

$$|u| \leq 1.$$

We want to bring the system from an arbitrary initial condition (x_0, y_0) to the origin $(0, 0)$ in minimum time.

1. *Show that the extremals are generated either with $u = 1$ or $u = -1$.*
2. *Show that for $u = 1$, we have the bundle of circles*
$$x^2 + y^2 = R_1^2,$$
and for $u = -1$, we have the bundle of circles
$$(x + 1)^2 + (y - 2)^2 = R_2^2.$$
3. *Discuss how to construct the optimal trajectory.*

PROBLEM 9.14 *In a horizontal plane with Cartesian coordinates (x, y), an airplane is flying with constant speed V for a duration T from the point $(-a, 0)$ to the point $(a, 0)$. Find the trajectory $y = y(x)$ such that the area enclosed by that trajectory and the x-axis is maximum. Note that the length of that trajectory is specified as $l = VT$.*

PROBLEM 9.15 *Consider the second-order linear time invariant system:*

$$\dot{x}_1 = x_2$$
$$\dot{x}_2 = -x_1 + u,$$

where the control u is subject to the constraint

$$|u| \leq 1.$$

We want to bring the system from a given initial condition $(x_1, x_2) = (a, b)$ to the origin $(0, 0)$ in minimum time.

1. *Show that the optimal trajectory is a sequence of circular arcs from the two bundles of circles:*
$$(x_1 - 1)^2 + x_2^2 = R_1^2$$
$$(x_1 + 1)^2 + x_2^2 = R_2^2, \qquad (9.224)$$
where R_1 and R_2 are constants.
2. *Show that the motion is clockwise on the circles (9.224).*
3. *Show that switching occurs every π units of time and sketch an optimal trajectory with one switch.*

10 Introduction to Differential Games

Game theory is an extension of control theory that deals with situations where several **players** exercise authority over a system. In general, each player may pursue its own objective. The players are not necessarily adversaries; they may choose to cooperate if it is to their advantage. In addition, the players may not know everything there is to know. They may act to learn what they do not know, or exploit the situation based on what they do know. This is a typical exploration–exploitation trade-off.

Game theory has applications to economics (several Nobel Prizes were awarded in this field), international diplomacy, guidance and pursuit evasion, warfare, and sports. Game theory is also applicable to control (see Figure 1.4), where the exogenous and endogenous inputs are two players in a game.

Section 10.1 presents a taxonomy of two-player games. Section 10.2 describes an example of a simple game of pursuit evasion in a two-player football scrimmage. Section 10.3 describes the Bellman–Isaacs equation. Sections 10.4 and 10.5 present modeling and features of the solution for the homicidal chauffeur game. Section 10.6 describes a game-theoretic view of proportional navigation. Sections 10.7, 10.8, and 10.9 present a summary of the key results in the chapter, bibliographic notes for further reading, and homework problems, respectively.

10.1 Taxonomy of Two-Player Games

We consider two-player games, in which the two players have control actions u and v, respectively.

DEFINITION 10.1 *In a **zero-sum game**, the players have the same objective, $J(u, v)$. One attempts to minimize it, and the other attempts to maximize it. Hence, the two players decide how to act by solving the following optimization problems:*

$$\begin{aligned} \textit{for } u &: \min_u \max_v J(u, v) \\ \textit{for } v &: \max_v \min_u J(u, v). \end{aligned} \tag{10.1}$$

*In a **non-zero-sum game**, each player has its own objective and attempts to optimize it. In the two-player case, let the objectives be $J_u(u, v)$ and $J_v(u, v)$. Then, the players' actions are decided by solving the following optimization problems:*

$$\begin{aligned} \textit{for } u &: \min_u J_u(u, v) \\ \textit{for } v &: \max_v J_v(u, v). \end{aligned} \tag{10.2}$$

DEFINITION 10.2 *A zero-sum game is said to have value J^* if the underlying minimax and maximin problems of* (10.1) *yield the same outcome, in which case the common outcome is called the **value of the game**. Hence, when a zero-sum game has a value, we have*

$$J^* = \min_u \max_v J(u, v) = \max_v \min_u J(u, v). \tag{10.3}$$

DEFINITION 10.3 *In a **game of kind**, the outcomes are in a discretum, for example, {u wins, u loses, draw}. In a **game of degree**, the outcomes are in a continuum, for example, $\mathbb{R}$.*

Chess is a classic example of a game of kind. How close to the friendly ship does the enemy UAV come before it gets intercepted is an example of outcome in a continuum in a game of degree.

DEFINITION 10.4 *In a **static game**, u and v are real vectors, and $J(u, v)$ is a real function. In a **differential game**, $u(.)$ and $v(.)$ are functions of time, and $J(u, v)$ is a functional, for example,*

$$J(u, v) = K(x(t_f)) + \int_{t_0}^{t_f} L(x(t), u(t), v(t))\,\mathrm{d}t, \tag{10.4}$$

where

$$\begin{aligned} \dot{x}(t) &= f(x(t), u(t), v(t)), \\ x(t_0) &= x_0, \\ g(x(t_f)) &= 0 \end{aligned} \tag{10.5}$$

represent the dynamics, initial condition, and termination condition, respectively.

DEFINITION 10.5 *In a **sequential game**, one player decides first, then the other player decides, knowing what the first player decided. In a **synchronous game**, both players decide simultaneously, without knowing what the other player is going to do.*

Chess is a classic example of a sequential game. Rock-paper-scissors is a classic example of a synchronous game.

DEFINITION 10.6 *In a **full information game**, each player knows everything, including the state of the system, the objective function of the other player, and that the other player knows everything too. Anything short of that is a **partial information game**, and is characterized by an **information structure**; for example, u knows, v doesn't know, and u knows that v doesn't know.*

Partial information games are complex, difficult, and practical. They illustrate the old adage that "knowledge is power."

DEFINITION 10.7 *Differential games can also be classified by the form of input, u, that we wish to obtain.*

1. *If $u = u(t)$, we are searching for a **control**. This class of input is usually relatively computationally inexpensive to obtain.*
2. *If $u = u(x)$, we are searching for a **policy**. Policies are, in general, more computationally expensive to obtain than controls.*
3. *If $u = u(x, v(t))$, we are searching for a **strategy**. Strategies are, in general, more computationally expensive to obtain than policies.*
4. *If $u = u(x, v(x))$, we also say that we are searching for a **strategy**. This type of input is very computationally expensive to obtain.*

For example, chess is a strategic game: a good chess player decides its next move not only based on the current configuration of the board (i.e., x) but also based on an inference of what the strategy pursued by the opponent is.

DEFINITION 10.8 *In a game of kind, if, in a situation, a player can choose a strategy that guarantees a win regardless of what the other player does, then the player is said to **dominate** that situation.*

DEFINITION 10.9 *The simple **pursuit game** (or pursuit-evasion game) has two players, the pursuer u, and the evader v, each controlling the motion of a vehicle. The usual question one seeks to answer is: How should u best pursue v? That is, if at each instant in time u knows its own and v's relevant state variables, how should u choose the various inputs at its disposal? A pursuit game terminates when **capture** occurs, that is, the distance between u and v becomes less than a certain prescribed quantity ϵ. We also have to define what is meant by "best," for example, whether capture can be achieved at all, and whether there are secondary payoffs (e.g., accuracy, energy, and so on).*

In view of the preceding taxonomy, the pursuit games of interest in guidance, and in this chapter, are then zero-sum, differential, with terminal cost, synchronous, full-information games for which we seek (to the extent possible, dominant) strategies.

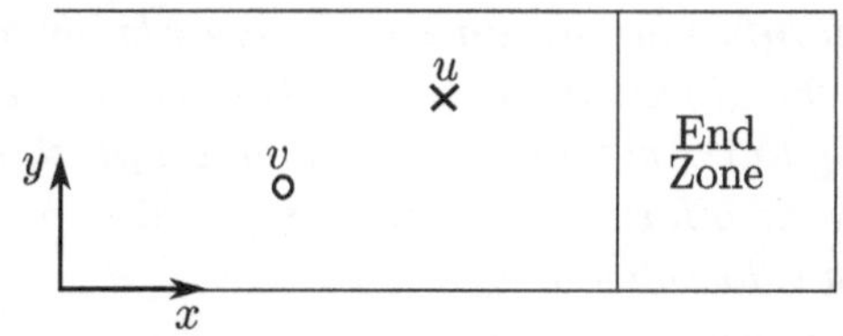

Figure 10.1. Setup for the football two-player scrimmage example.

10.2 Example of a Simple Pursuit Game: Two-Player Football Scrimmage

In this example, we have two players, u and v. Player u is the defender and wants to intercept the other player as far from the end zone as possible; we assume that player u travels at constant speed S_u. Player v, the ball carrier, wants to carry the ball as close to the end zone as possible; we assume that player v travels at constant speed S_v. The setup for the game is shown in Figure 10.1.

10.2.1 Modeling

Let x_u, y_u, x_v, and y_v be the Cartesian coordinates of u and v, and let u and v be their respective heading angles. We assume that both players follow unicycle kinematics:

$$\begin{aligned} \dot{x}_u &= S_u \cos u, \\ \dot{y}_u &= S_u \sin u, \\ \dot{x}_v &= S_v \cos v, \\ \dot{y}_v &= S_v \sin v. \end{aligned} \tag{10.6}$$

In addition, the ball carrier cannot leave the field:

$$0 \le y_v \le Y. \tag{10.7}$$

The termination condition for the game is

$$\begin{cases} x_u(t_f) - x_v(t_f) = 0 \\ y_u(t_f) - y_v(t_f) = 0. \end{cases} \tag{10.8}$$

The game has a zero-sum terminal objective:

$$J(u, v) = x_v(t_f). \tag{10.9}$$

10.2.2 Analysis

REMARK 10.1 *Wherever intercept occurs, each player must run straight toward the intercept point, otherwise they are wasting advantage. Note that the players are then running constant bearing trajectories (see Chapter 5) to the point of intercept.*

REMARK 10.2 *Intercept occurs at a point where the ratio of distance to initial conditions is the same as the speed ratio, $\gamma = S_u/S_v$. This is a consequence of Remark 10.1 and the law of similarity in triangles.*

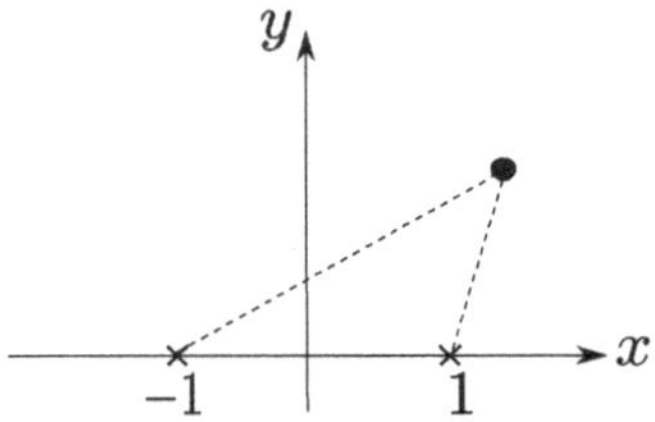

Figure 10.2. Setup for the Apollonius circle theorem.

10.2.3 The Apollonius Circle Theorem

The key result to solve this game is due to Apollonius of Perga, who was a Greek geometer and astronomer who lived in the second century B.C. His work in the field of conics influenced many later scholars, including Ptolemy, Johannes Kepler, Isaac Newton, and René Descartes, with many eventual applications to aerospace flight mechanics. It was Apollonius who gave the ellipse, the parabola, and the hyperbola the names by which we know them.

PROPOSITION 10.1 (Apollonius Circle Theorem) *In a plane, the locus of points such that the ratio of distances to two fixed points is constant is a circle.*

PROOF. Without loss of generality, let $(-1, 0)$ and $(1, 0)$ be the Cartesian coordinates of the fixed points. The geometry is shown in Figure 10.2. Then, the condition that the ratio of distances to the two fixed points be a constant (smaller than or equal to 1 without loss of generality) is expressed as

$$\frac{\sqrt{(x-1)^2+y^2}}{\sqrt{(x+1)^2+y^2}} = \gamma \leq 1. \tag{10.10}$$

This implies

$$(x-1)^2 + y^2 = \gamma^2\left((x+1)^2 + y^2\right), \tag{10.11}$$

or

$$x^2 - 2x + 1 + y^2 = \gamma^2(x^2 + 2x + 1 + y^2). \tag{10.12}$$

Grouping terms,

$$(1-\gamma^2)x^2 - 2(1+\gamma^2)x + (1-\gamma^2)y = \gamma^2 - 1, \tag{10.13}$$

or

$$x^2 - \frac{2(1+\gamma^2)}{1-\gamma^2}x + y^2 = -1. \tag{10.14}$$

This yields

$$\left(x - \frac{1+\gamma^2}{1-\gamma^2}\right)^2 - \left(\frac{1+\gamma^2}{1-\gamma^2}\right)^2 + y^2 = -1, \tag{10.15}$$

or

$$\left(x - \frac{1+\gamma^2}{1-\gamma^2}\right)^2 + y^2 = \left(\frac{1+\gamma^2}{1-\gamma^2}\right)^2 - 1. \tag{10.16}$$

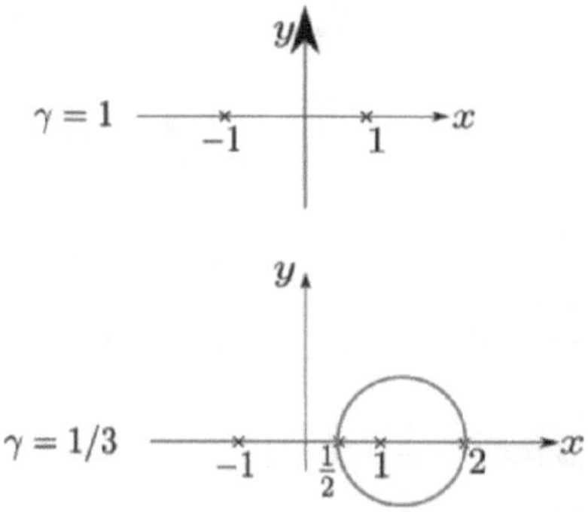

Figure 10.3. Examples for the Apollonius circle theorem.

The right-hand side of (10.16) can be rewritten as

$$\frac{(1+\gamma^2)^2-(1-\gamma^2)^2}{(1-\gamma^2)^2}=\frac{1+2\gamma^2+\gamma^4-(1-2\gamma^2+\gamma^4)}{(1-\gamma^2)^2}$$

$$=\frac{4\gamma^2}{(1-\gamma^2)^2}. \tag{10.17}$$

Hence, (10.16) becomes

$$\left(x-\frac{1+\gamma^2}{1-\gamma^2}\right)^2+y^2=\frac{4\gamma^2}{(1-\gamma^2)^2}, \tag{10.18}$$

which has the form

$$(x-x_c)^2+(y-y_c)^2=R^2 \tag{10.19}$$

and is therefore the equation of a circle centered at

$$(x_c, y_c)=\left(\frac{1+\gamma^2}{1-\gamma^2}, 0\right) \tag{10.20}$$

and with radius

$$R=\frac{2\gamma}{|1-\gamma^2|}, \tag{10.21}$$

intercepting the x axis at

$$\frac{1+\gamma^2}{1-\gamma^2}\pm\frac{2\gamma}{1-\gamma^2}=\begin{cases}\frac{1+\gamma}{1-\gamma}\\ \frac{1-\gamma}{1+\gamma}.\end{cases} \tag{10.22}$$

Examples are shown in Figure 10.3 for $\gamma = 1$ and $\gamma = 1/3$.

REMARK 10.3 *Changing γ to $1/\gamma$ yields a result that is a mirror image with respect to the y axis. It is equivalent to switching the roles of the two fixed points.*

REMARK 10.4 *Assume that the two fixed points are the initial positions of two vehicles moving with speed ratio γ. The Apollonius circle determines two disjoint regions in the plane such that each vehicle is initially in its region. Then, each vehicle is dominant*

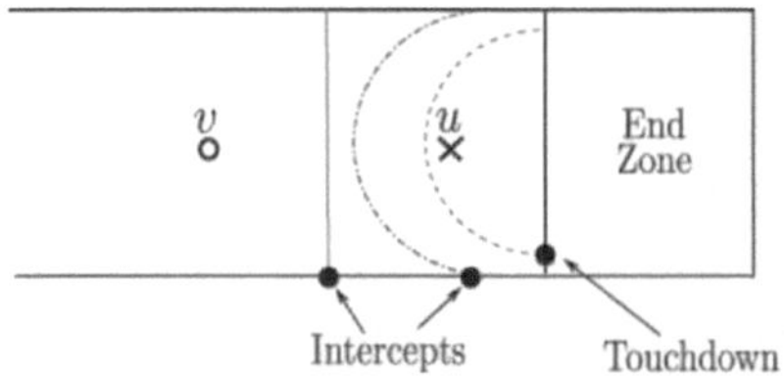

Figure 10.4. Solution to the football two-player scrimmage problem.

in its region in the sense that for every point in its region, the vehicle can reach that point ***before*** *the other vehicle,* ***regardless of what the other vehicle does****. Hence, the Apollonius circle determines the* ***dominance boundary*** *for the game of kind where a player wins if it arrives at a given destination before the other player.*

10.2.4 Solution to the Football Two-Player Scrimmage Problem

Player v heads to the point on the Apollonius circle closest to the end zone, as shown in Figure 10.4. If player v can head straight to the end zone without penetrating the dominance region of u, then v is guaranteed a touchdown.

10.3 The Bellman–Isaacs Equation

In this section, we give a result that extends Bellman's equation of Chapter 9 to the class of differential games of interest in this chapter. Consider a differential game specified as follows:

$$\begin{aligned} \dot{x} &= f(x, u, v, t), \\ x(t_0) &= x_0, \\ x &\in \mathbb{R}^n, x \in C^{*1}, \\ u &\in \mathbb{R}^m, u \in C^{*0}, u \in U, \\ v &\in \mathbb{R}^p, v \in C^{*0}, v \in V, \\ J(u(.), v(.); x_0, t_0) &= K(x_0, t_0, x_f, t_f) + \int_{t_0}^{t_f} L(x, u, v, t)\, \mathrm{d}t, \end{aligned} \tag{10.23}$$

where x is the nth-order, piecewise differentiable state vector, u is the mth-order, piecewise continuous control of player u, v is the pth-order, piecewise continuous control of player v, t_0 is the initial time, x_0 is the initial state, U and V characterize possible amplitude constraints on the controls, and $J(.,.;.,.)$ is the zero-sum objective that u minimizes and v maximizes.

Assume that the preceding differential game of degree has a value, and let the value be

$$\begin{aligned} J^*(x_0, t_0) &= \min_{u \in U} \max_{v \in V} \int_{t_0}^{t_f} L(x, u, v, t)\, \mathrm{d}t \\ &= \max_{v \in V} \min_{u \in U} \int_{t_0}^{t_f} L(x, u, v, t)\, \mathrm{d}t, \end{aligned} \tag{10.24}$$

subject to (10.23). Then the value of the game satisfies

$$\min_u \max_v \left\{ \left(\frac{\mathrm{d}J^*}{\mathrm{d}t} \right)_{u,v,t} + L(x, u, v, t) \right\} = 0, \tag{10.25}$$

which is called the **Bellman–Isaacs functional equation** and should be compared with (9.157).

Moreover, if the value is a continuously differentiable function of its arguments, then a development similar to that preceding (9.159) yields

$$\frac{\partial J^*}{\partial t} = \min_u \max_v \left\{ L(x, u, v, t) + \left(\frac{\partial J^*}{\partial x} \right)^T f(x, u, v, t) \right\}, \tag{10.26}$$

which is a more common form of the Bellman–Isaacs equation and should be compared with (9.159).

If the value function is twice continuously differentiable, then (9.196) still holds and the necessary conditions for optimality can be written in terms of the Hamiltonian,

$$H(x, p, u, v, t) = p^T f(x, u, v, t) - L(x, u, v, t), \tag{10.27}$$

as

$$\begin{aligned}
\dot{x} &= \frac{\partial H}{\partial p}(x, p, u, v, t), \\
\dot{p} &= -\frac{\partial H}{\partial x}(x, p, u, v, t), \\
u &= \operatorname*{argmax}_{\xi \in U} H(x, p, \xi, v, t), \\
v &= \operatorname*{argmin}_{\xi \in V} H(x, p, u, \xi, t), \\
\delta K + \left[p^T \delta x - H \delta t \right]_{t_0}^{t_f} &= 0 \text{ for all admissible variations,}
\end{aligned} \tag{10.28}$$

which should be compared to (9.68) and (9.69).

10.4 The Homicidal Chauffeur: Modeling

In the next two sections, we discuss a particular pursuit-evasion differential game that has been extensively studied in the literature and has potential application to flight mechanics in contested environments. The reader may recognize that this game is simplistic in its modeling assumptions. Despite this simplicity, however, this game is surprisingly rich in that it helps illustrate many features found in more sophisticated games. Hence, the purpose here is not to give a detailed solution of the game but rather to showcase the features of its solution, in nature, process, and phenomenology, that foreshadow more complex games.

The homicidal chauffeur game is formulated as follows. In a fixed plane, a pursuer P and evader E, both modeled as points, move under the following assumptions. The pursuer moves with constant speed and can change its heading angle by

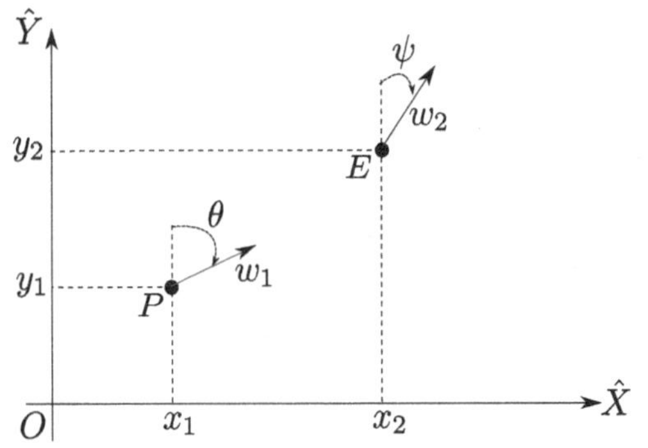

Figure 10.5. Setup for the homicidal chauffeur.

choosing instantaneously the radius of curvature of its trajectory above a given threshold. The evader moves with constant speed and can choose instantaneously its heading angle without restriction. Capture happens when the distance between the pursuer and evader, that is, the range, reaches a given threshold. We consider the game of degree version where the zero-sum objective is the time at which capture happens: the pursuer and evader attempt to minimize it and maximize it, respectively. Finally, it is typically assumed that the pursuer has an advantage in speed but a disadvantage in maneuverability.

A dynamic model of the motions of the pursuer and evader is as follows (see Figure 10.5). Let (x_1, y_1) be the Cartesian coordinates of P with respect to a fixed frame $\{O, \hat{X}, \hat{Y}\}$. Let w_1 denote the speed of P, and θ denote its heading angle, counted clockwise from the $\hat{Y}$ axis. Also, let R denote the minimum radius of curvature of P's trajectory. Then the motion of P is modeled by the following equations:

$$
\begin{aligned}
\dot{x}_1 &= w_1 \sin\theta, \\
\dot{y}_1 &= w_1 \cos\theta, \\
\dot{\theta} &= \frac{w_1}{R}\phi, \\
|\phi| &\le 1, \qquad (10.29)
\end{aligned}
$$

where ϕ is the ratio of minimum radius of curvature to actual radius of curvature. Similarly, let (x_2, y_2) be the Cartesian coordinates of E with respect to the fixed frame $\{O, \hat{X}, \hat{Y}\}$. Let w_2 denote the speed of E and ψ denote its heading angle. Then, the motion of E is modeled by the following equations:

$$
\begin{aligned}
\dot{x}_2 &= w_2 \sin\psi \\
\dot{y}_2 &= w_2 \cos\psi. \qquad (10.30)
\end{aligned}
$$

Capture happens when the range reaches a given threshold R_c.

Equations (10.29) and (10.30) are a system of five differential equations with one inequality constraint that models the dynamics of the homicidal chauffeur game. Although they serve the purpose of modeling, it is expedient in differential games to reduce the dynamic order of the problem as much as possible. In the case at hand, the dynamic order can be reduced from five to two, as follows.

Introduce the new, orthonormal, moving vectrix $(\hat{x}, \hat{y})^T$ such that the $\hat{y}$ axis is always aligned with the velocity vector of P. Hence, the vectrices are related by

$$\begin{bmatrix} \hat{x} \\ \hat{y} \end{bmatrix} = \begin{bmatrix} \cos\theta & -\sin\theta \\ \sin\theta & \cos\theta \end{bmatrix} \begin{bmatrix} \hat{X} \\ \hat{Y} \end{bmatrix}, \tag{10.31}$$

and the angular velocity vector of the moving vectrix is

$$\omega = -\dot{\theta}\hat{z} = -\frac{w_1}{R}\phi\hat{z}, \tag{10.32}$$

where the unit vector $\hat{z} = \hat{x} \times \hat{y}$ is such that the vectrix $(\hat{x}, \hat{y}, \hat{z})^T$ is orthonormal direct.

REMARK 10.5 (Key Idea for Order Reduction) *A reduction of the dynamic order of* (10.29) *and* (10.30) *is obtained by noting that the absolute velocity of E should be obtained by adding to the absolute velocity of P the velocity of E relative to P, the key being to resolve all these velocities in the moving vectrix.*

By definition of the moving vectrix, the absolute velocity of P, expressed in the moving vectrix, is $w_1\hat{y}$.

The velocity of E relative to P, expressed in the moving vectrix, is obtained as follows. Let $\vec{R}$ denote the position vector of E relative to P, and let (x, y) denote the components of that vector, resolved in the moving vectrix. Then, the vector $\vec{r}$ and its time derivative, expressed in the moving vectrix, have the form

$$\begin{aligned} \vec{r} &= x\hat{x} + y\hat{y}, \\ \dot{\vec{r}} &= \dot{x}\hat{x} + \dot{y}\hat{y} + \omega \times \vec{r}, \\ &= \left(\dot{x} + \frac{w_1}{R}\phi y\right)\hat{x} + \left(\dot{y} - \frac{w_1}{R}\phi x\right)\hat{y}. \end{aligned} \tag{10.33}$$

The absolute velocity of E, expressed in the fixed vectrix, has the form $w_2(\sin\psi\hat{X} + \cos\psi\hat{Y})$. Because the fixed and moving vectrices are related by (10.31), the absolute velocity of E, expressed in the moving vectrix, has the form

$$w_2\left[(\cos\theta\sin\psi - \sin\theta\cos\psi)\hat{x} + (\sin\theta\sin\psi + \cos\theta\cos\psi)\hat{y}\right], \tag{10.34}$$

or equivalently,

$$w_2\left[\sin(\psi - \theta)\hat{x} + \cos(\psi - \theta)\hat{y}\right]. \tag{10.35}$$

Therefore, the key idea for order reduction yields the vector equation

$$w_1\hat{y} + \left(\dot{x} + \frac{w_1}{R}\phi y\right)\hat{x} + \left(\dot{y} - \frac{w_1}{R}\phi x\right)\hat{y} = w_2\left[\sin(\psi - \theta)\hat{x} + \cos(\psi - \theta)\hat{y}\right]. \tag{10.36}$$

Resolving along $\hat{x}$ and $\hat{y}$, we obtain the two scalar equations, together with the still applicable inequality condition

$$\begin{aligned} \dot{x} &= -\frac{w_1}{R}\phi y + w_2\sin(\psi - \theta), \\ \dot{y} &= \frac{w_1}{R}\phi x - w_1 + w_2\cos(\psi - \theta), \\ |\phi| &\le 1, \end{aligned} \tag{10.37}$$

which achieve the desired order reduction. Note that in this reduced-order model, the control variable of the evader is no longer ψ but rather $\psi - \theta$, where θ is a state of the full-order model but not of the reduced one.

The next step in modeling, which is also typical in differential games, is to nondimensionalize (10.37). Here, we do so by choosing as new length scale R, and as new time scale $\frac{R}{w_1}$. Hence,

$$x = \tilde{x}R,$$
$$y = \tilde{y}R,$$
$$\frac{\mathrm{d}}{\mathrm{d}t} = \frac{w_1}{R}\frac{\mathrm{d}}{\mathrm{d}\tau}, \tag{10.38}$$

where $(\tilde{x}, \tilde{y})$ are the dimensionless coordinates of the position of the evader relative to the pursuer resolved in the moving vectrix, and τ is the dimensionless time. In terms of these dimensionless quantities, the first two lines of (10.37) become

$$\frac{\mathrm{d}\tilde{x}}{\mathrm{d}\tau} = -\phi\tilde{y} + \frac{w_2}{w_1}\sin(\psi - \theta)$$
$$\frac{\mathrm{d}\tilde{y}}{\mathrm{d}\tau} = \phi\tilde{x} - 1 + \frac{w_2}{w_1}\cos(\psi - \theta). \tag{10.39}$$

To rewrite these nondimensional equations in the standard form, define

$$u = \phi,$$
$$v = \psi - \theta,$$
$$\gamma = \frac{w_2}{w_1}. \tag{10.40}$$

Moreover, forgo the use of the tilde notation and τ. Then, the nondimensional equations for the homicidal chauffeur game take the elegant form

$$\dot{x} = -uy + \gamma \sin v,$$
$$\dot{y} = ux - 1 + \gamma \cos v,$$
$$|u| \leq 1, \tag{10.41}$$

where the dimensionless capture radius is the quantity $\rho = R_c/R$, specified in terms of the dimensional capture radius and the minimum radius of curvature of the pursuer's trajectory.

In the preceding nondimensional equations, the control variable of the pursuer is the bounded quantity u, that of the evader is the unbounded quantity v, the pursuer has unit speed, and the evader has speed γ. It is typically assumed that $\gamma < 1$, in other words, that the pursuer has an advantage in speed but a disadvantage in maneuverability.

10.5 The Homicidal Chauffeur: Features of the Solution

The equations defining the homicidal chauffeur game, in nondimensional form, are repeated here for convenience, as follows:

$$
\begin{aligned}
\dot{x} &= -uy + \gamma \sin v, \\
\dot{y} &= ux - 1 + \gamma \cos v, \\
x(0), y(0) \text{ given}, &\qquad \text{satisfying } \sqrt{x(0)^2 + y(0)^2} > \rho, \\
|u| &\leq 1, \\
\sqrt{x(t_f)^2 + y(t_f)^2} &= \rho, \\
J(u(.), v(.); (x(0), y(0))) &= \int_0^{t_f} \mathrm{d}t,
\end{aligned}
\tag{10.42}
$$

where x and y are the coordinates of the evader relative to the pursuer in a frame where the y axis is always aligned with the velocity vector of the pursuer, u quantifies the turn rate of the pursuer, $\gamma < 1$ is the constant speed of the evader, v represents the difference of heading angles between the evader and pursuer and is the evader's control variable, ρ is the capture radius, and t_f is the final time.

In (10.42), the first two lines are state equations, the third specifies the initial conditions located outside the capture disk, the fourth specifies the turn rate limitation of the pursuer, the fifth specifies that the final time is defined by capture, and the sixth specifies that the zero-sum objective is the final time.

Because the problem defined by (10.42) is scleronomic, the value function is stationary of the form $J^*(x, y)$. Using the notation J_x^* and J_y^* for the partial derivatives of the value function, the Bellman–Isaacs equation (10.26) yields

$$
\min_{u:|u|\leq 1} \max_{v} \left[J_x^*(-uy + \gamma \sin v) + J_y^*(ux - 1 + \gamma \cos v) \right] = -1. \tag{10.43}
$$

Performing the preceding optimizations, we obtain the optimal controls as

$$
\begin{aligned}
u &= \operatorname{sign}\left(J_x^* y - J_y^* x \right) \\
\frac{\sin v}{J_x^*} &= \frac{\cos v}{J_y^*} = \frac{1}{\sqrt{J_x^{*2} + J_y^{*2}}},
\end{aligned}
\tag{10.44}
$$

which must hold in any region of the state space where the value function is continuously differentiable. The co-state equations are

$$
\begin{bmatrix} \dot{J}_x^* \\ \dot{J}_y^* \end{bmatrix} = \begin{bmatrix} 0 & -u \\ u & 0 \end{bmatrix} \begin{bmatrix} J_x^* \\ J_y^* \end{bmatrix}, \tag{10.45}
$$

which can be solved in terms of trigonometric functions when $u = \pm 1$, that is, $J_x^* y - J_y^* x \neq 0$ (see Problem 2.1, part 1).

We eliminate the possibility of singular controls in the homicidal chauffeur problem as follows. Assume that, on a time interval, the quantity $J_x^* y - J_y^* x$ is identically zero so that, as per (10.44), the control u is undetermined. Differentiating with respect to time yields

$$\dot{J}_x^* y + J_x^* \dot{y} - \dot{J}_y^* x - J_y^* \dot{x} \equiv 0. \tag{10.46}$$

Using (10.45) and collecting terms, we obtain

$$J_x^*(-1 + \gamma \cos v) - J_y^* \gamma \sin v \equiv 0. \tag{10.47}$$

Then, using (10.42) and collecting terms again, we obtain

$$J_x^* \equiv 0, \tag{10.48}$$

which, together with (10.44), yields

$$J_y^* \equiv 0. \tag{10.49}$$

Then, the Bellman–Isaacs equation (10.26) implies the contradiction $0 = -1$. Therefore, singular controls are not possible in the homicidal chauffeur problem.

When capture happens, this can only be at particular locations of the capture circle characterized as follows. Let capture happen at the point on the capture circle with coordinates $(x, y) = (\rho \sin s, \rho \cos s)$, where the quantity s is the azimuth angle at which capture happens, measured clockwise from the y axis. Then, it can be shown that the azimuth angle must satisfy

$$|s| < \arccos \gamma, \tag{10.50}$$

which agrees with our expectation that capture should happen close to the y axis, because this is the direction of the pursuer's motion. Moreover, it can be shown that capture happens at all when and only when

$$\rho > \sqrt{1 - \gamma^2} + \gamma \arcsin \gamma - 1. \tag{10.51}$$

As mentioned previously, the homicidal chauffeur game is surprisingly rich for a game that can be so simply stated. Equations (10.50) and (10.51) are sufficient to generate the extremals and produce a solution in the small. The solution in the large is obtained as a sequence of extremals connected through switching conditions specified by the necessary conditions. This is where complexity arises. To date, more than 20 qualitatively different solutions have been reported, featuring phenomena such as barriers, switching loci, switching envelopes, equivocal lines, dispersal lines and more.

10.6 A Game-Theoretic View of Proportional Navigation

This section provides a justification, stemming from the theory of differential games, for using proportional navigation as a strategy for homing guidance. Based on a dynamic model suggested in Remark 5.5, we pose the problem of homing guidance, for a missile with perfect autopilot dynamics, as a differential game with time varying cost functional, where capture is enforced by a target set condition. We derive necessary conditions for optimality, through which we show that the optimal guidance law uses a lateral missile acceleration that is proportional to the turn rate of the line of sight, that is, proportional navigation.

We account for the relative motions of the missile and target perpendicular to the nominal line of sight as follows (see Figure 10.6). In a fixed plane with Cartesian

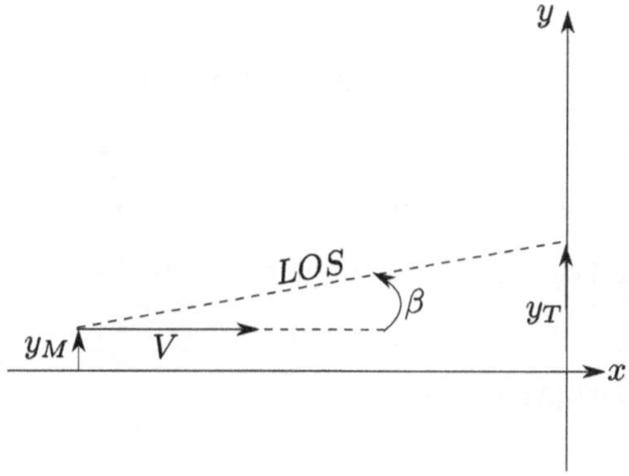

Figure 10.6. Setup for proportional navigation as a strategy for homing guidance.

coordinates, assume that the nominal line of sight is parallel to the x axis. The missile has a velocity that is nominally constant and parallel to the x axis and can add to this nominal velocity a small component that is parallel to the y axis. The target can only move along the y axis. Hence, the coordinates of the missile are

$$(V(t - t_f), y_M), \tag{10.52}$$

and those of the target are

$$(0, y_T), \tag{10.53}$$

where V is the closing speed, t_f is the time of intercept, and y_M and y_T are the displacement of the missile and target perpendicular to the nominal line of sight, respectively.

The dynamics of the engagement are modeled by two double integrators:

$$\begin{aligned} \ddot{y}_M &= u \\ \ddot{y}_T &= v, \end{aligned} \tag{10.54}$$

where the inputs u and v are the lateral accelerations of the missile and target, respectively. Assume that the line-of-sight angle is small. Then, it is given by

$$\beta = \frac{y_T - y_M}{V(t_f - t)}. \tag{10.55}$$

To pose a differential game problem, we first rewrite (10.54) in standard state-space form by defining

$$x_1 = y_M, \quad x_2 = y_T, \quad x_3 = \dot{y}_M, \quad x_4 = \dot{y}_T. \tag{10.56}$$

Then, (10.54) becomes

$$\begin{bmatrix} \dot{x}_1 \\ \dot{x}_2 \\ \dot{x}_3 \\ \dot{x}_4 \end{bmatrix} = \begin{bmatrix} 0 & 0 & 1 & 0 \\ 0 & 0 & 0 & 1 \\ 0 & 0 & 0 & 0 \\ 0 & 0 & 0 & 0 \end{bmatrix} \begin{bmatrix} x_1 \\ x_2 \\ x_3 \\ x_4 \end{bmatrix} + \begin{bmatrix} 0 \\ 0 \\ u \\ v \end{bmatrix}. \tag{10.57}$$

The zero-sum cost functional is

$$J(u(.), v(.)) = \frac{1}{2} \int_{t_0}^{t_f} (t_f - t) \left(u^2(t) r_M - v^2(t) r_T \right) \mathrm{d}t, \tag{10.58}$$

where t_0 is a given initial time and $r_M > 0$ and $r_T > 0$ are constants. Note that the signs of the coefficients of u^2 and v^2 are opposite, to make this a game rather than an LQR problem. Also note that the cost functional contains a discount factor $(t - t_f)$. When t is close to t_f, we are in a situation of cheap control (see Remark 9.22), hence the discount factor gives more urgency to times close to nominal intercept. The boundary conditions are

$$\begin{aligned} x_1(t_f) &= x_{1f},\ x_2(t_f) = x_{2f}, \\ x_3(t_f) &= x_{3f},\ x_4(t_f) = x_{4f}, \end{aligned} \tag{10.59}$$

where x_{3f} and x_{4f} are free, but x_{1f} and x_{2f} satisfy a target set condition that enforces interception:

$$g(x(t_f)) = y_T(t_f) - y_M(t_f) = \begin{bmatrix} -1 & 1 & 0 & 0 \end{bmatrix} \begin{bmatrix} x_1(t_f) \\ x_2(t_f) \\ x_3(t_f) \\ x_4(t_f) \end{bmatrix} = 0. \tag{10.60}$$

Equations (10.57)–(10.60) specify a pursuit-evasion differential game for which we demonstrate that the optimal u is proportional to $\dot{\beta}$.

The Hamiltonian (10.27) is

$$H(x, p, u, v, t) = \begin{bmatrix} p_1 & p_2 & p_3 & p_4 \end{bmatrix} \left(\begin{bmatrix} 0 & 0 & 1 & 0 \\ 0 & 0 & 0 & 1 \\ 0 & 0 & 0 & 0 \\ 0 & 0 & 0 & 0 \end{bmatrix} \begin{bmatrix} x_1 \\ x_2 \\ x_3 \\ x_4 \end{bmatrix} + \begin{bmatrix} 0 \\ 0 \\ u \\ v \end{bmatrix} \right)$$
$$-\frac{1}{2}(t_f - t)(u^2 r_M - v^2 r_T). \tag{10.61}$$

The state equations (10.28) (line 1) are

$$\begin{bmatrix} \dot{x}_1 \\ \dot{x}_2 \\ \dot{x}_3 \\ \dot{x}_4 \end{bmatrix} = \frac{\partial H}{\partial p} = \begin{bmatrix} 0 & 0 & 1 & 0 \\ 0 & 0 & 0 & 1 \\ 0 & 0 & 0 & 0 \\ 0 & 0 & 0 & 0 \end{bmatrix} \begin{bmatrix} x_1 \\ x_2 \\ x_3 \\ x_4 \end{bmatrix} + \begin{bmatrix} 0 \\ 0 \\ u \\ v \end{bmatrix}. \tag{10.62}$$

The co-state equations (10.28) (line 2) are

$$\begin{bmatrix} \dot{p}_1 \\ \dot{p}_2 \\ \dot{p}_3 \\ \dot{p}_4 \end{bmatrix} = -\frac{\partial H}{\partial x} = \begin{bmatrix} 0 & 0 & 0 & 0 \\ 0 & 0 & 0 & 0 \\ -1 & 0 & 0 & 0 \\ 0 & -1 & 0 & 0 \end{bmatrix} \begin{bmatrix} p_1 \\ p_2 \\ p_3 \\ p_4 \end{bmatrix}. \tag{10.63}$$

The co-state boundary condition, from (10.28) (line 5) and (9.94), is that there exists a real number μ such that

$$\begin{bmatrix} p_1(t_f) \\ p_2(t_f) \\ p_3(t_f) \\ p_4(t_f) \end{bmatrix} = \mu \frac{\partial g}{\partial x(t_f)} = \begin{bmatrix} -\mu \\ \mu \\ 0 \\ 0 \end{bmatrix}. \tag{10.64}$$

Integrating (10.63) with boundary condition (10.64) yields

$$\begin{bmatrix} p_1 \\ p_2 \\ p_3 \\ p_4 \end{bmatrix} = \begin{bmatrix} 1 & 0 & 0 & 0 \\ 0 & 1 & 0 & 0 \\ t_f - t & 0 & 1 & 0 \\ 0 & t_f - t & 0 & 1 \end{bmatrix} \begin{bmatrix} -\mu \\ \mu \\ 0 \\ 0 \end{bmatrix} = \begin{bmatrix} -\mu \\ \mu \\ \mu(t - t_f) \\ \mu(t_f - t) \end{bmatrix}. \tag{10.65}$$

The optimality conditions for u and v, (10.28) (lines 3 and 4), yield

$$\begin{aligned} \frac{\partial H}{\partial u} &= 0 \\ \frac{\partial H}{\partial v} &= 0, \end{aligned} \tag{10.66}$$

respectively, which imply

$$\begin{aligned} u &= \frac{p_3}{(t_f - t)r_M} \\ v &= \frac{p_4}{(t - t_f)r_T}. \end{aligned} \tag{10.67}$$

Note that

$$\begin{aligned} \frac{\partial^2 H}{\partial u^2} &= (t - t_f)r_M < 0 \\ \frac{\partial^2 H}{\partial v^2} &= (t_f - t)r_T > 0, \end{aligned} \tag{10.68}$$

which ensure that u maximizes and v minimizes the Hamiltonian, respectively.

Equations (10.67) and (10.65) (lines 3 and 4) imply

$$\begin{aligned} u &= -\frac{\mu}{r_M} \\ v &= -\frac{\mu}{r_T}. \end{aligned} \tag{10.69}$$

Then, (10.62) and (10.69), together with the boundary condition (10.59), imply

$$\begin{aligned} x_1 &= x_{1f} + (t - t_f)x_{3f} - \frac{\mu}{r_M}\frac{(t - t_f)^2}{2}, \\ x_2 &= x_{2f} + (t - t_f)x_{4f} - \frac{\mu}{r_T}\frac{(t - t_f)^2}{2}. \end{aligned} \tag{10.70}$$

Now consider expression (10.55) for the line of sight angle. Using (10.56), (10.70), (10.59) and (10.60), we obtain

$$\begin{aligned} \beta &= \frac{x_2 - x_1}{V(t_f - t)}, \\ &= \frac{(x_{4f} - x_{3f})(t - t_f) + \mu\left(\frac{1}{r_M} - \frac{1}{r_T}\right)\frac{(t-t_f)^2}{2}}{V(t_f - t)}, \\ &= \frac{x_{3f} - x_{4f}}{V} + \mu\left(\frac{1}{r_T} - \frac{1}{r_M}\right)\frac{(t - t_f)}{2V}. \end{aligned} \tag{10.71}$$

Hence, the turn rate of the line of sight is

$$\dot{\beta} = \frac{\mu}{2V}\left(\frac{1}{r_T} - \frac{1}{r_M}\right). \tag{10.72}$$

Now, for a vehicle with nominal speed V undergoing a lateral acceleration u, the turn rate of the velocity vector is given by

$$\dot{\theta} = \frac{u}{V}. \tag{10.73}$$

Therefore, comparing (10.69) (line 1) with (10.72) and (10.73) makes it clear that the optimal guidance law must implement proportional navigation of the form

$$\dot{\theta} = \lambda\dot{\beta}, \tag{10.74}$$

where the navigation constant is

$$\lambda = \frac{2}{\left(1 - \frac{r_M}{r_T}\right)}, \tag{10.75}$$

assuming that $r_T \neq r_M$. Note that if $r_M = r_T$ or $\mu = 0$, then (10.72) implies that $\dot{\beta} = 0$, which is constant bearing guidance and is a particular case of proportional navigation, as discussed in Chapter 5.

10.7 Summary of Key Results

The key results in Chapter 10 are as follows:

1. Proposition 10.1 (the Apollonius circle theorem), which delineates dominance regions in the simplest pursuit game of kind
2. Equations (10.25) and (10.26) (the Bellman–Isaacs equations), which determine the value function and optimal policies in zero-sum differential games
3. Equations (10.27) and (10.28), which extend Pontryagin's maximum principle into a maximin principle for games

10.8 Bibliographic Notes for Further Reading

The material in Chapter 10 is standard and is well covered in several publications, including [40], [26], [7], and [27].

An early application of game theory to guidance was the realization that proportional navigation is the solution to certain pursuit-evasion differential games [37], using a setting slightly different from that in Section 10.6.

10.9 Homework Problems

PROBLEM 10.1 *Let $a \neq 0$ and $b \neq 0$ be real numbers. Prove that*

$$a\cos u + b\sin u \tag{10.76}$$

is maximized when

$$\frac{\cos u}{a} = \frac{\sin u}{b} = \frac{1}{\sqrt{a^2 + b^2}}. \tag{10.77}$$

PROBLEM 10.2 *Prove the claim made in Remark 10.1:*

1. *Based on* (10.6)–(10.9) *and the Bellman–Isaacs equation, derive the co-state dynamics and the optimal controls.*
2. *Show that the extremals are straight lines.*

PROBLEM 10.3 (The Dolichobrachistochrone [40]) *A differential game version of the classical brachistochrone problem can be posed as follows:*

Dynamics:

$$\dot{x} = \sqrt{y}\cos u + \frac{w}{2}(v+1)$$
$$\dot{y} = \sqrt{y}\sin u + \frac{w}{2}(v-1),$$

where $(x, y) \in \mathbb{R}^2$ *is the state vector,* $(u, v) \in \mathbb{R}^2$ *is the control vector, and* $w > 0$ *is a constant.*

Boundary conditions:

$$x(0) = x_0 > 0 \text{ given}, \quad y(0) = y_0 > 0 \text{ given},$$
$$x(t_f) = 0, \quad y(t_f) \text{ free}.$$

Zero sum performance index:

$$J = \int_0^{t_f} \mathrm{d}t.$$

Constraint:

$$|v| \leq 1.$$

Derive and sketch the extremals in the state space.

PROBLEM 10.4 (The War of Attrition and Attack [40]) *Consider a differential game posed as follows:*

Dynamics:

$$\dot{x}_1 = m_1 - c_1 v x_2$$
$$\dot{x}_2 = m_2 - c_2 u x_1,$$

where $(x_1, x_2) \in \mathbb{R}^2$ *is the state vector,* $(u, v) \in \mathbb{R}^2$ *is the control vector, and* $m_1 > 0$, $c_1 > 0$, $m_2 > 0$, *and* $c_2 > 0$ *are constants satisfying* $c_1 > c_2$.

Boundary conditions:

$$x_1(0) = x_{10} > 0 \text{ given}, \; x_2(0) = x_{20} > 0 \text{ given},$$
$$t_f \text{ given}, \; x_1(t_f) \text{ free}, \; x_2(t_f) \text{ free}.$$

Zero sum performance index:

$$J = \int_0^{t_f} ((1-v)x_2 - (1-u)x_1)\, \mathrm{d}t.$$

Constraints:

$$0 \le u \le 1,$$
$$0 \le v \le 1,$$
$$x_1 \ge 0, \ x_2 \ge 0.$$

Derive and sketch the extremals in the state space.

PROBLEM 10.5 (The Isotropic Rocket [40]) *Consider a differential game posed as follows:*

Dynamics:

$$\dot{x}_1 = x_3,$$
$$\dot{x}_2 = x_4,$$
$$\dot{x}_3 = F \sin u - kx_3,$$
$$\dot{x}_4 = F \cos u - kx_4,$$
$$\dot{x}_5 = w \sin v,$$
$$\dot{x}_6 = w \cos v,$$

where $(x_1, x_2) \in \mathbb{R}^2$ *are the Cartesian coordinates of a pursuer,* $(x_3, x_4) \in \mathbb{R}^2$ *are the Cartesian components of the pursuer's velocity,* $F > 0$ *and* $k > 0$ *are constants,* $(x_5, x_6) \in \mathbb{R}^2$ *are the Cartesian coordinates of an evader,* $w > 0$ *is constant, and* (u, v) *are the controls of the pursuer and evader, respectively.*

Boundary conditions:

$$x_i(0) \text{ given}, 1 \le i \le 6, \quad t_f \text{ free}.$$

Capture condition:

$$\sqrt{(x_1(t_f) - x_5(t_f))^2 + (x_2(t_f) - x_6(t_f))^2} \le l,$$

where $l > 0$ *is a given constant.*

Zero sum performance index:

$$J = \int_0^{t_f} \mathrm{d}t.$$

Assume that the initial conditions violate the capture condition. Derive and sketch the extremals in the Cartesian plane.

Epilogue

During this 10-chapter journey through *Fundamentals of Aerospace Navigation and Guidance*, the book endeavors to give an orderly account of the lay of this land. This orderliness should now be apparent through synergies between many items. By this we mean that, often, a newly introduced fundamental relies on past fundamentals, sheds new light on some of them, and foreshadows future ones. Let us give examples of such synergies.

1. Inequality (1.2) establishes that, in static systems, good navigation combined with good navigation-based guidance guarantees good guidance. This principle is extended to deterministic linear dynamic systems through inequality (7.54) under the umbrella of the deterministic separation principle. Then, it is further extended to systems with linear dynamics, quadratic cost, and Gaussian noise processes in Section 9.5.6 under the umbrella of the stochastic separation principle.
2. The controllability Gramian is introduced in Proposition 2.22 to settle the binary question of controllability of a state. However, after deriving optimization results in Section 8.4 and extending them to optimal control in Section 9.3, we show in Example 9.6 that the same controllability Gramian can also be used as a quantitative measure of controllability through (9.54).
3. The implicit function theorem, introduced in Section 6.3 to analyze the accuracy of ballistic shots, is also used in Proposition 8.6 to establish the existence of Lagrange multipliers in constrained optimization. This result is extended in Sections 8.4 and 9.3 to establish the existence of the time varying co-state vector. We notice the striking similarity between the interpretation of the Lagrange multipliers (8.45) and that of the co-state vector (9.196). Moreover, for Mayer optimal control problems (9.5)–(9.7), the co-state vector defined by (9.35), (9.37), and (9.39) (line 2) is nothing other than the adjoint vector defined in (2.118) and (2.121), with appropriate interpretation of boundary conditions.
4. Remarks 8.5 and 8.6 point out that a static unconstrained optimization problem can be solved iteratively through local quadratic approximation of the objective function. This idea is extended in Remark 8.12 to constrained optimization, using a quadratic approximation of the objective function and a linear approximation

of the constraint. The same idea is further extended in Section 9.5.5 to dynamic situations, motivating the linear quadratic regulator as a subsidiary problem in an iterative solution of an optimal control problem.

5. Section 5.6.2 suggests that jinking may be an effective maneuver to escape a proportionally navigated homing missile with first-order autopilot. Then, after introducing the maximum principle in Section 9.4, we show in Section 9.4.5 that the optimal escape maneuver against such a missile must indeed be a jink.
6. The key achievement of the maximum principle is to turn a dynamic optimization problem into a static one. In other words, instead of searching for the time history of a control such as in (9.66), we maximize the Hamiltonian at every time, such as in (9.69) (line 3). Now, a zero-sum differential game can be viewed as an optimal control problem with two opposing control authorities. This suggests that we should be able to transform a zero-sum differential game into a static game. And indeed, this is what the Bellman-Isaacs equation accomplishes in (10.28) (lines 3 and 4).
7. The constant bearing principle, stated in Proposition 5.1, suggests that proportional navigation should be an effective method for homing guidance because it attempts to achieve constant bearing through feedback. After deriving results on optimal control and differential games, we show in Section 10.6 that the optimal guidance strategy in a simple pursuit-evasion differential game must, in fact, be a form of proportional navigation.

The preceding are just some of the numerous synergistic connections between items in this book. In the authors' opinion, the resulting epiphanies are worthwhile intellectual rewards for studying these fundamentals.

Navigation and guidance are enabling technologies for aerospace missions. In that respect, they have significantly contributed to many success stories in aerospace history. These include civilian flight (see Figure E.1), military aircraft (see Figure E.2), unmanned aerial vehicles (see Figure E.3), orbital flight (see Figure E.4), lunar manned exploration (see Figure E.5), Mars robotic exploration (see Figure E.6), and an exosolar probe (see Figure E.7).

At the end of our journey through *Fundamentals of Aerospace Navigation and Guidance*, we should ask ourselves the following two questions: Is there an overarching principle that ties together all these fundamentals? And what is one to do with the fundamentals learned?

In answer to the first question, it is the authors' opinion that all the results in this book rely on the principle of local simplicity, which states that "physics is simple only when analyzed locally" [54]. We use this principle numerous times throughout the text when we assume that certain quantities are small. For instance, the assumption that both the navigation and guidance errors are small leads to linear dynamic models. Also, the assumption that the statistical distributions of errors are Gaussian really stems, as per the central limit theorem, from assuming that these errors are the superposition of a large number of small, independent, and identically distributed contributions. Finally, the use of differential calculus methods to obtain necessary or sufficient conditions for optimality stems from postulating a small perturbation of an optimal solution – these methods are exploited in optimization, optimal control, and differential games.

Figure E.1. Airbus A380. Image courtesy of P.loos.

Figure E.2. Lockheed SR-71 Blackbird. Image courtesy of NASA.

Figure E.3. General Atomics MQ-9 Reaper. Image courtesy of U.S. Air Force Photo/Lt. Col. Leslie Pratt.

Figure E.4. International Space Station. Image courtesy of NASA.

Figure E.5. Apollo 11 astronaut Buzz Aldrin walks on the surface of the Moon near the leg of the lunar module Eagle. Image courtesy of NASA.

Figure E.6. An artist's concept of a NASA Mars Exploration Rover, such as Spirit or Opportunity, on the surface of Mars. Image courtesy of NASA.

Figure E.7. An artist's concept of Voyager 1 entering interstellar space. Image courtesy of NASA.

In answer to the second question, recall the adage; "Give a person a fish, and you give that person a meal. Teach a person to fish, and you feed that person for a lifetime." Through the 10 chapters of *Fundamentals of Aerospace Navigation and Guidance*, we have together caught many fish and, in the process, developed proficiency at fishing. It is the authors' opinion that the true value of this book is to be demonstrated when the reader, having acquired these fundamentals, puts them to use in engineering aerospace vehicles.

APPENDIX A

Useful Definitions and Mathematical Results

A.1 Results from Topology

DEFINITION A.1 *Let $x^* \in \mathbb{R}^n$ and $\delta > 0$. The* ***open ball*** *around x^* with radius δ is the set*

$$\mathring{B}(x^*, \delta) = \left\{x \in \mathbb{R}^n : ||x - x^*|| < \delta\right\}. \quad \text{(A.1)}$$

The ***closed ball*** *around x^* with radius δ is the set*

$$\bar{B}(x^*, \delta) = \left\{x \in \mathbb{R}^n : ||x - x^*|| \leq \delta\right\}. \quad \text{(A.2)}$$

DEFINITION A.2 *Let $x^* \in \mathbb{R}^n$ and $X \subseteq \mathbb{R}^n$. We say that x^* is* ***in the interior*** *of X, or equivalently, that X is a* ***neighborhood*** *of x^*, if there exists an open ball around x^* that is completely contained in X, that is,*

$$\exists \delta > 0 : \mathring{B}(x^*, \delta) \subset X. \quad \text{(A.3)}$$

DEFINITION A.3 *Let $X \subseteq \mathbb{R}^n$. The* ***interior*** *of X, notated $\mathring{X}$, is the set of points of which X is a neighborhood, or equivalently, the set of points that are in the interior of X. In other words,*

$$\mathring{X} = \left\{x \in \mathbb{R}^n : \exists \delta > 0 : \mathring{B}(x, \delta) \subset X\right\}. \quad \text{(A.4)}$$

Note that the interior of a set is always a subset of the set, that is,

$$\forall X \subseteq \mathbb{R}^n, \mathring{X} \subseteq X. \quad \text{(A.5)}$$

DEFINITION A.4 *Let $X \subseteq \mathbb{R}^n$. We say that X is* ***open*** *if it is equal to its interior, or equivalently, if it is a neighborhood of each of its points, or equivalently, if $X = \mathring{X}$.*

DEFINITION A.5 *Let $X \subset \mathbb{R}^n$. We say that X has* ***measure zero*** *if its interior is the empty set, that is, $\mathring{X} = \emptyset$.*

Note that, in $\mathbb{R}$, the interval $X = [0, 1]$ does not have measure zero. However, in $\mathbb{R}^2$, the line segment $X = [0, 1] \times \{0\} = \left\{(x, 0) : x \in [0, 1]\right\}$ does have measure zero.

DEFINITION A.6 *Let $x^* \in \mathbb{R}^n$ and $X \subseteq \mathbb{R}^n$. We say that x^* **adheres** to X if every open ball around x^* has a nonempty intersection with X, that is,*

$$\forall \delta > 0, \mathring{B}(x^*, \delta) \cap X \neq \emptyset. \tag{A.6}$$

DEFINITION A.7 *Let $X \subseteq \mathbb{R}^n$. The **closure** of X, or **adherence** of X, notated $\bar{X}$, is the set of points that adhere to X. In other words,*

$$\bar{X} = \left\{x \in \mathbb{R}^n : \forall \delta > 0, \mathring{B}(x, \delta) \cap X \neq \emptyset\right\}. \tag{A.7}$$

Note that the closure of a set is always a superset of the set, that is,

$$\forall X \subset \mathbb{R}^n, \bar{X} \supseteq X. \tag{A.8}$$

DEFINITION A.8 *Let $X \subseteq \mathbb{R}^n$. We say that X is **closed** if it is equal to its closure, that is, if all the points that adhere to X belong to X also, or in other words, $X = \bar{X}$.*

DEFINITION A.9 *Let $X \subset \mathbb{R}^n$. The **boundary** of X, notated ∂X, is the set of points that adhere to X without being in the interior of X. In other words,*

$$\partial X = \bar{X} \setminus \mathring{X}. \tag{A.9}$$

DEFINITION A.10 *Let $X \subseteq \mathbb{R}^n$. We say that X is **bounded** if there exists a closed ball around the origin that contains X, that is,*

$$\exists \delta > 0 : \bar{B}(0, \delta) \supset X. \tag{A.10}$$

DEFINITION A.11 *Let $X \subseteq \mathbb{R}^n$. We say that X is **compact** if it is both closed and bounded.*

DEFINITION A.12 *Let $f : \mathbb{R}^n \to \mathbb{R}^m : x \mapsto f(x)$, and $x^* \in \mathbb{R}^n$. We say that the function f is **continuous** at x^* if its value at x^* can be approximated arbitrarily closely by approximation of x^*, that is,*

$$\forall \epsilon > 0, \exists \delta > 0 : \forall x \in \mathbb{R}^n, ||x - x^*|| < \delta \implies ||f(x) - f(x^*)|| < \epsilon. \tag{A.11}$$

PROPOSITION A.1 *If $X \subset \mathbb{R}^n$ is compact and $f : X \to \mathbb{R}$ is continuous on X, then f achieves both its minimum and maximum on X.*

DEFINITION A.13 *Let $X \subset \mathbb{R}$ and $b \in \mathbb{R}$. We say that b is an **upper bound** (respectively, **lower bound**) for X if $\forall x \in X, x \leq b$ (respectively, $\forall x \in X, x \geq b$). We denote as $X^\uparrow$ the set of all the upper bounds of X and as $X^\downarrow$ the set of all its lower bounds. In other words,*

$$X^\uparrow = \{b \in \mathbb{R} : \forall x \in X, x \leq b\}$$
$$X^\downarrow = \{b \in \mathbb{R} : \forall x \in X, x \geq b\}. \tag{A.12}$$

DEFINITION A.14 *Let $X \subset \mathbb{R}$. The **supremum** (resp. **infimum**) of X is, if it exists, its least upper bound (resp. greatest lower bound). Let f be a real function or functional. The supremum (resp. infimum) of f is the supremum (resp. infimum) of its image – in* *other words, the supremum (resp. infimum) of the set of real values that f achieves. We use the notation* $\sup X$, $\sup f$ *for supremum and* $\inf X$, $\inf f$ *for infimum.*

Note that both the supremum and infimum of a set adhere to the set. The existence of a supremum or infimum is secured by the following foundational result.

PROPOSITION A.2 (Fundamental Axiom of Analysis) *If $X \subset \mathbb{R}$ has an upper (resp. lower) bound, then it has a supremum (resp. infimum). In other words,*

$$\begin{aligned} (X \subset \mathbb{R} \wedge X^{\uparrow} \neq \emptyset) &\implies (\exists x^* \in X^{\uparrow} : \forall x \in X^{\uparrow}, x^* \leq x) \\ (X \subset \mathbb{R} \wedge X^{\downarrow} \neq \emptyset) &\implies (\exists x^* \in X^{\downarrow} : \forall x \in X^{\downarrow}, x^* \geq x) . \end{aligned} \tag{A.13}$$

A.2 Results from Linear Algebra

DEFINITION A.15 *A **real vector space** $\mathcal{V}$ is a set of elements called "vectors," equipped with two operations: addition $+ : \mathcal{V} \times \mathcal{V} \to \mathcal{V}$ and scaling: $\mathbb{R} \times \mathcal{V} \to \mathcal{V}$, that satisfy the following axioms:*

$$\begin{aligned} \forall u, v \in \mathcal{V}, \exists! u + v \in \mathcal{V}, \\ \forall u, v \in \mathcal{V}, u + v = v + u, \\ \forall u, v, w \in \mathcal{V}, (u + v) + w = u + (v + w), \\ \exists 0 \in \mathcal{V} : \forall v \in \mathcal{V}, v + 0 = v, \\ \forall v \in \mathcal{V}, \exists (-v) \in \mathcal{V} : v + (-v) = 0, \\ \forall x \in \mathbb{R}, \forall v \in \mathcal{V}, \exists! xv \in \mathcal{V}, \\ \forall x, y \in \mathbb{R}, \forall v \in \mathcal{V}, (xy)v = x(yv), \\ \forall x \in \mathbb{R}, \forall u, v \in \mathcal{V}, x(u + v) = xu + xv, \\ \forall x, y \in \mathbb{R}, \forall v \in \mathcal{V}, (x + y)v = xv + yv, \\ \forall v \in \mathcal{V}, 1v = v. \end{aligned} \tag{A.14}$$

Note that the zero vector defined in (A.14) (line 4) and the additive inverse in (A.14) (line 5) are unique.

DEFINITION A.16 *A **subspace** of a real vector space is a subset that is itself a real vector space.*

DEFINITION A.17 *Let $\mathcal{V}$ be a real vector space and $V = \{v_1, v_2, \ldots, v_n\} \subset \mathcal{V}$. We say that V is **linearly independent** if*

$$\forall x_1, x_2, \ldots, x_n \in \mathbb{R}, x_1 v_1 + x_2 v_2 + \cdots + x_n v_n = 0 \implies x_1 = x_2 = \cdots = x_n = 0. \tag{A.15}$$

An infinite set is linearly independent if all its finite subsets are linearly independent.

DEFINITION A.18 *The **dimension** of a real vector space is the maximum number of elements in a linearly independent subset.*

DEFINITION A.19 *Let $\mathcal{V}$ be a real vector space and $V = \{v_1, v_2, \ldots, v_n\} \subset \mathcal{V}$. We say that V is* ***generating*** *if*

$$\forall v \in \mathcal{V}, \exists x_1, x_2, \ldots, x_n \in \mathbb{R} : v = x_1 v_1 + x_2 v_2 + \cdots + x_n v_n. \tag{A.16}$$

DEFINITION A.20 *Let $\mathcal{V}$ be a real vector space and $V \subset \mathcal{V}$. We say that V is a* ***basis*** *for $\mathcal{V}$ if V is both linearly independent and generating.*

Note that, in a real vector space, all bases have a number of elements equal to the dimension. Also, the dimension is the minimum number of elements in a generating subset. Finally, every vector is expressed uniquely as a linear combination of the basis vectors, as in (A.16) – the coefficients of that linear combination are called the **coordinates** of the vector with respect to the basis.

DEFINITION A.21 *Let $\mathcal{V}$ and $\mathcal{U}$ be real vector spaces and $\mathcal{A} : \mathcal{V} \to \mathcal{U}$. We say that the mapping $\mathcal{A}$ is* ***linear*** *if*

$$\forall v_1, v_2 \in \mathcal{V}, \forall x_1, x_2 \in \mathbb{R}, \mathcal{A}(x_1 v_1 + x_2 v_2) = x_1 \mathcal{A}(v_1) + x_2 \mathcal{A}(v_2). \tag{A.17}$$

DEFINITION A.22 *Let $\mathcal{V}$ and $\mathcal{U}$ be real vector spaces and $\mathcal{A} : \mathcal{V} \to \mathcal{U}$ be linear. The* ***range*** *or* ***range space*** *of $\mathcal{A}$ is the set*

$$\mathcal{R}(\mathcal{A}) = \{u \in \mathcal{U} : \exists v \in \mathcal{V} : u = \mathcal{A}(v)\}. \tag{A.18}$$

The ***null space*** *or* ***nullspace*** *or* ***kernel*** *of $\mathcal{A}$ is the set*

$$\mathcal{N}(\mathcal{A}) = \{v \in \mathcal{V} : \mathcal{A}(v) = 0\}. \tag{A.19}$$

Note that if $\mathcal{A} : \mathcal{V} \to \mathcal{U}$ is linear, then its range space is a subspace of $\mathcal{U}$ and its null space is a subspace of $\mathcal{V}$. This allows the following definition:

DEFINITION A.23 *Let $\mathcal{V}$ and $\mathcal{U}$ be real vector spaces and $\mathcal{A} : \mathcal{V} \to \mathcal{U}$ be linear. The* ***rank*** *of $\mathcal{A}$ is the dimension of its range space. The* ***nullity*** *of $\mathcal{A}$ is the dimension of its null space.*

PROPOSITION A.3 (Fundamental Theorem of Linear Algebra) *Let $\mathcal{V}$ and $\mathcal{U}$ be real vector spaces, and $\mathcal{A} : \mathcal{V} \to \mathcal{U}$ be linear. Then the sum of its rank and nullity equals the dimension of $\mathcal{V}$.*

DEFINITION A.24 *A* ***matrix*** *is a rectangular array of numbers denoted by*

$$A = \begin{bmatrix} a_{11} & a_{12} & \ldots & a_{1n} \\ a_{21} & a_{22} & \ldots & a_{2n} \\ \vdots & \vdots & \vdots & \vdots \\ a_{m1} & a_{m2} & \ldots & a_{mn} \end{bmatrix}. \tag{A.20}$$

If A has m rows and n columns, we say that A is an m by n matrix, written $m \times n$. If $m = n$, we say that A is a ***square matrix*** *of order n. For a square matrix A of order n, the diagonal elements are the entries a_{ii}, $1 \le i \le n$. A square matrix is called* ***diagonal*** *if its only nonzero entries are on its diagonal.*

The connection between linear mappings and matrices is as follows. Let $\mathcal{V}$ and $\mathcal{U}$ be real vector spaces of dimensions n and m, respectively, with known ordered bases. Let $\mathcal{A} : \mathcal{V} \to \mathcal{U}$ be linear. Then, to $\mathcal{A}$, we associate a matrix such as (A.20) as follows: the jth column of A contains the coordinates, with respect to the basis of $\mathcal{U}$, of the image, by $\mathcal{A}$, of the jth vector of the basis of $\mathcal{V}$. If $v \in \mathcal{V}$ has coordinates $x \in \mathbb{R}^n$ and $u = \mathcal{A}(v)$, then u has coordinates $y \in \mathbb{R}^m$ satisfying $y = Ax$.

DEFINITION A.25 *The **range or range space of a rectangular matrix** $A \in \mathbb{R}^{m\times n}$, $\mathcal{R}(A)$, is given by*

$$\mathcal{R}(A) = \{y \in \mathbb{R}^m : \exists x \in \mathbb{R}^n : y = Ax\}. \tag{A.21}$$

DEFINITION A.26 *The **null space or nullspace or kernel of a rectangular matrix** $A \in \mathbb{R}^{m\times n}$, $\mathcal{N}(A)$, is given by*

$$\mathcal{N}(A) = \{x \in \mathbb{R}^n : Ax = 0\}. \tag{A.22}$$

Note that if $A \in \mathbb{R}^{m\times n}$, then its range space is a subspace of $\mathbb{R}^m$ and its null space is a subspace of $\mathbb{R}^n$.

DEFINITION A.27 *Let $A \in \mathbb{R}^{m\times n}$. The **rank** of A is the dimension of its range space. The **nullity** of A is the dimension of its null space.*

A remarkable corollary of the fundamental theorem of linear algebra is that the rank of a matrix is equal to both the number of its linearly independent columns and the number of its linearly independent rows.

DEFINITION A.28 *If $A = [a_{ij}]$ is an $m \times n$ real matrix, then the **transpose** of A, denoted A^T, is the $n \times m$ real matrix defined by $A^T = [a_{ji}]$.*

DEFINITION A.29 *A real square matrix A is called **symmetric** if $A = A^T$.*

PROPOSITION A.4 *If A is a real symmetric matrix, then all its eigenvalues and eigenvectors are real (see Definition 2.10). Also, none of its eigenvalues is defective (see Definitions 2.11 and 2.12). Moreover, its eigenvectors associated with different eigenvalues are orthogonal in the following sense: if λ_i and λ_j, $\lambda_i \neq \lambda_j$, are eigenvalues of A associated with the eigenvectors x_i and x_j, respectively, then*

$$x_i^T x_j = 0. \tag{A.23}$$

As a consequence of Proposition A.4, if A is a real symmetric matrix of order n, then it can be factored as

$$A = X\Lambda X^T, \tag{A.24}$$

where the square matrix X is **orthogonal**, that is, it satisfies

$$X^T X = I, \tag{A.25}$$

and the square matrix Λ is diagonal. Specifically, the matrix X contains, columnwise, the eigenvectors of A, normalized to magnitude 1, and the matrix Λ has on its diagonal the corresponding eigenvalues of A.

DEFINITION A.30 *If A is a symmetric matrix, then the function $g : \mathbb{R}^n \to \mathbb{R}$ defined by*

$$g(x) = x^T Ax, \tag{A.26}$$

where

$$x = \begin{bmatrix} x_1 \\ x_2 \\ \vdots \\ x_n \end{bmatrix}, \tag{A.27}$$

*is called a **real quadratic form** in the n variables $x_1, x_2, \ldots, x_n$.*

DEFINITION A.31 *The **sign of a symmetric matrix** (**positive definite** [> 0], **positive semidefinite** [≥ 0], **negative definite** [< 0], **negative semidefinite** [≤ 0], or **indefinite** [$\gtrless 0$]) is defined by the sign of its quadratic form. For example, $A = A^T$ is positive definite if*

$$\forall x \neq 0, x^T Ax > 0. \tag{A.28}$$

The sign of a symmetric $n \times n$ matrix A in a linear subspace $\mathcal{S} = \{x \in \mathbb{R}^n : Cx = 0\}$, where matrix C is $m \times n$ and full rank, is defined as follows. Let S be an $n \times (n - m)$ matrix that contains a columnwise basis of $\mathcal{S}$, that is, $CS = 0$. Then, the sign of A in $\mathcal{S}$ is, by definition, the sign of $S^T AS$.

Note that the sign of a symmetric matrix is determined by the sign of its eigenvalues. For instance, a symmetric matrix is positive definite (resp. positive semidefinite) if and only if all its eigenvalues are strictly positive (resp. positive or zero).

A.3 Taylor's Theorem

THEOREM A.1 (Taylor's Theorem) *Let $f : \mathbb{R} \to \mathbb{R}$, $f \in C^{*n+1}$ on $[a, x]$. Then,*

$$f(x) = f(a) + f'(a)(x - a) + \frac{f''(a)}{2!}(x - a)^2 + \cdots + \frac{f^{(n)}(a)}{n!}(x - a)^n + R_n(x, a), \tag{A.29}$$

where

$$R_n(x, a) = \int_a^x \frac{(x - \tau)^n}{n!} f^{(n+1)}(\tau)\,\mathrm{d}\tau. \tag{A.30}$$

Moreover, if $f \in C^{n+1}$ on $[a, x]$, then there exists $c \in [a, x]$ such that

$$R_n(x, a) = \frac{f^{(n+1)}(c)}{(n+1)!}(x - a)^{n+1}. \tag{A.31}$$

A.4 Newton's Method

Consider the system of equations

$$y = g(x, t), \tag{A.32}$$

where $t \in \mathbb{R}$, $y \in \mathbb{R}^n$, and $g : \mathbb{R}^n \times \mathbb{R} \to \mathbb{R}^n$ are known and $x \in \mathbb{R}^n$ is unknown.

Assume that we proceed iteratively to find a solution for the system of equations (A.32), and let x^k be the kth iterate, that is, the candidate solution at the kth iteration. Assume that x^k is not a solution of (A.32), that is, $y \neq g(x^k, t)$. However, assume that x^{k+1} is a solution, that is, $y = g(x^{k+1}, t)$. Performing a first-order Taylor series expansion around x^k, and using the same notation for Jacobian as in Chapter 2, we obtain

$$y = g(x^k, t) + \left(\frac{\partial g}{\partial x}\right)_{x^k}^T (x^{k+1} - x^k), \tag{A.33}$$

where we have neglected the higher-order terms. We recognize that (A.33) is a *linear* equation for x^{k+1}. If the Jacobian matrix is nonsingular, we can solve this equation as

$$x^{k+1} = x^k + \left(\frac{\partial g}{\partial x}\right)_{x^k}^{-T} \left(y - g(x^k, t)\right). \tag{A.34}$$

Equation (A.34) defines **Newton's iteration**. The preceding discussion is quite informal and does not precisely answer the question of when this iteration is guaranteed to converge to a solution. This question is settled by the following result.

PROPOSITION A.5 (Newton's Theorem) *Let g be a differentiable function of x in the system of equations (A.32). Assume that the system (A.32) has a solution x^* such that the Jacobian matrix, $\left(\frac{\partial g}{\partial x}\right)_{x^*}$, is nonsingular. Then, there exists a neighborhood of x^* such that whenever one chooses an initial condition x^0 from that neighborhood (i.e., whenever $||x^0 - x^*||$ is small enough), iteration (A.34) is guaranteed to converge toward x^*.*

EXAMPLE A.1 *Figure A.1 illustrates Newton's iteration.*

A.5 The Implicit Function Theorem

PROPOSITION A.6 (Implicit Function Theorem) *Let F be a real m-vector function of n real variables, that is,*

$$F : \mathbb{R}^n \to \mathbb{R}^m, \quad m < n. \tag{A.35}$$

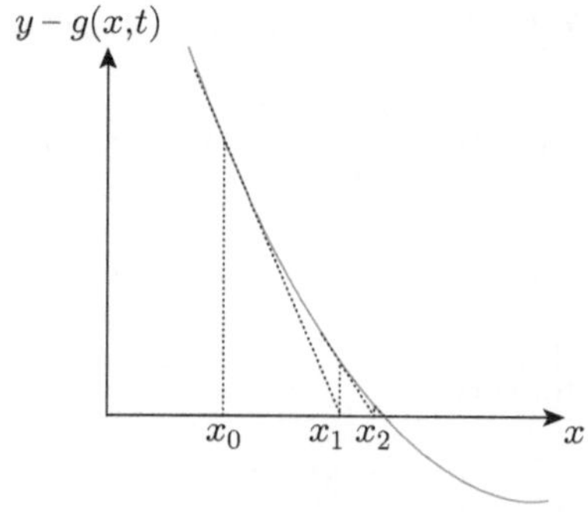

Figure A.1. Newton's iteration.

Assume that $x^0 \in \mathbb{R}^n$ *satisfies* $F(x^0) = 0$ *and that* F *is of class* $C^p, p \geq 1$, *in a neighborhood of* x^0. *Let*

$$x^0 = \begin{bmatrix} x_1^0 \\ x_2^0 \end{bmatrix}, \tag{A.36}$$

where $x_1^0 \in \mathbb{R}^m$ *and* $x_2^0 \in \mathbb{R}^{n-m}$.

Assume that the $m \times m$ *matrix of partial derivatives evaluated at* x^0,

$$\left[\frac{\partial F}{\partial x_1}\right]_{x^0}, \tag{A.37}$$

is nonsingular. (Note that this is a crucial assumption.) Then, there exists an open ball around x_2^0, $B(x_2^0, \delta)$, *and a function* ϕ *of class* C^p,

$$\phi : B(x_2^0, \delta) \to \mathbb{R}^m : x_2 \mapsto x_1 = \phi(x_2), \tag{A.38}$$

that makes the implicit relation $F(x) = 0$ *explicit on that neighborhood, that is, such that*

$$x_1^0 = \phi(x_2^0) \tag{A.39}$$

$$F(\phi(x_2), x_2) \equiv 0 \text{ on } B(x_2^0, \delta). \tag{A.40}$$

Moreover, its matrix of partial derivatives at x_2^0, $\left(\frac{\partial \phi}{\partial x_2}\right)_{x_2 = x_2^0}$, *satisfies the linear algebraic equation:*

$$\left(\frac{\partial F}{\partial x_1}\right)_{x^0}^T \left(\frac{\partial \phi}{\partial x_2}\right)_{x^0}^T + \left(\frac{\partial F}{\partial x_2}\right)_{x^0}^T = 0. \tag{A.41}$$

REMARK A.1 *Note that formula* (A.41) *is easily remembered by setting the total differential on the left-hand side of* (A.40) *to zero and using the chain rule. Also note that it is possible to evaluate the matrix of partial derivatives of* ϕ *at* x_2^0 *without solving for* ϕ *explicitly.*

REMARK A.2 (Local Inverse Theorem) *The implicit function theorem is sometimes called the **local inverse theorem**. To see why, consider the function*

$$f : \mathbb{R}^n \to \mathbb{R}^n : x \mapsto y = f(x). \tag{A.42}$$

We want to know if f can be inverted, and particularly, whether $x = \phi(y)$. Let $x_1 = x$, $x_2 = y$, and $g(x_1, x_2) = f(x) - y$. The implicit function theorem states that if $\frac{\partial g}{\partial x_1} = \frac{\partial f}{\partial x}$ is nonsingular, then the inverse exists locally. Moreover,

$$\frac{\partial \phi}{\partial x_2} = \frac{\partial \phi}{\partial y} = -(-I)\left(\frac{\partial f}{\partial x}\right)^{-1} = \left(\frac{\partial f}{\partial x}\right)^{-1}. \tag{A.43}$$

Bibliography

[1] Federal Aviation Administration. *FAR/AIM 2013: Federal Aviation Regulations/Aeronautical Information Manual.* Aviation Supplies and Academics, 2013.

[2] J. D. Anderson. *Introduction to Flight.* McGraw-Hill, 2012.

[3] K. J. Åström and B. Wittenmark. *Adaptive Control.* Dover, 2008.

[4] K. E. Atkinson. *An Introduction to Numerical Analysis.* John Wiley, 2008.

[5] Y. Bar-Shalom, X. R. Li, and T. Kirubarajan. *Estimation with Applications to Tracking and Navigation: Theory, Algorithms and Software.* John Wiley, 2001.

[6] R. B. Barrar. *An Analytic Proof That The Hohmann-Type Transfer Is The True Minimum Two-Impulse Transfer.* System Development Corporation, 1962.

[7] T. Başar and G. J. Olsder. *Dynamic Noncooperative Game Theory.* Society for Industrial and Applied Mathematics, 1998.

[8] R. H. Battin. *Astronautical Guidance.* McGraw-Hill, 1964.

[9] R. H. Battin. *An Introduction to the Mathematics and Methods of Astrodynamics.* AIAA, 1999.

[10] J. Z. Ben Asher. *Advances in Missile Guidance Theory.* AIAA, 1998.

[11] D. S. Bernstein and E. G. Gilbert. Optimal periodic control: The π test revisited. *IEEE Transactions on Automatic Control*, 25:673–84, 1980.

[12] D. P. Bertsekas. *Nonlinear Programming.* Athena Scientific, 1999.

[13] R. W. Brockett. *Finite Dimensional Linear Systems.* John Wiley, 1970.

[14] R. Brooks. A robust layered control system for a mobile robot. *IEEE Journal of Robotics and Automation*, 2:14–23, 1986.

[15] A. E. Bryson and Y. C. Ho. *Applied Optimal Control.* Hemisphere, 1975.

[16] E. F. Camacho and C. Bordons. *Model Predictive Control.* Springer, 2008.

[17] C. T. Chen. *Linear Systems Theory and Design.* Oxford University Press, 1998.

[18] H. D. Curtis. *Orbital Mechanics for Engineering Students.* Elsevier, 2010.

[19] J. M. A. Danby. *Fundamentals of Celestial Mechanics.* Willman-Bell, 1992.

[20] S. DiCairano, H. Park, and I. V. Kolmanovsky. Model predictive control approach for guidance of spacecraft rendezvous and proximity maneuvering. *International Journal of Robust and Nonlinear Control*, 22(12):1398–1427, 2012.

[21] C. T. J. Dodson and T. Poston. *Tensor Geometry: The Geometric Viewpoint and Its Uses*. Springer, 1991.

[22] B. Elias. *Pilotless Drones: Background and Considerations for Congress Regarding Unmanned Aircraft Operations in the National Airspace System*. Congressional Research Service, 2010.

[23] J. A. Farrell and M. Barth. *The Global Positioning System and Inertial Navigation*. McGraw-Hill, 1998.

[24] R. P. Feynman. *The Character of Physical Law*. Modern Library, 1994.

[25] M. W. Fossier. The development of radar homing missiles. *Journal of Guidance*, 7(6):641–51, 1984.

[26] A. Friedman. *Differential Games*. Dover, 2006.

[27] W. M. Getz and M. Pachter. Two-target pursuit-evasion differential games in the plane. *Journal of Optimization Theory and Application*, 34:383–403, 1981.

[28] E. G. Gilbert. Vehicle cruise: improved fuel economy by periodic control. *Automatica*, 12(2):159–66, 1976.

[29] E. G. Gilbert and M. G. Parsons. Periodic control and the optimality of aircraft cruise. *AIAA Journal of Aircraft*, 13(10):828–30, 1976.

[30] D. T. Greenwood. *Principles of Dynamics*. Prentice Hall, 1988.

[31] M. S. Grewal, L. R. Weill, and A. P. Andrews. *Global Positioning Systems, Inertial Navigation and Integration*. John Wiley, 2001.

[32] G. Grimmett. *Probability and Random Processes*. Oxford University Press, 2001.

[33] W. Haeussermann. Developments in the field of automatic guidance and control of rockets. *Journal of Guidance and Control*, 4(3):225–39, 1981.

[34] A. J. Healey, D. B. Marco, and R. B. McGhee. Autonomous underwater vehicle control coordination using a tri-level hybrid software architecture. In *IEEE Conference on Robotics and Automation*, pp. 2149–59, Minneapolis, MN, April 1996.

[35] J. P. Hespanha. *Linear Systems Theory*. Princeton University Press, 2009.

[36] P. Hill and C. Peterson. *Mechanics and Thermodynamics of Propulsion*. Prentice Hall, 1991.

[37] Y. C. Ho, A. E. Bryson, and S. Baron. Differential games and optimal pursuit-evasion strategies. *IEEE Transactions on Automatic Control*, 10(4):385–89, 1965.

[38] R. M. Howe. Guidance. In *Systems Engineering Handbook*, pp. 19-1–19-25. McGraw-Hill, 1965.

[39] B. Hyun, P. T. Kabamba, and A. R. Girard. Optimally-informative path planning for dynamic Bayesian classification. *Optimization Letters*, 6(8):1627–42, 2012.

[40] R. Isaacs. *Differential Games: A Mathematical Theory with Applications to Warfare and Pursuit, Control and Optimization*. Dover, 1965.

[41] A. Isidori. *Nonlinear Control Systems*. Springer, 1995.

[42] P. T. Kabamba. A note on navigation accuracy. *SIAM Review*, 35(3):481–87, 1993.

[43] T. Kailath. *Linear Systems*. Prentice Hall, 1980.

[44] H. K. Khalil. *Nonlinear Systems*. Prentice Hall, 2001.

[45] V. Krishnan. *Probability and Random Processes*. John Wiley, 2006.

[46] H. Kwakernaak and R. Sivan. *Linear Optimal Control Systems*. John Wiley, 1972.

[47] R. D. Launius. Spaceflight: the development of science, surveillance, and commerce in space. In *Proceedings of IEEE, Special Centennial Issue*, 100:1785–1818, 2012.

[48] E. Lavretsky and K. Wise. *Robust and Adaptive Control: With Aerospace Applications*. Springer, 2012.

[49] C. F. Lin. *Modern Navigation, Guidance and Control Processing*. Prentice Hall, 1991.

[50] D. Mackenzie. *Inventing Accuracy – A Historical Sociology of Nuclear Missile Guidance*. MIT Press, 1990.

[51] Complex Magazine. The 50 greatest juke moves in football history. 2010. http://www.complex.com/sports/2010/08/the-50-greatest-juke-moves-in-football-history/.

[52] J. S. Meditch. *Optimal Estimation and Control*. McGraw-Hill, 1969.

[53] J. S. Meditch. On the problem of optimal thrust programming for a lunar soft landing. *IEEE Transactions on Automatic Control*, 9(4):477–84, 1964.

[54] C. W. Misner, K. S. Thorne, and J. A. Wheeler. *Gravitation*. W. H. Freeman, 1973.

[55] K. S. Narendra and A. M. Annaswamy. *Stable Adaptive Systems*. Dover, 1991.

[56] J. Newhouse. *War and Peace in the Nuclear Age*. New York: Knopf, 1989.

[57] NIST. *NIST/SEMATECH e-Handbook of Statistical Methods*. 2012. http://www.itr.niat.gov/div898/handbook/.

[58] U.S. Department of Defense and National Information Service Department of Transportation. *2001 Federal Radionavigation Systems*. DOT-VNTSC-RSPA-01-3.1/DOD-4650.5. 2001.

[59] Office of the Secretary of Defense. *Unmanned Aircraft Systems Roadmap, 2005–2030*. U.S. Government Printing Office, 2005.

[60] G. R. Pitman. *Inertial Guidance*. John Wiley, 1962.

[61] L. S. Pontryagin, V. G. Boltyanskii, R. V. Gamkrelidze, and E. F. Mishchenko. *The Mathematical Theory of Optimal Processes*. Wiley Interscience, 1962.

[62] S. S. Rao. *Optimization Theory and Applications*. John Wiley, 1983.

[63] A. P. Sage and C. C. White. *Optimal Systems Control*. Prentice Hall, 1977.

[64] S. Sastry. *Nonlinear Systems: Analysis, Stability and Control*. Springer, 1999.

[65] C. E. Shannon. A mathematical theory of communication. *Bell System Technical Journal*, 27(3):379–423, 1948.

[66] G. M. Siouris. *Missile Guidance and Control Systems*. Springer, 2004.

[67] M. Skolnik. *Radar Handbook*. McGraw-Hill Professional, 2008.

[68] D. Skoog, F. J. Holler, and S. Couch. *Principles of Instrumental Analysis*. Brooks/Cole, 2006.

[69] J. J. E. Slotine and W. Li. *Applied Nonlinear Control*. Prentice Hall, 1991.

[70] J. L. Speyer. Non-optimality of the steady-state cruise. *AIAA Journal*, 14(11):1604–10, 1976.

[71] R. F. Stengel. *Optimal Control and Estimation*. Dover, 1994.

[72] J. Strickland. *Missile Flight Simulation*. lulu.com, 2012.

[73] T. Takehira, N. X. Vinh, and P. T. Kabamba. Analytical solution of missile terminal guidance. *AIAA Journal of Guidance, Control, and Dynamics*, 21(2):342–48, 1998.

[74] A. van den Bos. *Parameter Estimation for Scientists and Engineers*. John Wiley, 2007.

[75] N. X. Vinh. *Flight Mechanics of High-Performance Aircraft*. Cambridge University Press, 1993.

[76] W. Von Braun, F. I. Ordway, and F. Durant. *Space Travel: A History: An Update of History of Rocketry and Space Travel*. HarperCollins, 1985.

[77] H. G. Wang and T. C. Williams. Strategic inertial navigation systems. *IEEE Control Systems Magazine*, 28(1):65–85, 2008.

[78] J. R. Wertz, D. F. Everett, and J. J. Puschell, eds. *Space Mission Engineering: The New SMAD*. Microcosm Press, 2011.

[79] Y. Yildiz and I. V. Kolmanovsky. Stability properties and cross coupling performance of the control allocation scheme CAPIO. *AIAA Journal of Guidance, Control and Dynamics*, 34(4):1190–96, 2011.

Index